CAMBRIDGE MONOGRAPHS ON PHYSICS

The classical thermodynamics of deformable materials

The classical thermodynamics of deformable materials

A. G. McLELLAN
Professor of Physics, University of Canterbury, Christchurch, New Zealand

CAMBRIDGE UNIVERSITY PRESS
CAMBRIDGE
LONDON NEW YORK NEW ROCHELLE
MELBOURNE SYDNEY

CAMBRIDGE UNIVERSITY PRESS
Cambridge, New York, Melbourne, Madrid, Cape Town, Singapore,
São Paulo, Delhi, Dubai, Tokyo, Mexico City

Cambridge University Press
The Edinburgh Building, Cambridge CB2 8RU, UK

Published in the United States of America by Cambridge University Press, New York

www.cambridge.org
Information on this title: www.cambridge.org/9780521180122

First published 1980
First paperback edition 2010

A catalogue record for this publication is available from the British Library

Library of Congress Cataloguing in Publication data

McLellan, Alister George, 1919–
The classical thermodynamics of deformable materials
(Cambridge monographs on physics)
Includes bibliographical references and index
1. Deformations (Mechanics) 2. Strains and
stresses. 3. Thermodynamics. I. Title.
TA417.6.M32 620.1′123 76–2277

ISBN 978-0-521-21237-3 Hardback
ISBN 978-0-521-18012-2 Paperback

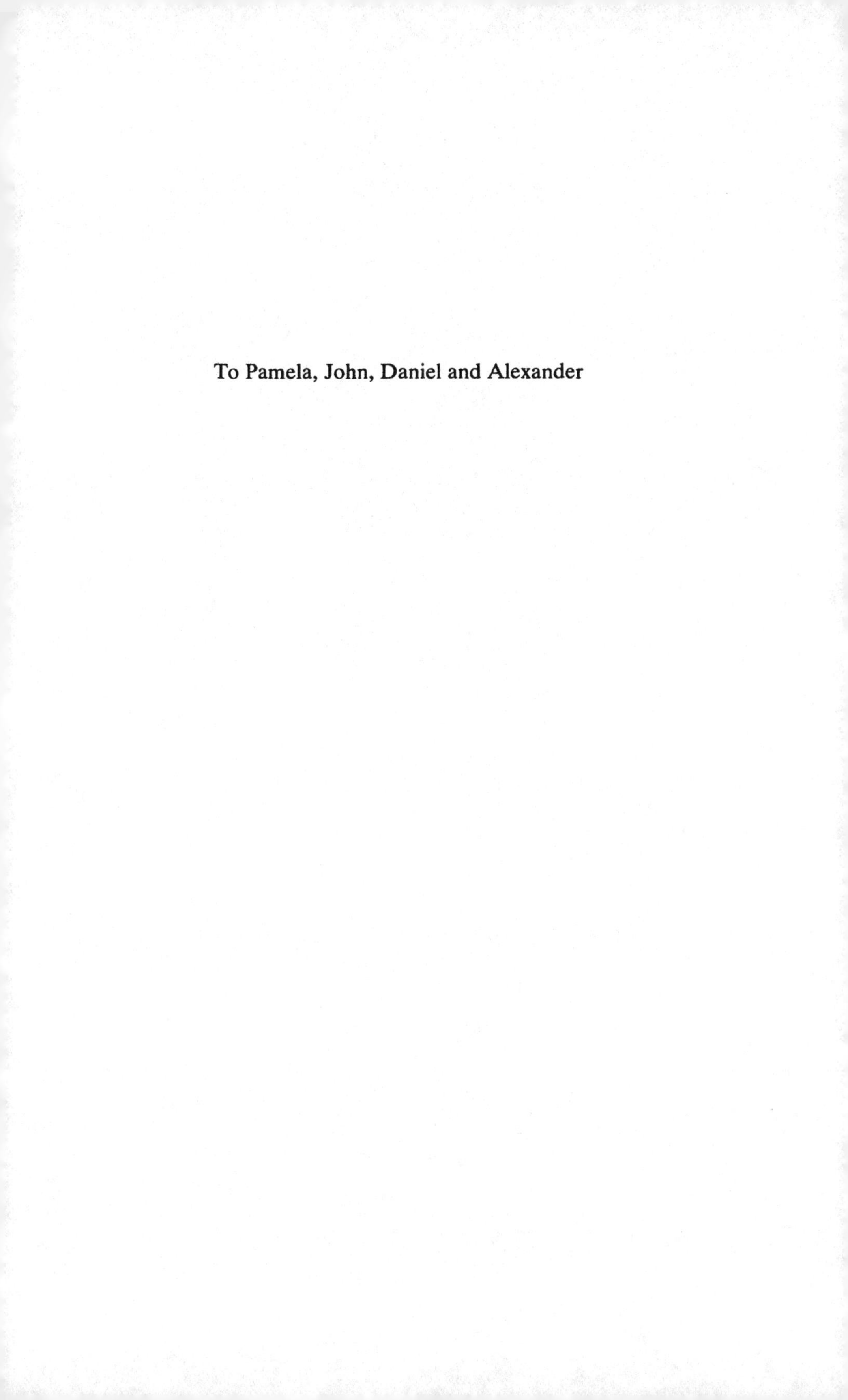

To Pamela, John, Daniel and Alexander

Contents

Preface

The study of the elastic properties of solids (and liquids) is of great importance in the science and technology of materials. In such studies the elastic behaviour must properly be treated thermodynamically, although many books on elasticity acknowledge this only by assuming that there is an elastic energy function. In this book the theory of elastic materials is developed from a consistently thermodynamic point of view.

The thermodynamics of solids under strain is of importance in many fields, such as metallurgy, solid state physics, solid state chemistry and the earth sciences. In the earth sciences it is of interest when the orientation and the chemical or phase composition of crystalline material must be considered in relation to the environment within the earth – where non-hydrostatic stress situations must be common.

In physics and chemistry the application of stresses to homogeneous phases is an important experimental tool in studies of the elastic, optical and electrical properties of crystals as well as of solid state reactions involving diffusion.

This book has been written to give a basic treatment of the thermodynamics of solids under strain. The knowledge of mathematics required may be termed elementary, that is, undergraduate mathematics involving the techniques of partial differentiation, integration and the elementary manipulation of matrices and determinants. In fact the required mathematical properties of special matrices such as symmetric or orthogonal matrices are derived and described in the early chapters. The mathematical description of finite strains is given and extensively used, since, in non-hydrostatic situations, infinitesimal strain theory is not correct even as a limit. Complicated tensorial descriptions are avoided by adhering throughout to a laboratory Cartesian co-ordinate axis system. There is no loss of generality in so doing. From a mathematical point of view the text should be readable by any trained worker in the fields mentioned above. It could be read by a senior undergraduate student in physics, chemistry or metallurgy.

The description of stresses is approached from first principles. For example, it is shown how the existence of a stress tensor depends on the elastic forces arising from forces of very short range. Energy and work performed in deformations are considered in relation to several systems of strain co-ordinates. One of these systems is new and is based on the unique factorisation of a non-singular matrix into the product of two matrices, one orthogonal and the other triangular with positive diagonal elements. This system enables the strain co-ordinates of infinitesimal strain theory to be correctly used as finite strain co-ordinates, and it is of unique importance in the treatment of thermodynamic stability of solids under stress.

The thermodynamic basis of systems under non-hydrostatic stresses is then introduced, and the fundamental assumptions of classical thermodynamics are stated. A complete account is then given of the multivariate thermodynamics required for the derivation of thermodynamic relations and the conditions of equilibrium and stability. Relations involving elastic moduli, specific heats and thermal expansion coefficients are deduced and symmetry schemes for these quantities in the different crystal classes derived. A simple method of writing down the thermodynamic functions for the crystal classes is based on a treatment of integrity bases of crystal groups recently given by the author.

The equilibrium conditions are derived for mechanical and thermal equilibrium. A treatment of chemical equilibrium is given for the diffusion of gases and liquids in a stressed solid as well as for the case of a stressed crystal in contact with a liquid solution.

Stability of equilibrium is discussed. By the use of the new strain co-ordinates mentioned above, the difficulties of the so-called trivial instabilities of a compressed solid are avoided. A new stability condition thus results. Stability against solid state phase transitions is considered and a thermodynamic theory of such phase transitions is given. It is shown how thermodynamic properties at a phase transition may be classified in terms of the rate of change of entropy at the limits of stability. In particular, a method of deriving new limiting thermodynamic relations at a transition is described.

Professor W. S. Fyfe, FRS, first aroused my interest in the field of non-hydrostatic thermodynamics, when he told me of the problems and controversies arising in the application of this subject in geology. I thank him for this and for the many discussions we have had.

I am grateful to the Commonwealth Scholarship Commission in the United Kingdom for the award of a Commonwealth Visiting Professorship at the University of Hull, where I was able to complete this book; to the University of Hull and to Professor Cole for their hospitality; to the University of Canterbury, where I enjoy a happy working life, for leave;

to Michael Woolfson and to everyone at the Press who have helped with this book; and finally and especially to my colleagues – Brian Wybourne for his encouragement, and Archie Ross for his careful, painstaking and useful criticisms.

Christchurch, New Zealand A. G. McL.

Part 1

The mathematical foundations of finite strain theory

1

Introduction

The physical properties of materials deformed by the application of external forces and magnetic and electric fields are of interest to physicists, chemists, earth scientists, metallurgists and biophysicists. The state of strain of solid material can be affected by changes in temperature, phase, or chemical composition, as well as by the application of magnetic, electric or gravitational fields. Conversely, applied stress can affect the properties of a solid. For example, Gibbs (1906) considered the conditions of thermodynamic chemical equilibrium of a stressed solid in contact with a liquid containing the solid in solution. He showed that, at equilibrium, the chemical potential of the dissolved solid must depend on the state of stress of the solid. Gibbs also discussed the equilibrium concentration of a fluid diffusing and absorbed into a stressed solid.

In fact, practically every property of a solid may be affected by its state of strain. To name just a few we may mention the conduction of heat, and electricity, and the transmission of light. Some properties may change merely in degree, but many will change in nature because of the production of anisotropic effects by applied stress. For instance, a stress-free transparent solid may transmit light isotropically, i.e., the optical effects may be described by a single refractive index for each frequency. If stress is applied, the solid may exhibit the phenomena of polarisation and double refraction, properties which cannot be described by a single refractive index. This behaviour is used in the well-known technique for studying the distribution of stresses in engineering structures such as bridges – models of the structures are made, using a suitable transparent material, and the distribution of the stress is made visible in the resulting optical effects. Another example occurs in the study of the crystal field spectra of transition metal ions, where the application of a suitable stress can reduce the symmetry of the crystal and thus lower the degeneracies of some of the crystal field energy levels. This effect is observed in the resulting splitting of the spectra, and useful information as to the nature of the crystal field is often obtained.

The effect of stress on crystalline substances is further of considerable interest in mineralogy and solid state physics. For instance, there are many phase transitions which are produced by minor movements of the lattice ions. A good example of this is the α–β quartz transition at 575 °C in which the low temperature α form is trigonal, while above the transition temperature the β form is hexagonal. The change is produced in the quartz by small rotations and displacements of the SiO_4 tetrahedral units of the silica structures.

The α–β quartz transition takes place rapidly and reversibly at the transition temperature, and this temperature has been shown to be dependent not only on the pressure but on the non-hydrostatic stress. The theory of such a dependence is of interest in mineralogy.

Silica occurs commonly in three low pressure crystalline forms: quartz, tridymite and cristobalite; and each of these has a high and low temperature modification, e.g., α tridymite changes into β tridymite at a temperature somewhere between 120 °C and 160 °C, while α cristobalite changes into β cristobalite between 200 °C and 275 °C.

The structural difference between the three forms of silica bears no relation to that between the low and high temperature forms (α, β). The three minerals have their tetrahedra linked together according to different schemes, whereas, as described above, the change from the α to β form does not alter the way in which the tetrahedra are linked. A high pressure form of silica, coesite, is also known and has been produced at pressures in the region of 20 kilobars, and possibly this too has α and β forms.

The simple chemical silica thus has a complicated and interesting phase story, and it is important to understand the thermodynamics of these phase transitions in as wide a context as possible, including the effects of non-hydrostatic stresses.

Many other minerals of relatively simple composition have similar phase forms, as may be seen by glancing through *Crystal Structures of Minerals* by Bragg & Claringbull (1965). Calcium carbonate, $CaCO_3$, has the two well-known forms calcite (hexagonal) and aragonite (rhombic), and the transition between these forms has been studied at pressures between 5 and 10 kilobars and temperatures between 300 °C and 600 °C. It is not known for certain, experimentally, whether the non-hydrostatic stresses affect the transition temperature. The material $MgSiO_3$ has three crystalline forms, proto-, ortho- and clino-enstatite. The ortho–clino transition appears to be due to minor displacements of atoms and ions and hence to be a rapid reversible transformation which could be affected by shearing stresses. It is important to study the thermodynamics of non-hydrostatic stress in order to be able to understand such solid–solid transitions.

Another interesting phenomenon is twinning in crystals and, here, shearing stresses can affect the production or disappearance of twins in crystals. Bragg & Bragg (1933) state that

> a twinned crystal consists of two individuals which are symmetrically united. The one may be derived from the other by reflection across a plane or by rotation around an axis. These operations of reflection or rotation must naturally not be symmetry operations of either individual, since in that case there could be no distinction between them, but the planes or axes are always net planes or zone axes of each crystal.

α-Quartz twins in several ways, but here only Dauphiné twinning will be described. In this type of twinning one individual is related to the other by 180° rotation about the c-axis.

Thomas & Wooster (1951) showed that detwinning of such twins could be brought about by applying non-hydrostatic stresses at elevated temperatures. They showed that application of shear stresses in some directions produced rapid detwinning while it was possible to choose directions at which the twins would continue to exist. Here again a thermodynamic theory is needed to explain this effect.

In the above mineralogical phenomena the mechanical co-ordinates required are limited to those describing the mechanical strain, but, in such processes as magnetostriction, magnetic co-ordinates are required. Conversely, in a magnetic material it is necessary to consider the mechanical co-ordinates, describing shape and size of the specimen, as well as the magnetic co-ordinates. That this is so arises in thermodynamics from the need to evaluate all forms of work done on the specimen.

The thermodynamic theory should also be able to relate the effects of temperature on such quantities as the elastic constants and heat capacities of chemically inert substances under non-hydrostatic stresses.

Where the emphasis is on the study of the properties of materials under stress, as in this book, we can often restrict ourselves to cases where the material is under homogeneous conditions, just as in thermodynamics the pressure is usually taken as uniform. Hence there is no reason to use anything but Cartesian co-ordinates, and, in what follows, finite and infinitesimal deformations will be described in Cartesian co-ordinates throughout. This leads to considerable simplification and we lose nothing of physical importance. The description of deformations will be given according to one mathematical method, and no loss of generality occurs because, if necessary, transformations could be made to other possible mathematical methods or co-ordinate systems.

In this book we will develop, from first principles, the description of deformations, as well as that of the forces acting, and the expression of internal forces, in terms of stress tensors. The thermodynamics of materials under strain will be developed from the classical Gibbsian thermodynamic point of view. For example, we will make the most important basic assumption that the equilibrium state of a substance can be defined uniquely in terms of state variables.

In classical hydrostatic thermodynamics, the state variables usually chosen are T and p. It is usually implied that the system is made up of *fluid* phases only and that the time taken to come to equilibrium is sufficiently small that the states studied are truly at equilibrium with regard to all processes which may occur. However, in practice, with glasses and crystalline solids, etc., the use of classical thermodynamics can be meaningful and useful, so long as the states studied can be in equilibrium against all the processes taking times much less than that of an experiment, while other processes are so slow that no noticeable effects of these occur in the duration of an experiment. It may be said that the use of classical thermodynamics depends on the existence of two characteristic times t_s and t_l (s for short, and l for long), such that $t_s \ll t \ll t_l$, where t is the time over which the experiment occurs, and the possible processes are such that their relaxation times fall into two well separated ranges – one range less than t_s and the other range more than t_l.

In the case of crystals under strain then, it is reasonable to apply classical thermodynamics to non-hydrostatic situations, just as it is commonplace to apply it to hydrostatic cases. The application of shearing stresses may alter the rate of some processes, or even introduce new processes, but so long as the above criterion is satisfied, this approach is justified and is very useful.

Classical thermodynamics in this sense is thus a simple limiting or idealised case. For example, the state of a part of the interior of the earth will, over millions of years, have been subject to such processes as slow viscous flow, crystallisation and production of defects and microcrystals. During this time, the state variables T and p may have returned to their values at an earlier epoch, although the physical description may be entirely different. In this extreme case the use of classical thermodynamics would be inappropriate over the whole period. The development of irreversible thermodynamics is necessary for such systems, and even there care must be taken. One of the arguments against the theory of continental drift was based on a viscosity value for the earth derived from tidal measurements, i.e., measurements involving processes whose relaxation times are of the order of 24 hours, whereas the viscosity appropriate for continental drift is that arising from processes whose relaxation times range up to the millions of years required for appreciable drift to occur. These two viscosities could thus be completely unrelated, or at least the long term viscosity be much less than the short term.

In this book each chapter is prefaced by a summary of the contents, and at the end of each chapter a list of the important definitions and equations is given.

2

Mathematical description of homogeneous deformations

The mathematical description of deformations is given in terms of a transformation from a reference state. That is, the transformation is that between a point $\boldsymbol{x}$ in the reference state and the position $\boldsymbol{X}$ of that point in the deformed state. The transformation of an infinitesimal line element is discussed and employed to define the symmetric finite strain tensor $\boldsymbol{\eta}$. The transformation of an infinitesimal element of area is discussed, as is the well-known transformation of a volume element $\mathrm{d}v$ in the reference state to $\mathrm{d}V$ in the deformed state, and it is shown that $\mathrm{d}V = J\,\mathrm{d}v$, where J is the determinant of the matrix in the relation between the related line elements $\mathrm{d}\boldsymbol{X}$ and $\mathrm{d}\boldsymbol{x}$. J is the Jacobian of the transformation $\boldsymbol{x} \rightarrow \boldsymbol{X}$.

The discussion then moves to the special case of homogeneous deformations, where the transformation is linear: $\boldsymbol{X} = \boldsymbol{D}\boldsymbol{x}$, where $\boldsymbol{D}$ is a non-singular real transformation matrix. The unique factorisation of $\boldsymbol{D}$ into the product $\boldsymbol{RA}$ where $\boldsymbol{A}$ is a positive-definite real symmetric matrix, and $\boldsymbol{R}$ is a real orthogonal matrix, is described. It is not proved, but the proof is indicated and given as a problem in example 14.

The properties of real symmetric matrices are discussed, especially emphasising the use of their eigenvalues and eigenvectors. The properties of rigid rotations and related orthogonal matrices are also discussed.

2.1. The description of deformations

The method used here may be called an isomorphic mapping of space. That is, we consider a correspondence between points $\boldsymbol{x}$ in a reference or initial space and points $\boldsymbol{X}$ in a space mapped from this initial space. This is represented by the transformation

$$\boldsymbol{x} \rightarrow \boldsymbol{X}, \quad \text{or} \quad \boldsymbol{X}(\boldsymbol{x}). \tag{2.1.1}$$

The space or region mapped out by $\boldsymbol{X}$ will be of different shape, size and position to that mapped out by $\boldsymbol{x}$, as illustrated in fig. 1(a) for two dimensions. Equation (2.1.1) may also be written

$$\boldsymbol{X} = \boldsymbol{u} + \boldsymbol{x}. \tag{2.1.2}$$

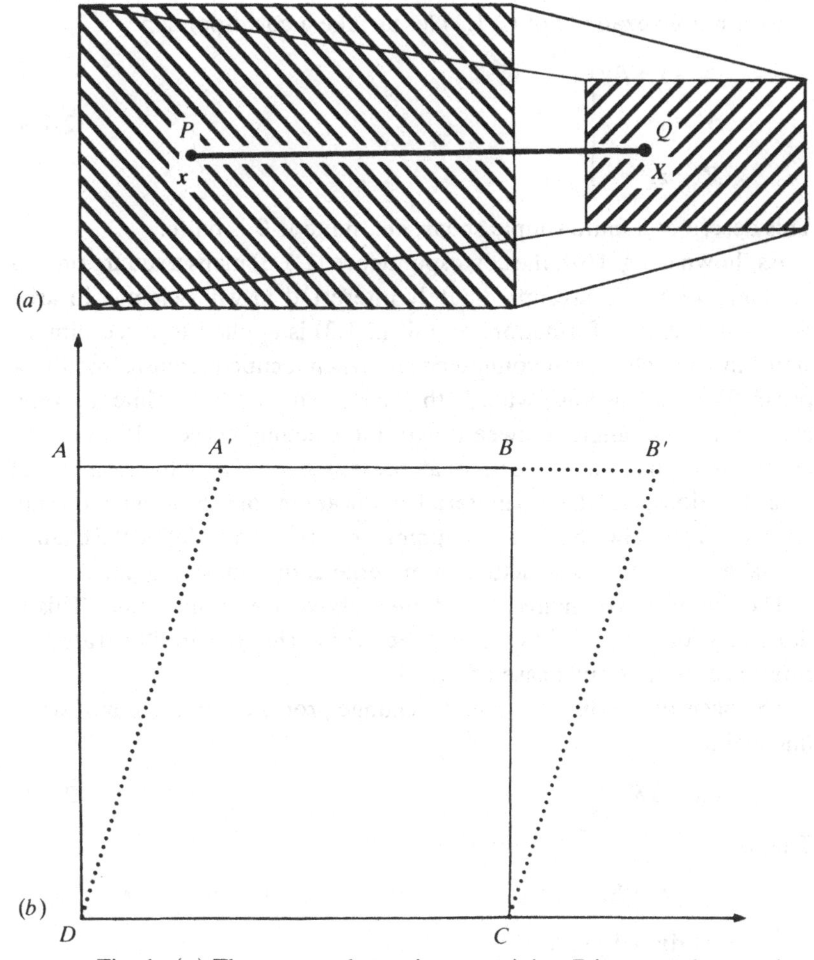

Fig. 1. (*a*) The rectangular region containing P is mapped on to that containing Q. (*b*) A finite simple shear. The space $ABCD$ is mapped on to the space $A'B'CD$.

Here $\boldsymbol{X}$, $\boldsymbol{x}$, $\boldsymbol{u}$ are vectors in three-dimensional space and the representation used for them is by the same fixed Cartesian co-ordinate axis system; for example X_α, x_α, u_α $(\alpha = 1, 2, 3)$ are the components of these vectors referred to this axis system. Sometimes, for convenience, the suffix notation may be dispensed with temporarily, and X, Y, Z taken, for instance, as the components of $\boldsymbol{X}$. We shall say that the point $\boldsymbol{X}$ is the image of the point $\boldsymbol{x}$, and we can obviously extend the use of this term to apply to surfaces and other regions. For example, the image surface Σ of σ, a surface in the reference space, is the surface traced out by the images of points on σ.

As a simple example of (2.1.1) let us take a correspondence

$$\left.\begin{aligned} X &= x + \beta y \\ Y &= y \\ Z &= z. \end{aligned}\right\} \tag{2.1.3}$$

This describes a finite simple shear, i.e., finite if β is finite.

As shown in fig. 1(*b*), the correspondence (2.1.3) maps a square on to a parallelogram. We are immediately interested in what change in area occurs in this transformation, and, if (2.1.3) is applied to three dimensions, in what change in volume occurs when a cube is transformed to a prism. We may also ask what is the change in length of a line, and the change in the angle between two intersecting curves. It must be emphasised that the discussion at present is confined to geometrical considerations, and that no material media are in consideration, although later it will be shown how such mappings may be used to define the change in shape of a material structure or material continuous medium.

The dimensional changes mentioned above are easily found. This is done first for the general case and then these changes are illustrated by reference to the simple case of (2.1.3).

The basis of the discussion is the change produced in the infinitesimal line element

$$\mathrm{d}\boldsymbol{x} \to \mathrm{d}\boldsymbol{X}. \tag{2.1.4}$$

That is

$$\boldsymbol{x} \to \boldsymbol{X},$$

$$\boldsymbol{x} + \mathrm{d}\boldsymbol{x} \to \boldsymbol{X} + \mathrm{d}\boldsymbol{X},$$

as in fig. 2.

The relation for (2.1.4) is, for $\alpha = 1, 2, 3$,

$$\begin{aligned} \mathrm{d}X_\alpha &= \frac{\partial X_\alpha}{\partial x_\beta}\mathrm{d}x_\beta \\ &= \frac{\partial X_\alpha}{\partial x_1}\mathrm{d}x_1 + \frac{\partial X_\alpha}{\partial x_2}\mathrm{d}x_2 + \frac{\partial X_\alpha}{\partial x_3}\mathrm{d}x_3, \end{aligned} \tag{2.1.5}$$

where the summation convention for repeated Greek indices is used here and throughout. Equation (2.1.5) follows from the fact that each component X_α is a function of the three components of $\boldsymbol{x}$.

To find the effects of this transformation on the length of a line element and the angle between two line elements, let us consider two infinitesimal

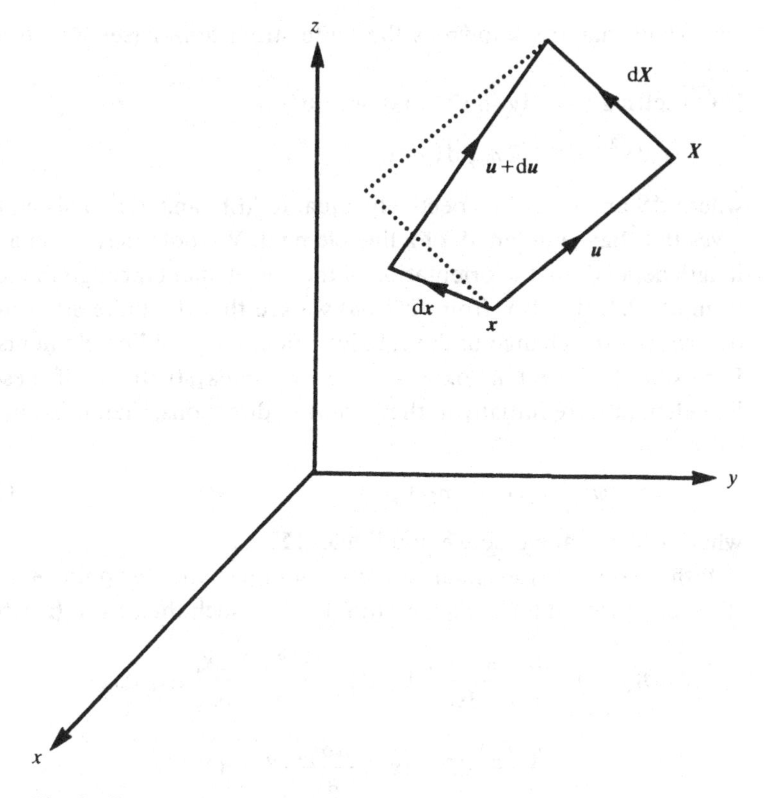

Fig. 2. The increments in $\boldsymbol{X}$ and $\boldsymbol{u}$ induced by the change d$\boldsymbol{x}$ in $\boldsymbol{x}$.

line elements, d$\boldsymbol{x}$, d$\boldsymbol{y}$, at $\boldsymbol{x}$, and their images, d$\boldsymbol{X}$, d$\boldsymbol{Y}$, at $\boldsymbol{X}$. The scalar product

$$\begin{aligned}(\mathrm{d}\boldsymbol{X}\cdot\mathrm{d}\boldsymbol{Y}) = \mathrm{d}X_\alpha\,\mathrm{d}Y_\alpha &= \frac{\partial X_\alpha}{\partial x_\beta}\frac{\partial X_\alpha}{\partial x_\gamma}\,\mathrm{d}x_\beta\,\mathrm{d}y_\gamma \\ &= (\mathrm{d}\boldsymbol{x}\cdot\mathrm{d}\boldsymbol{y}) + 2\eta_{\beta\gamma}\,\mathrm{d}x_\beta\,\mathrm{d}y_\gamma,\end{aligned} \tag{2.1.6a}$$

where

$$2\eta_{\beta\gamma} = \frac{\partial u_\beta}{\partial x_\gamma} + \frac{\partial u_\gamma}{\partial x_\beta} + \frac{\partial u_\alpha}{\partial x_\beta}\frac{\partial u_\alpha}{\partial x_\gamma}. \tag{2.1.6b}$$

Equation (2.1.6b) may be readily derived from (2.1.6a) on substituting

$$\frac{\partial X_\alpha}{\partial x_\beta} = \frac{\partial u_\alpha}{\partial x_\beta} + \delta_{\alpha\beta}, \tag{2.1.7}$$

a relation which easily follows from (2.1.2). Here $\delta_{\alpha\beta}$ is the Kronecker delta, which is unity when $\alpha = \beta$ and zero when $\alpha \neq \beta$. $\boldsymbol{\eta}$ is a very important quantity known as the finite strain tensor (see also example 14).

On setting $\mathrm{d}\boldsymbol{x} = \mathrm{d}\boldsymbol{y}$ in (2.1.6a), we obtain

$$\mathrm{d}S^2 = \mathrm{d}s^2 + 2\eta_{\beta\gamma}\,\mathrm{d}x_\beta\,\mathrm{d}x_\gamma, \tag{2.1.8}$$

where $\mathrm{d}S$ and $\mathrm{d}s$ are respectively equal to $|\mathrm{d}\boldsymbol{X}|$ and $|\mathrm{d}\boldsymbol{x}|$. This equation gives the change in length of a line element. We note that the change in length depends on the orientation of the line element through the second term of (2.1.8). Also, from (2.1.6a) we see that the finite strain tensor determines the change in the relative orientations of line elements at $\boldsymbol{x}$. For example, if we take $\mathrm{d}\boldsymbol{x} = (0, 0, \mathrm{d}s)$; $\mathrm{d}\boldsymbol{y} = (\mathrm{d}s_1, 0, 0)$, i.e., if these two line elements are initially in the x_3 and x_1 directions, then it is simple to show that

$$\cos\psi = \eta_{13}/(\eta_{11}\;\eta_{33})^{\frac{1}{2}}, \tag{2.1.9}$$

where ψ is the angle between $\mathrm{d}\boldsymbol{X}$ and $\mathrm{d}\boldsymbol{Y}$.

If there is an infinitesimal variation $\delta\boldsymbol{u}$ in $\boldsymbol{u}$ so that the point $\boldsymbol{X}$ goes to $\boldsymbol{X} + \delta\boldsymbol{u}$, there will be a change $\delta(\mathrm{d}S^2)$ in $\mathrm{d}S^2$ such that, from (2.1.6a),

$$\begin{aligned}
\delta(\mathrm{d}S^2) &= \frac{\partial X_\alpha}{\partial x_\beta}\frac{\partial(\delta u_\alpha)}{\partial x_\gamma}\,\mathrm{d}x_\beta\,\mathrm{d}x_\gamma + \frac{\partial(\delta u_\alpha)}{\partial x_\beta}\frac{\partial X_\alpha}{\partial x_\gamma}\,\mathrm{d}x_\beta\,\mathrm{d}x_\gamma \\
&= \frac{\partial(\delta u_\alpha)}{\partial x_\gamma}\,\mathrm{d}X_\alpha\,\mathrm{d}x_\gamma + \frac{\partial(\delta u_\alpha)}{\partial x_\beta}\,\mathrm{d}x_\beta\,\mathrm{d}X_\alpha \\
&= \frac{\partial(\delta u_\alpha)}{\partial x_\gamma}\frac{\partial x_\gamma}{\partial X_\mu}\,\mathrm{d}X_\alpha\,\mathrm{d}X_\mu + \frac{\partial(\delta u_\alpha)}{\partial x_\beta}\frac{\partial x_\beta}{\partial X_\mu}\,\mathrm{d}X_\mu\,\mathrm{d}X_\alpha \\
&= \left\{\frac{\partial(\delta u_\alpha)}{\partial X_\mu} + \frac{\partial(\delta u_\mu)}{\partial X_\alpha}\right\}\mathrm{d}X_\alpha\,\mathrm{d}X_\mu,
\end{aligned} \tag{2.1.10}$$

where use has been made of (2.1.5) and the corresponding reciprocal relation giving the components of $\mathrm{d}\boldsymbol{x}$ in terms of those of $\mathrm{d}\boldsymbol{X}$. Relation (2.1.10) shows that the infinitesimal variation of $\boldsymbol{u}$ is equivalent, as one would expect, to an infinitesimal transformation $\boldsymbol{X} \to \boldsymbol{X} + \delta\boldsymbol{u}$, from the reference state defined by $\boldsymbol{X}$, the terms quadratic in $\delta\boldsymbol{u}$ being neglected.

An important variation of (2.1.10) is obtained by transforming the $\mathrm{d}X_\alpha\,\mathrm{d}X_\mu$ back to $\mathrm{d}x_\beta\,\mathrm{d}x_\gamma$, to obtain

$$\delta(\mathrm{d}S^2) = \left\{\frac{\partial(\delta u_\alpha)}{\partial X_\mu} + \frac{\partial(\delta u_\mu)}{\partial X_\alpha}\right\}\frac{\partial X_\alpha}{\partial x_\beta}\frac{\partial X_\mu}{\partial x_\gamma}\,\mathrm{d}x_\beta\,\mathrm{d}x_\gamma,$$

which, from (2.1.6b), by definition

$$= 2\delta\eta_{\beta\gamma}\,\mathrm{d}x_\beta\,\mathrm{d}x_\gamma, \tag{2.1.11}$$

and so

$$\delta\eta_{\beta\gamma} = \frac{1}{2}\left\{\frac{\partial(\delta u_\alpha)}{\partial X_\mu} + \frac{\partial(\delta u_\mu)}{\partial X_\alpha}\right\}\frac{\partial X_\alpha}{\partial x_\beta}\frac{\partial X_\mu}{\partial x_\gamma}, \tag{2.1.12}$$

which is a result of some importance when considering the virtual work done in deformations.

Finally in this section we shall evaluate the finite strain tensor $\boldsymbol{\eta}$ for our simple example, (2.1.3), of a simple homogeneous shear. For this case (2.1.5) and (2.1.6) become

$$\left.\begin{aligned} \mathrm{d}X &= \mathrm{d}x + \beta\,\mathrm{d}y \\ \mathrm{d}Y &= \mathrm{d}y \\ \mathrm{d}Z &= \mathrm{d}z, \end{aligned}\right\} \tag{2.1.13}$$

and

$$\mathrm{d}S^2 = \mathrm{d}s^2 + 2\beta\,\mathrm{d}x\,\mathrm{d}y + \beta^2\,dy^2 \tag{2.1.14}$$

and hence $2\eta_{xy} = 2\eta_{yx} = \beta$, $2\eta_{yy} = \beta^2$, all other components of $\boldsymbol{\eta}$ being zero.

2.2. The transformation of a volume element

In discussing this topic we shall need some simple properties of determinants, which we shall introduce first. We shall consider determinants of a 3×3 matrix only. We can write a determinant

$$\begin{vmatrix} a_1 & b_1 & c_1 \\ a_2 & b_2 & c_2 \\ a_3 & b_3 & c_3 \end{vmatrix} = \delta_{\alpha\beta\gamma} a_\alpha b_\beta c_\gamma. \tag{2.2.1}$$

$\delta_{\alpha\beta\gamma}$ is the well-known Levi–Civita pseudo-tensor which is zero unless α, β, γ are all different, 1 if α, β, γ is an even permutation of 1, 2, 3, and -1 if they are an odd permutation. Equation (2.2.1) then follows from the fundamental definition of a determinant. We shall be using a determinant written using a common symbol with two suffices, for example the determinant

$$\Delta = \begin{vmatrix} A_{11} & A_{12} & A_{13} \\ A_{21} & A_{22} & A_{23} \\ A_{31} & A_{32} & A_{33} \end{vmatrix}; \tag{2.2.2}$$

then it can be seen that

$$\delta_{\mu\nu\omega}\,\Delta = \delta_{\alpha\beta\gamma} A_{\alpha\mu} A_{\beta\nu} A_{\gamma\omega}, \tag{2.2.3a}$$

or

$$\delta_{\alpha\beta\gamma}\,\Delta = \delta_{\mu\nu\omega} A_{\alpha\mu} A_{\beta\nu} A_{\gamma\omega}, \tag{2.2.3b}$$

where in the first expression summation is over α, β, γ, and in the second over μ, ν, ω. That (2.2.3) is true can be seen from two well-known properties of determinants, (i) that a determinant changes sign if two adjacent rows or columns are interchanged, (ii) if two rows or columns are identical then the determinant vanishes – with (ii) following naturally from (i).

Now we can consider the change of a volume element. An infinitesimal volume element in the space of $\boldsymbol{x}$ can be defined by a triplet, $\mathrm{d}\boldsymbol{a}$, $\mathrm{d}\boldsymbol{b}$, $\mathrm{d}\boldsymbol{c}$, of infinitesimal line elements at $\boldsymbol{x}$, (see fig. 3). In the mapping these will transform to $\mathrm{d}\boldsymbol{A}$, $\mathrm{d}\boldsymbol{B}$, $\mathrm{d}\boldsymbol{C}$, at $\boldsymbol{X}$. The volume element is given by

$$\begin{aligned}\mathrm{d}\boldsymbol{A}\cdot\mathrm{d}\boldsymbol{B}\wedge\mathrm{d}\boldsymbol{C} &= \delta_{\alpha\beta\gamma}\,\mathrm{d}A_\alpha\,\mathrm{d}B_\beta\,\mathrm{d}C_\gamma \\ &= \delta_{\alpha\beta\gamma}\frac{\partial X_\alpha}{\partial x_\mu}\frac{\partial X_\beta}{\partial x_\nu}\frac{\partial X_\gamma}{\partial x_\omega}\mathrm{d}a_\mu\,\mathrm{d}b_\nu\,\mathrm{d}c_\omega. \qquad (2.2.4)\end{aligned}$$

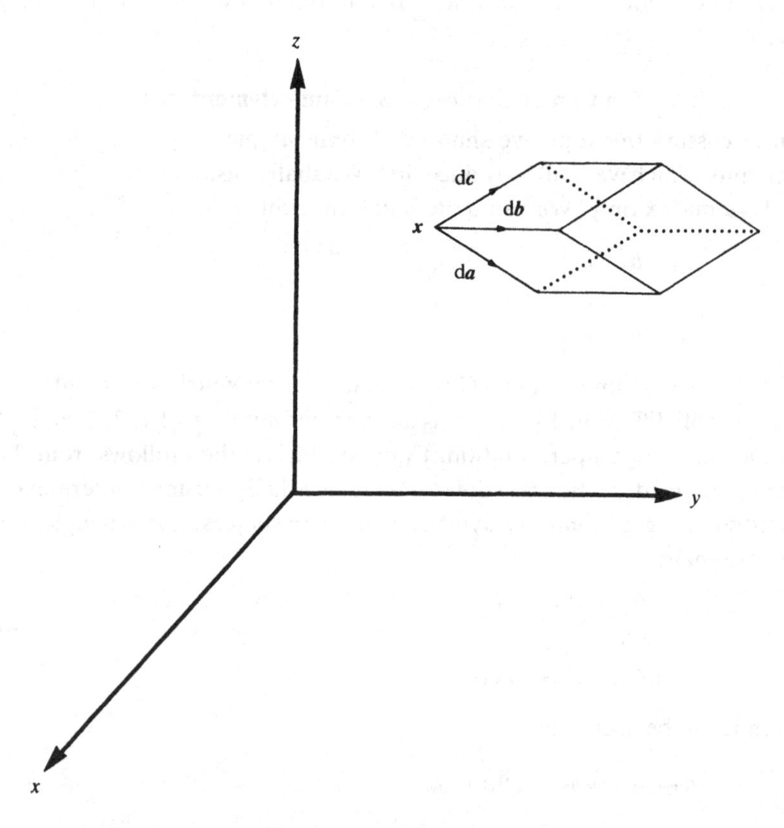

Fig. 3. The infinitesimal volume element.

If we define the Jacobian of the transformation as

$$J = \det(\partial \boldsymbol{X}/\partial \boldsymbol{x}) = \begin{vmatrix} \dfrac{\partial X_1}{\partial x_1} & \dfrac{\partial X_1}{\partial x_2} & \dfrac{\partial X_1}{\partial x_3} \\ \dfrac{\partial X_2}{\partial x_1} & \dfrac{\partial X_2}{\partial x_2} & \dfrac{\partial X_2}{\partial x_3} \\ \dfrac{\partial X_3}{\partial x_1} & \dfrac{\partial X_3}{\partial x_2} & \dfrac{\partial X_3}{\partial x_3} \end{vmatrix}, \tag{2.2.5}$$

then, from (2.2.3a), we see that (2.2.4) can be written

$$\begin{aligned} \mathrm{d}\boldsymbol{A}\cdot\mathrm{d}\boldsymbol{B}\wedge\mathrm{d}\boldsymbol{C} &= J\delta_{\mu\nu\omega}\,\mathrm{d}a_\mu\,\mathrm{d}b_\nu\,\mathrm{d}c_\omega \\ &= J\,\mathrm{d}\boldsymbol{a}\cdot\mathrm{d}\boldsymbol{b}\wedge\mathrm{d}\boldsymbol{c}. \end{aligned} \tag{2.2.6}$$

From simple vector theory we know that if we have such a triplet of infinitesimal line elements as above, then we can choose the order of $\mathrm{d}\boldsymbol{a}$, $\mathrm{d}\boldsymbol{b}$, $\mathrm{d}\boldsymbol{c}$ so that the triple vector product $\mathrm{d}\boldsymbol{a}\cdot\mathrm{d}\boldsymbol{b}\wedge d\boldsymbol{c}$ is positive and equal to the volume element. Thus, if J is positive, the left-hand side of (2.2.6) is also positive and equal to the volume $\mathrm{d}V$ of the transformed element. If this is so, then we can write (2.2.6) as

$$\mathrm{d}V = J\,\mathrm{d}v, \tag{2.2.7}$$

our desired transformation law for volume.

Why should J be positive? First of all consider the possibility that it is zero at some point. Then the matrix $(\partial\boldsymbol{X}/\partial\boldsymbol{x})$ would be singular and it would not be possible to invert the relation $\boldsymbol{x}\to\boldsymbol{X}$ at this point, to find the relation $\boldsymbol{X}\to\boldsymbol{x}$. Thus we would not have an isomorphic mapping, and if we restrict ourselves to such mappings then J must not be zero. Hence, J must be either always positive or always negative everywhere in the region, if the mapping is described by continuous derivatives. For transformations which describe the deformations of a material body it is easy to see that J will always be positive, and in what follows we shall assume this. The reader could illustrate this point by considering possible examples of homogeneous transformations for which J is negative.

Finally, we apply (2.2.7) to the example of a homogeneous simple shear, (2.1.3), for which we see that $J=1$, and thus this is a case of an isovolumetric transformation.

2.3. The transformation of a surface element

An infinitesimal surface element is defined by two infinitesimal line elements $\mathrm{d}\boldsymbol{a}$ and $\mathrm{d}\boldsymbol{b}$ at $\boldsymbol{x}$. The surface element is a pseudo-vector

$$\mathrm{d}\boldsymbol{\sigma} = \boldsymbol{n}\,\mathrm{d}\sigma = \mathrm{d}\boldsymbol{a}\wedge\mathrm{d}\boldsymbol{b}, \tag{2.3.1}$$

where $\boldsymbol{n}$ is a unit vector whose direction is determined by the vector product on the right-hand side, and $d\sigma$ is the modulus of the vector product, i.e., the area of the infinitesimal parallelogram generated by $d\boldsymbol{a}$ and $d\boldsymbol{b}$. Under the transformation $\boldsymbol{x} \rightarrow \boldsymbol{X}$, this parallelogram is transformed to that described by $d\boldsymbol{A}$ and $d\boldsymbol{B}$, the transformations of $d\boldsymbol{a}$ and $d\boldsymbol{b}$, and the transformed surface element can be described by

$$d\boldsymbol{\Sigma} = \boldsymbol{N}\, d\Sigma = d\boldsymbol{A} \wedge d\boldsymbol{B}, \tag{2.3.2a}$$

where $\boldsymbol{N}$ is a unit vector parallel to the vector product $d\boldsymbol{A} \wedge d\boldsymbol{B}$, and $d\Sigma$ is the area of this surface element.

In components, we can write

$$d\Sigma_\alpha = \delta_{\alpha\beta\gamma}\, dA_\beta\, dB_\gamma, \tag{2.3.2b}$$

and using (2.1.5), (2.3.2b) becomes

$$d\Sigma_\alpha = \delta_{\alpha\beta\gamma} \frac{\partial X_\beta}{\partial x_\nu} \frac{\partial X_\gamma}{\partial x_\omega}\, da_\nu\, db_\omega. \tag{2.3.3}$$

Now, consider the identity (2.2.3a) applied to the determinant J of (2.2.5). Multiply the resulting identity by $\partial x_\mu / \partial X_{\alpha'}$. Since

$$\frac{\partial X_\alpha}{\partial x_\mu} \frac{\partial x_\mu}{\partial X_{\alpha'}} = \frac{\partial X_\alpha}{\partial X_{\alpha'}} = \delta_{\alpha\alpha'}, \tag{2.3.4}$$

we obtain

$$\delta_{\alpha\beta\gamma} \frac{\partial X_\beta}{\partial x_\nu} \frac{\partial X_\gamma}{\partial x_\omega} = J \delta_{\mu\nu\omega} \frac{\partial x_\mu}{\partial X_\alpha}, \tag{2.3.5}$$

where the prime has been dropped from α on both sides. Hence on using (2.3.5), (2.3.3) may be written

$$d\Sigma_\alpha = J \delta_{\mu\nu\omega} \frac{\partial x_\mu}{\partial X_\alpha}\, da_\nu\, db_\omega \tag{2.3.6}$$

$$= J(\partial x_\mu / \partial X_\alpha)\, d\sigma_\mu. \tag{2.3.7}$$

Thus, comparing (2.3.7) and (2.1.5), one sees that the transformation of a pseudo-vector describing a surface element is entirely different from that of an infinitesimal line element.

In fact, if we define the matrix $\boldsymbol{D}$ such that

$$D_{\alpha\beta} = \partial X_\alpha / \partial x_\beta, \tag{2.3.8}$$

i.e.,

$$dX_\alpha = D_{\alpha\beta}\, dx_\beta, \tag{2.3.9}$$

we may write (2.3.7) as

$$\mathrm{d}\boldsymbol{\Sigma} = J(\boldsymbol{D}^{\mathrm{t}})^{-1}\,\mathrm{d}\boldsymbol{\sigma}, \tag{2.3.10}$$

on noting that (2.3.4) expresses the fact that the reciprocal of the matrix $\boldsymbol{D}$ is the matrix whose $\mu\alpha$ element is $\partial x_\mu/\partial X_\alpha$. $\boldsymbol{D}^{\mathrm{t}}$ is the transpose of $\boldsymbol{D}$ (see (2.4.2)).

The transformation of the pseudo-vector $\mathrm{d}\boldsymbol{\sigma}$ is thus clearly related to the reciprocal of the transformation of the line element. It may be helpful to write out the transformation (2.3.10) in full, using x, y, z instead of the suffix notation, to obtain

$$\left.\begin{aligned}
\mathrm{d}\Sigma_x &= J\left(\frac{\partial x}{\partial X}\,\mathrm{d}\sigma_x + \frac{\partial y}{\partial X}\,\mathrm{d}\sigma_y + \frac{\partial z}{\partial X}\,\mathrm{d}\sigma_z\right)\\
\mathrm{d}\Sigma_y &= J\left(\frac{\partial x}{\partial Y}\,\mathrm{d}\sigma_x + \frac{\partial y}{\partial Y}\,\mathrm{d}\sigma_y + \frac{\partial z}{\partial Y}\,\mathrm{d}\sigma_z\right)\\
\mathrm{d}\Sigma_z &= J\left(\frac{\partial x}{\partial Z}\,\mathrm{d}\sigma_x + \frac{\partial y}{\partial Z}\,\mathrm{d}\sigma_y + \frac{\partial z}{\partial Z}\,\mathrm{d}\sigma_z\right).
\end{aligned}\right\} \tag{2.3.11}$$

If the reader were to consider special cases, including those for which $J = 1$, of transformations of surface elements which, before transformation, are normal to the co-ordinate axes, this would help to illustrate (2.3.10) and (2.3.11).

In the example (2.1.3), the reciprocal transformation is easily seen to be described by

$$\left.\begin{aligned}
x &= X - \beta Y\\
y &= Y\\
z &= Z.
\end{aligned}\right\} \tag{2.3.12}$$

Thus, since $J = 1$ in this case,

$$\left.\begin{aligned}
\mathrm{d}\Sigma_x &= \mathrm{d}\sigma_x\\
\mathrm{d}\Sigma_y &= \mathrm{d}\sigma_y - \beta\,\mathrm{d}\sigma_x\\
\mathrm{d}\Sigma_z &= \mathrm{d}\sigma_z,
\end{aligned}\right\} \tag{2.3.13}$$

for which the transformation matrix is the transpose of that in (2.3.12).

2.4. Homogeneous transformations, rigid rotations and a unique factorisation of $\boldsymbol{D}$

A homogeneous transformation is one in which the $\partial X_\alpha/\partial x_\beta$ are independent of position. In that case (2.1.5) can be integrated directly to obtain

$$X_\alpha = D_{\alpha\beta} x_\beta, \quad \text{i.e.,} \quad \boldsymbol{X} = \boldsymbol{D}\boldsymbol{x}, \tag{2.4.1}$$

where the matrix $\boldsymbol{D}$ has as elements the constant coefficients, the $D_{\alpha\beta}$. No constants of integration have been included in (2.4.1), which is equivalent to saying that the transformations being considered do not include a uniform translational displacement. For transformations which describe real deformations, it is clear that $\boldsymbol{D}$ will be a real non-singular matrix. If J were zero, then we see from (2.2.7) that the transformation would involve that of a finite volume element to one of zero volume. Such unrealistic transformations will be excluded from consideration.

There is a very useful mathematical theorem which will enable us to show that the matrix $\boldsymbol{D}$ can be factorised, so that the transformation can be regarded as a unique product of a symmetric transformation and a rigid rotation.

This theorem is a consequence of a general theorem due to Autonne (see for example Macduffee (1946), p. 77) which states that:

Every non-singular matrix is uniquely expressible as a product of a unitary matrix and a positive-definite Hermitian matrix.

That the described decomposition is unique should be particularly noted. A Hermitian matrix $\boldsymbol{H}$ is one which is equal to the conjugate complex of its transpose. That is, the transpose $\boldsymbol{H}^t$ and the Hermitian conjugate $\boldsymbol{H}^\dagger$ of $\boldsymbol{H}$ are defined respectively as

$$H^t_{\alpha\beta} = H_{\beta\alpha} \tag{2.4.2}$$

$$H^\dagger_{\alpha\beta} = H^{t*}_{\alpha\beta} = H^*_{\beta\alpha}, \tag{2.4.3}$$

where the asterisk denotes the complex conjugate. Thus $\boldsymbol{H}^\dagger$ is the complex conjugate of $\boldsymbol{H}^t$. A matrix is said to be symmetric if it is equal to its transpose, $\boldsymbol{H}^t = \boldsymbol{H}$, and is said to be Hermitian if it is equal to its Hermitian conjugate ($\boldsymbol{H}^\dagger = \boldsymbol{H}$). A real symmetric matrix, for example, is Hermitian.

A unitary matrix $\boldsymbol{U}$ is one which is defined by

$$\boldsymbol{U}\boldsymbol{U}^\dagger = \boldsymbol{U}^\dagger\boldsymbol{U} = \boldsymbol{E}, \tag{2.4.4}$$

$\boldsymbol{E}$ being the unit matrix. An important example of a unitary matrix is a real orthogonal matrix $\boldsymbol{R}$ which describes a proper rotation if its determinant is $+1$, and an improper rotation if its determinant is -1. Thus from (2.4.4), the transpose of $\boldsymbol{R}$ is its reciprocal ($\boldsymbol{R}^t\boldsymbol{R} = \boldsymbol{R}\boldsymbol{R}^t = \boldsymbol{E}$).

For 1×1 matrices the above theorem is equivalent to the polar factorisation of a complex number z into $r \exp(\mathrm{i}\theta)$ and hence the decomposition described in the theorem is often called 'polar factorisation'.

When the theorem of Autonne is restricted to the real domain it is simple to prove (and this is left as an exercise for the reader) the following statement:

Every real non-singular matrix is uniquely expressible as a product of a real orthogonal matrix and a positive-definite real symmetric matrix.

A positive-definite real symmetric matrix $\boldsymbol{A}$ is one for which the quadratic form $t_\alpha A_{\alpha\beta} t_\beta > 0$, for arbitrary choices of the real variables t_α, except where they are all zero.

Thus the homogeneous transformation described by the real non-singular matrix $\boldsymbol{D}$ may be described uniquely as the product of a symmetric transformation and a rigid rotation, by the decomposition

$$\boldsymbol{D} = \boldsymbol{R}\boldsymbol{A}, \tag{2.4.5}$$

where $\boldsymbol{R}$ is the real orthogonal matrix and $\boldsymbol{A}$ the real symmetric positive-definite matrix into which $\boldsymbol{D}$ uniquely decomposes.

Since this is a very important theorem in the theory of deformations, the properties of a symmetric transformation $\boldsymbol{A}$ will be discussed further.

First of all, since $\boldsymbol{A}$ is a Hermitian matrix all of its eigenvalues are real, and its eigenvectors are, or may be chosen to be, orthogonal to each other.

The eigenvalue equation of a matrix $\boldsymbol{H}$ is

$$\boldsymbol{H}\boldsymbol{v} = \lambda \boldsymbol{v}, \quad \text{i.e.,} \quad H_{\alpha\beta} v_\beta = \lambda v_\alpha, \tag{2.4.6}$$

where λ is an eigenvalue and $\boldsymbol{v}$ a corresponding eigenvector. An $n \times n$ matrix has n eigenvalues, and to each distinct eigenvalue there corresponds at least one eigenvector. Hermitian and real orthogonal $n \times n$ matrices have n linearly independent eigenvectors. These properties of a matrix will be discussed more generally in section 2.5.

If $\boldsymbol{H}$ is Hermitian, its eigenvalues are all real. To prove this, take the scalar product of (2.4.6) with $\boldsymbol{v}^*$, the complex conjugate of $\boldsymbol{v}$. In suffix notation, we then have

$$v_\alpha^* H_{\alpha\beta} v_\beta = \lambda v_\alpha^* v_\alpha = v_\alpha^* H_{\beta\alpha}^* v_\beta, \tag{2.4.7}$$

where the last term arises from substituting $H_{\beta\alpha}^*$ for $H_{\alpha\beta}$ in the first term. Therefore

$$\lambda v_\alpha^* v_\alpha = (H_{\beta\alpha} v_\alpha)^* v_\beta = \lambda^* v_\beta^* v_\beta \tag{2.4.8}$$

on using the equation which is the complex conjugate of (2.4.6). Thus, since

$$v_\alpha^* v_\alpha = v_\beta^* v_\beta = |\boldsymbol{v}|^2 = |v_1|^2 + |v_2|^2 + |v_3|^2 \tag{2.4.9}$$

and is non-zero except for the trivial case $\boldsymbol{v} = 0$, then $\lambda = \lambda^*$ and hence is real.

Thus $\boldsymbol{A}$, a real symmetric matrix, has real eigenvalues.

Further, if

$$\boldsymbol{A}\boldsymbol{v} = \lambda \boldsymbol{v} \quad \text{and} \quad \boldsymbol{A}\boldsymbol{w} = \mu \boldsymbol{w}, \tag{2.4.10}$$

where $\lambda \neq \mu$, we show that $\boldsymbol{w} \cdot \boldsymbol{v} = 0$. To prove this we take the scalar product of $\boldsymbol{w}$ with the first equation of (2.4.10) and then, using the fact that $\boldsymbol{A}$ is symmetric, we obtain successively

$$w_\alpha A_{\alpha\beta} v_\beta = \lambda w_\alpha v_\alpha = w_\alpha A_{\beta\alpha} v_\beta = \mu w_\beta v_\beta. \tag{2.4.11}$$

The last step in this set of equalities is the result of using the second equation of (2.4.10). Since $\lambda \neq \mu$ we must have

$$\boldsymbol{w} \cdot \boldsymbol{v} = w_\alpha v_\alpha = 0. \tag{2.4.12}$$

If, as often happens, there is more than one eigenvector corresponding to the same eigenvalue, for instance if

$$\boldsymbol{A}\boldsymbol{v} = \lambda \boldsymbol{v}, \ \boldsymbol{A}\boldsymbol{w} = \lambda \boldsymbol{w}, \tag{2.4.13}$$

then

$$\boldsymbol{A}(\alpha \boldsymbol{v} + \beta \boldsymbol{w}) = \lambda(\alpha \boldsymbol{v} + \beta \boldsymbol{w}) \tag{2.4.14}$$

where α and β are any numbers. Thus, any linear combination of multiple eigenvectors is also an eigenvector, and we can choose such eigenvectors to be orthogonal to each other as well as to eigenvectors corresponding to different eigenvalues. For example, if in (2.4.13), $\boldsymbol{v} \cdot \boldsymbol{w} \neq 0$, we can choose $\boldsymbol{w}' = \alpha \boldsymbol{v} + \boldsymbol{w}$, so that

$$\boldsymbol{v} \cdot \boldsymbol{w}' = \alpha |\boldsymbol{v}|^2 + (\boldsymbol{v} \cdot \boldsymbol{w}) = 0 \tag{2.4.15}$$

by taking $\alpha = -(\boldsymbol{v} \cdot \boldsymbol{w})/|\boldsymbol{v}|^2$. In this case λ is said to have a degeneracy of 2; the choosing of an orthogonal set of eigenvectors for greater degeneracies may be carried out by an obvious generalisation of the above method.

If $\boldsymbol{v}$ is an eigenvector corresponding to λ, then so is $\alpha \boldsymbol{v}$ where α is any number. Thus, it is easy to choose an orthonormal set of eigenvectors by taking $\boldsymbol{v}/|\boldsymbol{v}|$ for every eigenvector, so that the set of eigenvectors $\boldsymbol{v}^{(ij)}$, i.e., the jth eigenvector corresponding to λ_i the ith eigenvalue, satisfies

$$\boldsymbol{v}^{(ij)} \cdot \boldsymbol{v}^{(i'j')} = \delta_{ii'} \delta_{jj'}. \tag{2.4.16}$$

Does a real symmetric matrix $\boldsymbol{A}$ have real eigenvectors? Not necessarily, for if $\boldsymbol{v}$ is an eigenvector then so is $i\boldsymbol{v}$. If $\boldsymbol{v}$ is real, $i\boldsymbol{v}$ is a complex eigenvector. However, if $\boldsymbol{A}\boldsymbol{v} = \lambda\boldsymbol{v}$, we have

$$\boldsymbol{A}\boldsymbol{v}^* = \lambda\boldsymbol{v}^* \tag{2.4.17}$$

as λ is real. Hence, if $\boldsymbol{v}$ is not proportional to $\boldsymbol{v}^*$, we can take the real and imaginary parts of $\boldsymbol{v}$, and use them to make up the orthonormal set. If $\boldsymbol{v}$ is proportional to $\boldsymbol{v}^*$, then $\boldsymbol{v}^* + \boldsymbol{v}$ is a real eigenvector corresponding to λ.

Thus an orthonormal real set of eigenvectors can be chosen for a real symmetric matrix.

Any real vector can be expressed as a linear sum of such an orthonormal set. Suppose this set is written $\boldsymbol{u}$, $\boldsymbol{v}$, $\boldsymbol{w}$, corresponding to the eigenvalues λ, μ, ν (there is nothing to stop some or all of these eigenvalues being equal). Any vector can then be written

$$\boldsymbol{t} = (\boldsymbol{t}\cdot\boldsymbol{u})\boldsymbol{u} + (\boldsymbol{t}\cdot\boldsymbol{v})\boldsymbol{v} + (\boldsymbol{t}\cdot\boldsymbol{w})\boldsymbol{w} = t_x\boldsymbol{u} + t_y\boldsymbol{v} + t_z\boldsymbol{w}, \tag{2.4.18}$$

where $t_x = (\boldsymbol{t}\cdot\boldsymbol{u})$ etc. are the components of $\boldsymbol{t}$ referred to this axis system. So

$$\boldsymbol{A}\boldsymbol{t} = \lambda t_x\boldsymbol{u} + \mu t_y\boldsymbol{v} + \nu t_z\boldsymbol{w}, \tag{2.4.19}$$

and we can write the quadratic form as

$$\boldsymbol{t}\cdot\boldsymbol{A}\boldsymbol{t} = \lambda t_x(\boldsymbol{t}\cdot\boldsymbol{u}) + \cdots = \lambda t_x^2 + \mu t_y^2 + \nu t_z^2 \tag{2.4.20}$$

which is positive-definite if, and only if, λ, μ, ν are all > 0.

Hence a symmetric homogeneous transformation has three positive non-zero eigenvalues, as one would expect.

The above treatment enables a clear description of a symmetric homogeneous transformation to be given. If the orthonormal eigenvectors of $\boldsymbol{A}$ are chosen as a co-ordinate axis system, $\boldsymbol{A}$ will be a diagonal matrix

$$\boldsymbol{A} = \begin{bmatrix} \lambda & 0 & 0 \\ 0 & \mu & 0 \\ 0 & 0 & \nu \end{bmatrix}. \tag{2.4.21}$$

Referred to this axis system, a sphere of unit radius will be transformed to an ellipsoid,

$$\left(\frac{X}{\lambda}\right)^2 + \left(\frac{Y}{\mu}\right)^2 + \left(\frac{Z}{\nu}\right)^2 = 1, \tag{2.4.22}$$

whose semi-major axes are λ, μ, ν respectively. If $\mu = \nu$, for instance, the ellipsoid is one of revolution about the x-axis. A unit cube, whose edges are parallel to the eigenvectors, becomes a rectangular parallelepiped, whose edges are λ, μ, ν respectively. The volume of the cube is $\lambda\mu\nu$, and the ratio of the volume of the ellipsoid to that of the sphere is $\lambda\mu\nu:1$.

Surface elements normal to the x-axis are transformed to surface elements *still normal to the x-axis*, but multiplied in area by $\mu\nu$, as can easily be seen by referring to the example of the unit cube; analogous results hold for surface elements normal to the y- and z-axes. The reader should reconcile these results obtained in the principal axis system of $\boldsymbol{A}$, as it is called, with the results listed in the summary of transformations at the end of this chapter.

For completeness, something should be said of the properties of $\boldsymbol{R}$, a real orthogonal matrix. Such a matrix describes homogeneous transformations in which scalar products of vectors remain invariant. That is, if there is the following transformation

$$\boldsymbol{X} = \boldsymbol{R}\boldsymbol{x}, \qquad \boldsymbol{Y} = \boldsymbol{R}\boldsymbol{y},$$

or

$$X_\alpha = R_{\alpha\beta}x_\beta, \quad Y_\alpha = R_{\alpha\beta}y_\beta, \tag{2.4.23}$$

then $\boldsymbol{R}$ is an orthogonal matrix if

$$\boldsymbol{X}\cdot\boldsymbol{Y} = \boldsymbol{x}\cdot\boldsymbol{y}, \tag{2.4.24}$$

for arbitrary vectors $\boldsymbol{x}$ and $\boldsymbol{y}$. So

$$\boldsymbol{R}\boldsymbol{x}\cdot\boldsymbol{R}\boldsymbol{y} = R_{\alpha\beta}x_\beta R_{\alpha\gamma}y_\gamma = x_\nu y_\nu, \tag{2.4.25}$$

or

$$x_\beta(\boldsymbol{R}^{\mathrm{t}}\boldsymbol{R})_{\beta\gamma}y_\gamma = x_\nu y_\nu, \tag{2.4.26}$$

where $(\boldsymbol{R}^{\mathrm{t}}\boldsymbol{R})_{\beta\gamma}$ is the $\beta\gamma$ element of the product $\boldsymbol{R}^{\mathrm{t}}\boldsymbol{R}$.

Since (2.4.26) must hold for arbitrary vectors, let us choose $\boldsymbol{x}$ as a vector whose ith component is unity and all others zero, $\boldsymbol{y}$ as a vector whose jth component is unity and all others zero. Then on substituting in (2.4.26) and summing over β and γ the equation reduces to

$$(\boldsymbol{R}^{\mathrm{t}}\boldsymbol{R})_{ij} = \delta_{ij}. \tag{2.4.27}$$

Therefore

$$\boldsymbol{R}^{\mathrm{t}}\boldsymbol{R} = \boldsymbol{E}, \tag{2.4.28}$$

where $\boldsymbol{E}$ is the unit matrix. *That is, the reciprocal of an orthogonal matrix equals its transpose.*

What are the eigenvectors and eigenvalues of $\boldsymbol{R}$? Since a transformation described by a rotation obviously satisfies (2.4.24), it is clear that in this case, if $\boldsymbol{R}$ describes a rotation, the axis of rotation is an eigenvector corresponding to an eigenvalue of 1 (invariant axis). No other real eigenvectors can be seen, so we immediately consider the possibility that $\boldsymbol{R}$ may have complex eigenvectors and eigenvalues.

If we have

$$\boldsymbol{R}\boldsymbol{u} = \lambda \boldsymbol{u}; \qquad \boldsymbol{R}^{\mathrm{t}}\boldsymbol{u} = \lambda^{-1}\boldsymbol{u}, \tag{2.4.29}$$

where the second equation follows from the first by (2.4.28), and take the scalar product of the first equation with $\boldsymbol{u}^*$, then

$$u_\alpha^* R_{\alpha\beta} u_\beta = \lambda u_\alpha^* u_\alpha = (R_{\beta\alpha}^{\mathrm{t}} u_\alpha)^* u_\beta = (\lambda^{-1})^* u_\beta^* u_\beta, \tag{2.4.30}$$

hence $\lambda = (\lambda^{-1})^*$ or $|\lambda|^2 = 1$. *Thus the eigenvalues of* $\boldsymbol{R}$ *have unit modulus.* They may have the values ± 1, $\mathrm{e}^{\pm \mathrm{i}\theta}$, where the complex eigenvalues occur in pairs since $\boldsymbol{R}$ is real.

There are two types, $\boldsymbol{V}$ and $\boldsymbol{I}$, of real orthogonal 3×3 matrices. Type $\boldsymbol{V}$ has determinant $+1$, and eigenvalues of the form $+1$, $\mathrm{e}^{\mathrm{i}\theta}$, $\mathrm{e}^{-\mathrm{i}\theta}$, where θ is real. $\boldsymbol{V}$ describes a proper rotation. Type $\boldsymbol{I}$ has determinant -1, and eigenvalues of the form -1, $-\mathrm{e}^{\mathrm{i}\theta}$, $-\mathrm{e}^{-\mathrm{i}\theta}$. To every proper rotation matrix $\boldsymbol{V}$ there corresponds an orthogonal matrix $\boldsymbol{I}$ given by $\boldsymbol{I} = -\boldsymbol{E}\boldsymbol{V}$. $\boldsymbol{I}$, whose determinant is -1, represents a proper rotation followed by an inversion, $-\boldsymbol{E}$, in the origin and is termed an improper rotation.

To prove that $\boldsymbol{V}$ describes a proper rotation, consider the eigenvalue equation

$$\boldsymbol{V}\boldsymbol{t} = \mathrm{e}^{\mathrm{i}\theta}\boldsymbol{t}, \tag{2.4.31}$$

where

$$\boldsymbol{t} = \boldsymbol{u} + \mathrm{i}\boldsymbol{v}, \qquad \mathrm{e}^{\mathrm{i}\theta} = \cos\theta + \mathrm{i}\sin\theta, \tag{2.4.32}$$

$\boldsymbol{u}$ and $\boldsymbol{v}$ being real. On equating real and imaginary parts in (2.4.31), we obtain

$$\left.\begin{aligned} \boldsymbol{V}\boldsymbol{u} &= \boldsymbol{u}\cos\theta - \boldsymbol{v}\sin\theta \\ \boldsymbol{V}\boldsymbol{v} &= \boldsymbol{u}\sin\theta + \boldsymbol{v}\cos\theta. \end{aligned}\right\} \tag{2.4.33}$$

It is left as an exercise for the reader to prove that $\boldsymbol{t}\cdot\boldsymbol{t} = 0$, unless $\exp(2\mathrm{i}\theta) = 1$, and hence that $|\boldsymbol{u}| = |\boldsymbol{v}|$ and $\boldsymbol{u}\cdot\boldsymbol{v} = 0$; further that $\boldsymbol{u}\cdot\boldsymbol{w} = \boldsymbol{v}\cdot\boldsymbol{w} = 0$, where $\boldsymbol{V}\boldsymbol{w} = \boldsymbol{w}$, $\boldsymbol{w}$ being the eigenvector corresponding to the eigenvalue $+1$. That is, $\boldsymbol{u}$, $\boldsymbol{v}$ and $\boldsymbol{w}$ can all be normalised to unity to form a real orthonormal set.

Thus (2.4.33), together with $\boldsymbol{V}\boldsymbol{w} = \boldsymbol{w}$, shows that $\boldsymbol{V}$ corresponds to a proper rotation.

An arbitrary vector $\boldsymbol{p}$ may be referred to $\boldsymbol{u}$, $\boldsymbol{v}$, $\boldsymbol{w}$ as an axis system, i.e.,

$$\boldsymbol{p} = p_x\boldsymbol{u} + p_y\boldsymbol{v} + p_z\boldsymbol{w}, \tag{2.4.34}$$

and so, using (2.4.33),

$$\boldsymbol{V}\boldsymbol{p} = \begin{bmatrix} \cos\theta & \sin\theta & 0 \\ -\sin\theta & \cos\theta & 0 \\ 0 & 0 & 1 \end{bmatrix} \begin{bmatrix} p_x \\ p_y \\ p_z \end{bmatrix}. \tag{2.4.35}$$

Hence the 3×3 matrix in (2.4.35) gives the real representation of $\boldsymbol{V}$ referred to this real axis system.

It is clear that if an orthogonal matrix occurs as part of a transformation corresponding to a deformation of a material medium, it will describe a proper rotation.

Equation (2.1.3), with β a constant, is an example of a homogeneous strain, i.e., one in which the $\partial X_\alpha/\partial x_\beta$ are independent of position.

If we take $\beta = 2\tan\alpha$, it is a matter of simple algebra to show that the matrix of the coefficients of (2.1.3),

$$\boldsymbol{S} = \begin{bmatrix} 1 & \beta & 0 \\ 0 & 1 & 0 \\ 0 & 0 & 1 \end{bmatrix}, \tag{2.4.36}$$

may be expressed as a product

$$\boldsymbol{S} = \boldsymbol{R}\boldsymbol{A}, \tag{2.4.37}$$

where $\boldsymbol{R}$ is the proper real orthogonal matrix,

$$\boldsymbol{R} = \begin{bmatrix} c & s & 0 \\ -s & c & 0 \\ 0 & 0 & 1 \end{bmatrix}, \tag{2.4.38}$$

and

$$\boldsymbol{A} = \begin{bmatrix} c & s & 0 \\ s & c+2st & 0 \\ 0 & 0 & 1 \end{bmatrix}, \tag{2.4.39}$$

with $c = \cos\alpha$, $s = \sin\alpha$, $t = \tan\alpha$.

Thus the simple homogeneous shear described by (2.4.36) is equivalent to the symmetric shear of (2.4.39) followed by the rigid rotation of angle α about the z-axis of (2.4.38) (see fig. 4).

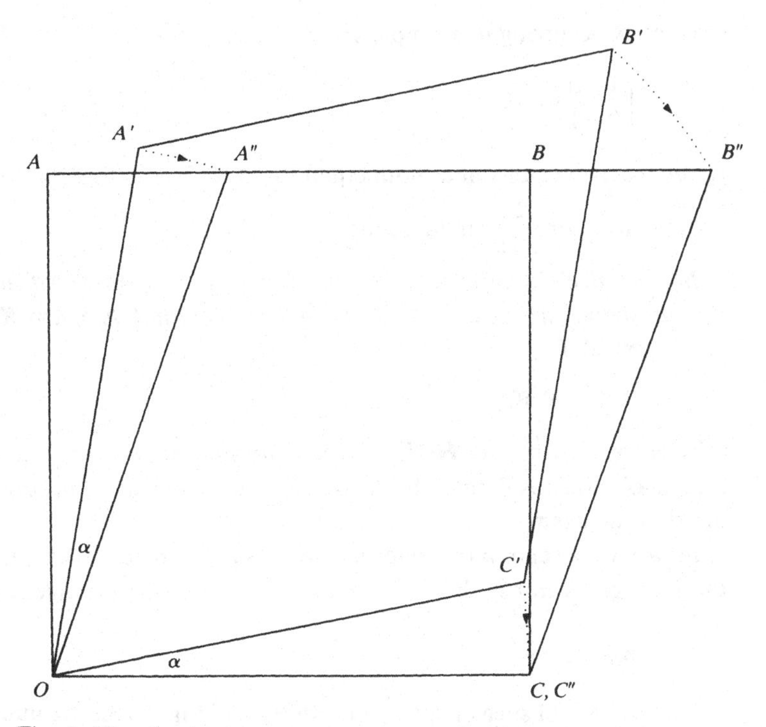

Fig. 4. The simple shear as a pure shear followed by a rotation. The square $OABC \rightarrow OA'B'C' \rightarrow OA''B''C''$; $\tan \alpha = \frac{1}{2}(AA''/OA)$.

2.5. A note on eigenvectors of matrices

n Eigenvalues can always be found for an $n \times n$ matrix, and some of these may be equal. *For each distinct eigenvalue at least one eigenvector can be found.* This follows from a well-known theorem (see e.g. Ferrar (1941), Theorem 11) which states: *A necessary and sufficient condition that a vector* $\boldsymbol{v}$*, whose components are not all zero, may be found which satisfies the* $\boldsymbol{n}$ *homogeneous equations*

$$\boldsymbol{B}\boldsymbol{v} = 0$$

$$\text{i.e.,} \qquad B_{\alpha\beta}v_\beta = 0 \quad (\alpha = 1, 2, \ldots, n) \tag{2.5.1}$$

where $\boldsymbol{B}$ *is an* $n \times n$ *matrix, is that the determinant* $|\boldsymbol{B}|$ *of* $\boldsymbol{B}$ *vanishes.* Hence, the eigenvalues of $\boldsymbol{D}$ are found by solving

$$|\boldsymbol{D} - \lambda \boldsymbol{E}| = 0, \tag{2.5.2}$$

to obtain n roots λ_i.

If two or more roots are equal, the above theorem ensures at least one eigenvector corresponding to these equal roots. As an example consider the matrix occurring in a simple shear,

$$\begin{bmatrix} 1 & a \\ 0 & 1 \end{bmatrix},$$

which has two eigenvalues both equal to +1. It is easy to show that only one eigenvector $\begin{bmatrix} 1 \\ 0 \end{bmatrix}$ can be found.

In general, n eigenvectors can be found for an n × n matrix if either (a) all the eigenvalues are distinct, or (b) the matrix is normal. A matrix $\boldsymbol{K}$ is said to be normal if

$$\boldsymbol{K}\boldsymbol{K}^{\dagger} = \boldsymbol{K}^{\dagger}\boldsymbol{K}. \tag{2.5.3}$$

A Hermitian matrix $\boldsymbol{H}(\boldsymbol{H}^{\dagger} = \boldsymbol{H})$ is a normal matrix; so is a unitary matrix $\boldsymbol{U}$, which is defined by $\boldsymbol{U}^{\dagger}\boldsymbol{U} = \boldsymbol{E}$, and of which a real orthogonal matrix is an example.

In case (b) a set of n orthonormal vectors $\boldsymbol{v}^{(ij)}$ can be found, i.e., $\boldsymbol{v}^{(ij)}$ is the jth eigenvector in this set corresponding to λ_i, the ith eigenvalue,

$$\boldsymbol{K}\boldsymbol{v}^{(ij)} = \lambda_i \boldsymbol{v}^{(ij)}. \tag{2.5.4}$$

The number d_i of such eigenvectors corresponding to λ_i is the number of times λ_i occurs in the characteristic equation (2.5.2) as a repeated root, i.e., (2.5.2) reduces to

$$(\lambda - \lambda_1)^{d_1}(\lambda - \lambda_2)^{d_2}(\lambda - \lambda_3)^{d_3} \cdots = 0. \tag{2.5.5}$$

The set is orthonormal in the complex sense

$$\boldsymbol{v}^{(ij)*} \cdot \boldsymbol{v}^{(i'j')} = \delta_{ii'}\delta_{jj'}, \tag{2.5.6}$$

as it may include complex vectors.

If the vectors $\boldsymbol{v}^{(ij)}$ are taken as the columns of a matrix $\boldsymbol{T}^{-1}$, and the complex conjugate vectors $\boldsymbol{v}^{(ij)*}$ are taken as the rows of a matrix $\boldsymbol{T}$, it may be seen from (2.5.6) that $\boldsymbol{T}$ is unitary since $\boldsymbol{T}^{\dagger}\boldsymbol{T} = \boldsymbol{E}$, and that under the transformation

$$\boldsymbol{T}\boldsymbol{K}\boldsymbol{T}^{-1} = \text{diagonal}(\lambda_1, \ldots, \lambda_1; \lambda_2, \lambda_2, \ldots; \lambda_3, \lambda_3, \ldots; \ldots), \tag{2.5.7}$$

i.e., $\boldsymbol{K}$ is diagonalised with each eigenvalue λ_i occurring along the principal diagonal (top left to bottom right) d_i times. Thus, normal matrices may be put into diagonal form by a unitary transformation. In case (a) above of a matrix with distinct roots this is not necessarily true unless the matrix is normal. As an example, the reader may consider the matrix

$$\boldsymbol{K}=\begin{bmatrix}1 & a\\ 0 & b\end{bmatrix}, \tag{2.5.8}$$

find its eigenvectors, discuss whether they are normal to each other and consider what happens as $b\to 1$.

Summary of important equations

A. *Transformations*

1. of a point,

$$\boldsymbol{x}\to\boldsymbol{X}, \quad \text{or} \quad \boldsymbol{X}(\boldsymbol{x}), \tag{2.1.1}$$

$$\boldsymbol{X}=\boldsymbol{u}+\boldsymbol{x}; \tag{2.1.2}$$

2. of a line element,

$$\mathrm{d}X_\alpha=\frac{\partial X_\alpha}{\partial x_\beta}\,\mathrm{d}x_\beta, \tag{2.1.5}$$

$$\mathrm{d}X_\alpha=D_{\alpha\beta}\,\mathrm{d}x_\beta; \tag{2.3.9}$$

3. of the length of a line element,

$$\mathrm{d}S^2=\mathrm{d}s^2+2\eta_{\beta\gamma}\,\mathrm{d}x_\beta\,\mathrm{d}x_\gamma; \tag{2.1.8}$$

4. of a volume element,

$$\mathrm{d}V=J\,\mathrm{d}v, \tag{2.2.7}$$

$$J=\det(\partial\boldsymbol{X}/\partial\boldsymbol{x}); \tag{2.2.5}$$

5. of a surface element,

$$\mathrm{d}\Sigma_\alpha=J(\partial x_\mu/\partial X_\alpha)\,\mathrm{d}\sigma_\mu, \tag{2.3.7}$$

that is,

$$\mathrm{d}\boldsymbol{\Sigma}=J(\boldsymbol{D}^{\mathrm{t}})^{-1}\,\mathrm{d}\boldsymbol{\sigma}, \tag{2.3.10}$$

where $\boldsymbol{D}^{\mathrm{t}}$ is the transpose of $\boldsymbol{D}$, that is $D^{\mathrm{t}}_{\alpha\beta}=D_{\beta\alpha}$.

B. *The finite strain tensor* $\boldsymbol{\eta}$

$$2\eta_{\alpha\beta}=\frac{\partial u_\alpha}{\partial x_\beta}+\frac{\partial u_\beta}{\partial x_\alpha}+\frac{\partial u_\mu}{\partial x_\alpha}\frac{\partial u_\mu}{\partial x_\beta}, \tag{2.1.6b}$$

$$2\boldsymbol{\eta}=\boldsymbol{D}^{\mathrm{t}}\boldsymbol{D}-\boldsymbol{E}. \qquad \text{(example 14)}$$

Examples

1.(*a*) $\boldsymbol{D}=\boldsymbol{RA}$ describes a homogeneous transformation, and $\boldsymbol{A}$ is the corresponding symmetric transformation. Prove that a volume element is transformed to the same value under both transformations, as is the angle between two line elements. Discuss the effect of these transformations on the area and orientation of a surface element.

(*b*) Verify the results of (*a*) for the case where $\boldsymbol{D}$ describes a homogeneous simple shear.

2. Consider a homogeneous symmetric transformation, $\boldsymbol{A}$. Compare, in terms of eigenvalues and eigenvectors, the matrix $\boldsymbol{A}$ with that defining the transformation of a surface element. Show, in fact, that both matrices have the same principal axes.

3.(*a*) The ellipsoid of a real symmetric matrix $\boldsymbol{S}$ is defined by the equation $x_\alpha S_{\alpha\beta} x_\beta = 1$. Show that the direction of $\boldsymbol{Sx}$ is that of the normal to the ellipsoid drawn from that point, on the surface of the ellipsoid, whose position vector is parallel to $\boldsymbol{x}$. Prove that the principal axes of $\boldsymbol{S}$ are those of the ellipsoid. Find the semi-axes of the ellipsoid in terms of the eigenvalues of $\boldsymbol{S}$.

(*b*) Discuss and contrast the ellipsoids for the transformation matrices of a line and surface element, in the case of a symmetric homogeneous transformation.

4. Show that a real proper orthogonal 3×3 matrix may be represented by the complex diagonal form

$$\begin{bmatrix} e^{i\theta} & 0 & 0 \\ 0 & e^{-i\theta} & 0 \\ 0 & 0 & 1 \end{bmatrix}$$

using the correct complex basis.

5.(*a*) Is the reciprocal of a symmetric matrix $\boldsymbol{A}$ also symmetric? Is $\boldsymbol{A}^2$ and, in general, $\boldsymbol{A}^n$ symmetric?

(*b*) If $\boldsymbol{A}$ and $\boldsymbol{B}$ are symmetric matrices, prove that $(\boldsymbol{AB})^t=\boldsymbol{BA}$. Hence, prove that $\boldsymbol{AB}$ is not symmetric unless $\boldsymbol{A}$ and $\boldsymbol{B}$ commute.

(*c*) If $\boldsymbol{A}$ and $\boldsymbol{B}$ commute prove that they have the same principal axes.

6. Consider the inhomogeneous transformation $X=ax^2$, $Y=y$, $Z=z$. Could this describe, satisfactorily, a deformation of a real material medium? If it would not be satisfactory for all values of x, could it be so for positive x?

7. Consider the following inhomogeneous two-dimensional transformation,

$$X=a(x^2-y^2), \qquad Y=2axy,$$

where a has the dimensions of reciprocal length. Determine the transformations of line and surface elements. Prove that, in this transformation, the angle between two line elements, at any point, is unaltered. Prove, also, that the transformed length of a line element is independent of its orientation, although it is dependent on its position. Over what region could this transformation describe a possible deformation of a real material medium?

Examples

8. Discuss the results of example 7 in terms of the Argand diagram for complex numbers. Discuss, generally, the corresponding results for the transformation described by

$$z \to Z(z) = X + \mathrm{i}Y,$$

where $Z(z)$ is an analytic function of $z = x + \mathrm{i}y$, and X and Y, the real and imaginary parts of Z, are functions of x and y.

9.(*a*) Prove that the trace (the sum of the diagonal elements) of a proper orthogonal 3×3 matrix is $1 + 2\cos\theta$, where θ is the angle of the rotation described by the matrix.

(*b*) Prove that the three eigenvalues of the symmetric shear, $\boldsymbol{A}$ of (2.4.39), are 1, $(1 \pm s)/c$. Verify that $\det(\boldsymbol{A}) = 1$, and is the product of the three eigenvalues. What are the corresponding eigenvectors?

10. The following are real symmetric matrices. Find the eigenvalues and a set of real orthonormal eigenvectors for each. Are they positive-definite matrices? Evaluate the strain tensors for the homogeneous transformations corresponding to ε times each matrix.

$$\begin{bmatrix} 3 & -1 & 1 \\ -1 & 5 & -1 \\ 1 & -1 & 3 \end{bmatrix}; \quad \begin{bmatrix} 51 & -5 & -25 \\ -5 & 11 & 5 \\ -25 & 5 & 21 \end{bmatrix}; \quad \begin{bmatrix} 3 & 1 & 1 \\ 1 & 5 & 1 \\ 1 & 1 & 3 \end{bmatrix};$$

$$\begin{bmatrix} 7 & 2 & 1 \\ 2 & 10 & 2 \\ 1 & 2 & 7 \end{bmatrix}.$$

Verify that the sum of the diagonal elements equals the sum of the eigenvalues, and that the sum of the squares of all the elements is equal to the sum of the squares of the eigenvalues. Can you prove these results for a general real symmetric matrix?

11. If $\boldsymbol{B}$ is any matrix, show that $\boldsymbol{B} + \boldsymbol{B}^{\mathrm{t}}$ is symmetric and that $\boldsymbol{B} - \boldsymbol{B}^{\mathrm{t}}$ is skew-symmetric, i.e., it equals minus its transpose. Show that the diagonal elements of a skew-symmetric matrix are all zero.

12.(*a*) If $\boldsymbol{S}$ is a symmetric matrix, prove that $S_{\alpha\beta}B_{\beta\alpha} = S_{\alpha\beta}B^{\mathrm{t}}_{\beta\alpha} = \frac{1}{2}S_{\alpha\beta}(\boldsymbol{B} + \boldsymbol{B}^{\mathrm{t}})_{\beta\alpha}$, where $\boldsymbol{B}$ is another matrix. Hence, if $\boldsymbol{B}$ is skew symmetric, prove that $S_{\alpha\beta}B_{\beta\alpha} = 0$.

(*b*) If $\boldsymbol{I}$ is a skew-symmetric matrix show that

$$I_{\alpha\beta}B_{\beta\alpha} = -I_{\alpha\beta}B^{\mathrm{t}}_{\beta\alpha} = \tfrac{1}{2}I_{\alpha\beta}(\boldsymbol{B} - \boldsymbol{B}^{\mathrm{t}})_{\beta\alpha},$$

i.e., the trace of $\boldsymbol{IB} = \frac{1}{2}$ trace of $\boldsymbol{I}(\boldsymbol{B} - \boldsymbol{B}^{\mathrm{t}})$.

13.(*a*) $\boldsymbol{S}$ is a real symmetric $n \times n$ matrix with eigenvalues λ^i and corresponding eigenvectors $\boldsymbol{v}^i$, $i = 1, \ldots, n$, where the $\boldsymbol{v}^i$ have been chosen to form an orthonormal set. Prove that

$$\boldsymbol{S} = \sum_{i=1}^{n} \lambda^i \boldsymbol{v}^i \boldsymbol{v}^i, \quad \text{i.e.,} \quad S_{\alpha\beta} = \sum_{i=1}^{n} \lambda^i v^i_\alpha v^i_\beta.$$

Hint: two matrices $\boldsymbol{A}$ and $\boldsymbol{B}$ are equal if $\boldsymbol{Ax} = \boldsymbol{Bx}$, for an arbitrary vector $\boldsymbol{x}$.

(*b*) Consider the matrix $\boldsymbol{R}$ whose columns are the eigenvectors $\boldsymbol{v}^i$. Prove that $\boldsymbol{R}^{\mathrm{t}}\boldsymbol{SR}$ is a diagonal matrix.

(c) If $\boldsymbol{v}$ is a unit vector, and if we form the symmetric matrix defined by

$$\boldsymbol{P}=\lambda \boldsymbol{v}\boldsymbol{v}, \quad \text{i.e.,} \quad P_{\alpha\beta}=\lambda v_\alpha v_\beta,$$

show that $\boldsymbol{P}$ has an eigenvalue λ corresponding to an eigenvector $\boldsymbol{v}$. Show, also, that all other eigenvalues of $\boldsymbol{P}$ are zero. Why is such a matrix called a projection operator?

14.(a) If $\boldsymbol{B}$ is a real matrix, show that $\boldsymbol{B}^t\boldsymbol{B}$ is symmetric and has non-negative eigenvalues.

(b) If $\boldsymbol{D}$ is a real non-singular matrix, show that a symmetric positive-definite matrix $\boldsymbol{A}$ exists such that $\boldsymbol{A}^2=\boldsymbol{D}^t\boldsymbol{D}$, by choosing $\boldsymbol{A}$ such that it has the same eigenvectors as $\boldsymbol{D}^t\boldsymbol{D}$ and positive eigenvalues which equal the square-roots of the corresponding eigenvalues of $\boldsymbol{D}^t\boldsymbol{D}$.

(c) Following (b), prove that $\boldsymbol{R}=\boldsymbol{D}\boldsymbol{A}^{-1}$ is a real orthogonal matrix, and hence that there is a unique factorisation of $\boldsymbol{D}$ into $\boldsymbol{RA}$.

(d) If $\boldsymbol{X}=\boldsymbol{A}\boldsymbol{x}$, $\boldsymbol{X}'=\boldsymbol{R}\boldsymbol{X}=\boldsymbol{R}\boldsymbol{A}\boldsymbol{x}=\boldsymbol{D}\boldsymbol{x}$, relate the result in (b) to the strain tensors for the transformations $\boldsymbol{x}\to\boldsymbol{X}'$ and $\boldsymbol{x}\to\boldsymbol{X}$, to show that the finite strain tensor $\boldsymbol{\eta}$ is given by $2\boldsymbol{\eta}=\boldsymbol{D}^t\boldsymbol{D}-\boldsymbol{E}=\boldsymbol{A}^2-\boldsymbol{E}$, where $\boldsymbol{E}$ is the unit tensor.

(e) Prove the results corresponding to (b) and (c), when one starts with the symmetric matrix $\boldsymbol{D}\boldsymbol{D}^t$.

15.(a) If $\boldsymbol{A}$ is a singular $n\times n$ matrix, show that a non-singular matrix $\boldsymbol{A}+\alpha\boldsymbol{E}$ can be chosen which has the same eigenvectors as $\boldsymbol{A}$, while its eigenvalues are those of $\boldsymbol{A}$ with α added to each.

(b) *(Attempt this only if interested in section 2.5)*
If $\boldsymbol{A}$ is a non-singular $n\times n$ normal matrix, which has an eigenvector $\boldsymbol{u}$ corresponding to an eigenvalue λ, prove that the sequence $\boldsymbol{u}, \boldsymbol{A}^\dagger\boldsymbol{u}, (\boldsymbol{A}^\dagger)^2\boldsymbol{u}, \ldots$, are all eigenvectors corresponding to λ. Suppose that $(A^\dagger)^{p+1}\boldsymbol{u}$ is the first member of this sequence that is a linear sum of the preceding members, i.e., that it is the first member that is dependent on the preceding members. Prove that $(A^\dagger)^p\boldsymbol{u}=\boldsymbol{u}_p$ satisfies $\boldsymbol{A}^\dagger\boldsymbol{u}_p=\lambda^*\boldsymbol{u}_p$, or, if $\boldsymbol{u}_p$ is regarded as a one-column matrix, $\boldsymbol{u}_p^\dagger\boldsymbol{A}=\lambda\boldsymbol{u}_p^\dagger$. (For example if $\boldsymbol{A}$ is Hermitian, the sequence consists only of $\boldsymbol{u}$, $\boldsymbol{A}\boldsymbol{u}$ and $\boldsymbol{u}^\dagger\boldsymbol{A}=\lambda\boldsymbol{u}^\dagger$.) Hence show that a unitary transformation can reduce $\boldsymbol{A}$ to the form $\begin{bmatrix}1 & 0\\ 0 & \boldsymbol{A}_1\end{bmatrix}$, where $\boldsymbol{A}_1$ is an $(n-1)\times(n-1)$ matrix on the diagonal, and the first row and column of the transformed $\boldsymbol{A}$ consists of 1 in the first (diagonal) place and zeros elsewhere.

Hence prove that a normal matrix has n eigenvectors which can be chosen orthonormal, and that it can be put in diagonal form by a unitary transformation. The reader should, perhaps, prove this for Hermitian, unitary and real orthogonal matrices, before tackling the general case of a normal matrix.

[*Hint*: since $\boldsymbol{u}, \boldsymbol{A}^\dagger\boldsymbol{u}, \ldots, (\boldsymbol{A}^\dagger)^p\boldsymbol{u}$, are an independent set they can be transformed to an orthogonal set which includes $\boldsymbol{u}_p$; using this it is easy to prove $\boldsymbol{A}^\dagger\boldsymbol{u}_p=\lambda^*\boldsymbol{u}_p$.]

3

Infinitesimal deformations

The theory of deformations is often greatly simplified by regarding them as infinitesimal – an approximation which, in practice, is often very satisfactory.

This brief chapter discusses the infinitesimal strain tensor as well as the properties of an infinitesimal rotation.

3.1. Infinitesimal transformations

As shown in chapter 2, homogeneous transformations can be described by a matrix $\boldsymbol{D}$, which may be expressed as $\boldsymbol{RA}$, where $\boldsymbol{R}$ is a real orthogonal matrix and $\boldsymbol{A}$ is a real symmetric positive-definite matrix. When the dimensionless derivatives $\partial u_\alpha/\partial x_\beta$ are all infinitesimal quantities, $\boldsymbol{D}$ is then a matrix whose elements differ infinitesimally from those of the unit matrix. Now

$$\boldsymbol{A} = \boldsymbol{R}^{\mathrm{t}}\boldsymbol{D}, \tag{3.1.1}$$

and, since $\boldsymbol{D}$ to a zeroth approximation is equal to the unit matrix, to this approximation

$$\boldsymbol{A} = \boldsymbol{R}^{\mathrm{t}}, \tag{3.1.2}$$

where of course $\boldsymbol{R}^{\mathrm{t}}$ is a real orthogonal matrix, whose determinant is $+1$ for the cases we will consider. It is easily seen that the only real orthogonal matrix which is also symmetric and positive-definite is the unit matrix. Thus $\boldsymbol{A}$ and $\boldsymbol{R}$ are infinitesimally close to the unit matrix if $\boldsymbol{D}$ is. Hence, if

$$\boldsymbol{D} = \boldsymbol{E} + \boldsymbol{d}; \qquad \boldsymbol{R} = \boldsymbol{E} + \boldsymbol{l}; \qquad \boldsymbol{A} = \boldsymbol{E} + \boldsymbol{\varepsilon}, \tag{3.1.3a}$$

then

$$\boldsymbol{d} = \boldsymbol{\varepsilon} + \boldsymbol{l} \tag{3.1.3b}$$

on neglecting the quadratic product of the infinitesimal rotation matrix $\boldsymbol{l}$ and $\boldsymbol{\varepsilon}$. Since $\boldsymbol{R}^{\mathrm{t}}\boldsymbol{R} = \boldsymbol{E}$, we have, to a first approximation,

$$\boldsymbol{E} + \boldsymbol{l} + \boldsymbol{l}^{\mathrm{t}} = \boldsymbol{E}. \tag{3.1.4}$$

Therefore $\boldsymbol{l} = -\boldsymbol{l}^{\mathrm{t}}$ and is skew-symmetric; $\boldsymbol{\varepsilon}$ is, of course, symmetric (not necessarily positive-definite). $\boldsymbol{\varepsilon}$ and $\boldsymbol{l}$ are the symmetric and skew-symmetric parts of $\boldsymbol{d}$ respectively, i.e., $\frac{1}{2}(\boldsymbol{d}+\boldsymbol{d}^{\mathrm{t}})$ and $\frac{1}{2}(\boldsymbol{d}-\boldsymbol{d}^{\mathrm{t}})$. A skew-symmetric matrix may be written, conventionally, as

$$\boldsymbol{l} = \delta\theta \begin{bmatrix} 0 & -n_z & n_y \\ n_z & 0 & -n_x \\ -n_y & n_x & 0 \end{bmatrix}, \tag{3.1.5}$$

and the operation of an infinitesimal rotation on an arbitrary vector $\boldsymbol{b}$ is given by

$$(\boldsymbol{E}+\boldsymbol{l})\boldsymbol{b} = \boldsymbol{b} + \boldsymbol{l}\boldsymbol{b}, \tag{3.1.6}$$

where

$$\boldsymbol{l}\boldsymbol{b} = \delta\theta \begin{bmatrix} n_y b_z - n_z b_y \\ n_z b_x - n_x b_z \\ n_x b_y - n_y b_x \end{bmatrix} = \delta\theta \boldsymbol{n} \wedge \boldsymbol{b}. \tag{3.1.7}$$

It may be seen from fig. 5 that this describes a rotation of angle $\delta\theta$ about an axis defined by a unit vector $\boldsymbol{n} = (n_x, n_y, n_z)$, where $n_x^2 + n_y^2 + n_z^2$ is chosen equal to 1. The sense of the rotation is that of a right-handed screw moving linearly in the direction of the unit vector $\boldsymbol{n}$.

The part $\boldsymbol{\varepsilon}$ of the infinitesimal transformation describes a symmetric infinitesimal homogeneous transformation and is the infinitesimal matrix

$$\varepsilon_{\alpha\beta} = \tfrac{1}{2}\{\partial u_\alpha/\partial x_\beta + \partial u_\beta/\partial x_\alpha\}. \tag{3.1.8}$$

This is identical with the strain-tensor (2.1.6b) of chapter 2 when terms quadratic in the components of the tensor $\partial \boldsymbol{u}/\partial \boldsymbol{x}$ are neglected.

Thus, in infinitesimal strain theory, the *infinitesimal strain tensor* $\boldsymbol{\varepsilon}$ describes both the transformation of the co-ordinates and that of the length of a line element, as follows:

$$\boldsymbol{u} = \boldsymbol{\varepsilon}\boldsymbol{x}, \tag{3.1.9a}$$

$$\mathrm{d}S^2 - \mathrm{d}s^2 = 2\varepsilon_{\alpha\beta}\,\mathrm{d}x_\alpha\,\mathrm{d}x_\beta. \tag{3.1.9b}$$

The transformation of volume elements becomes

$$\mathrm{d}V - \mathrm{d}v = \varepsilon_{\alpha\alpha}\,\mathrm{d}v, \tag{3.1.10}$$

since J to a first approximation becomes $1+\varepsilon_{\alpha\alpha}$, where $\varepsilon_{\alpha\alpha} = \varepsilon_{xx}+\varepsilon_{yy}+\varepsilon_{zz}$ is the trace of the strain tensor. Thus, if a continuous medium is deformed under such a transformation, (3.1.10) leads to the

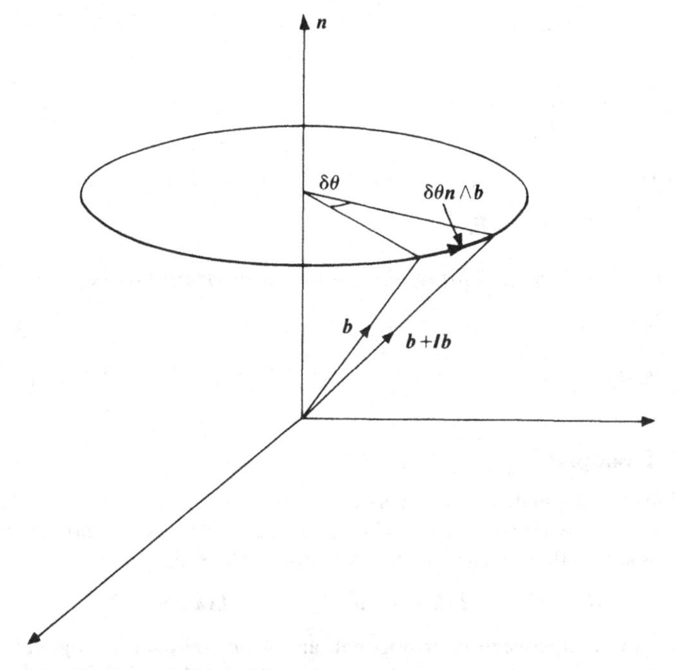

Fig. 5. An infinitesimal rotation about the axis $\boldsymbol{n}$.

following important result for the change in density,

$$\delta\rho/\rho = -\varepsilon_{\alpha\alpha}. \tag{3.1.11}$$

Since many deformations may, in practice, be described in terms of infinitesimal transformations, the above simplifications can be very useful and will be used later. They can be very important even in the context of finite deformations when an infinitesimal variation of the state of a system is being considered. This state can be taken as a reference state, and the variation from it can be described by an infinitesimal deformation.

Summary of important equations

A. *An infinitesimal rotation*

$$\boldsymbol{R} = \boldsymbol{E} + \boldsymbol{I}, \tag{3.1.3a}$$

where

$$\boldsymbol{I} = \delta\theta \begin{bmatrix} 0 & -n_z & n_y \\ n_z & 0 & -n_x \\ -n_y & n_x & 0 \end{bmatrix}, \tag{3.1.5}$$

represents an infinitesimal rotation of angle $\delta\theta$ about the unit vector $\boldsymbol{n}$ as axis.

B. *The infinitesimal strain tensor* $\boldsymbol{\varepsilon}$

$$\varepsilon_{\alpha\beta} = \frac{1}{2}\left(\frac{\partial u_\alpha}{\partial x_\beta} + \frac{\partial u_\beta}{\partial x_\alpha}\right), \tag{3.1.8}$$

that is,

$$\boldsymbol{\varepsilon} = \tfrac{1}{2}(\boldsymbol{D} + \boldsymbol{D}^{\mathrm{t}}) - \boldsymbol{E}.$$

C. *Volume and density change in infinitesimal strain theory*

$$\mathrm{d}V - \mathrm{d}v = \varepsilon_{\alpha\alpha}\,\mathrm{d}v, \tag{3.1.10}$$

$$\delta\rho/\rho = -\varepsilon_{\alpha\alpha}. \tag{3.1.11}$$

Examples

1.(*a*) Consider the matrix $\boldsymbol{L}$ which is the ordered product of two vectors $\boldsymbol{ku}$, i.e., $L_{\alpha\beta} = k_\alpha u_\beta$, where $|k| = |u| = 1$. Prove that its determinant is zero. Prove that $\boldsymbol{Lu} = \boldsymbol{k}$, and that its eigenvectors and eigenvalues are given by

$$\boldsymbol{Lk} = (\boldsymbol{k}\cdot\boldsymbol{u})\boldsymbol{k}; \qquad \boldsymbol{L}\{\boldsymbol{k} - (\boldsymbol{k}\cdot\boldsymbol{u})\boldsymbol{u}\} = 0; \qquad \boldsymbol{L}(\boldsymbol{k}\wedge\boldsymbol{u}) = 0.$$

Are these eigenvectors orthogonal, and if not may an orthogonal set be chosen? If $\boldsymbol{k} = \boldsymbol{u}$, $\boldsymbol{L}$ is a projection matrix; discuss its properties. If $\boldsymbol{k}\cdot\boldsymbol{u} = 0$, discuss the properties of $\boldsymbol{L}$.

(*b*) Prove that the determinant of $\boldsymbol{E} + \lambda\boldsymbol{L} = 1 + \lambda\,\mathrm{tr}(\boldsymbol{L}) = 1 + \lambda(\boldsymbol{k}\cdot\boldsymbol{u})$, where $\mathrm{tr}(\boldsymbol{L})$ denotes the trace of $\boldsymbol{L}$.

2. A homogeneous transformation is described by $\boldsymbol{X} = (\boldsymbol{E} + \boldsymbol{L})\boldsymbol{x}$. Investigate the properties of the transformation when $\boldsymbol{L} = \lambda\boldsymbol{ku}$, $|\boldsymbol{k}| = |\boldsymbol{u}| = 1$. Is this transformation $\boldsymbol{E} + \boldsymbol{L}$ ever a symmetric transformation? Consider the special cases $\boldsymbol{k}\cdot\boldsymbol{u} = 0$ and $\boldsymbol{k} = \boldsymbol{u}$, showing that the first case describes a simple shear and the second one a symmetric extension. Choose the most convenient axis system to express these transformations as simply as possible.

3. Show that the eigenvectors of the matrix $\boldsymbol{ku} + \boldsymbol{uk}$, where $|\boldsymbol{k}| = |\boldsymbol{u}| = 1$, are $\boldsymbol{k}\wedge\boldsymbol{u}$; $(\boldsymbol{k} + \boldsymbol{u})$; $(\boldsymbol{k} - \boldsymbol{u})$. Normalise these and verify that they are orthogonal.

4. If $\boldsymbol{D} = \boldsymbol{E} + \lambda\boldsymbol{ku}$, where $\boldsymbol{k}$ and $\boldsymbol{u}$ are unit vectors, show that $\boldsymbol{D}^{\mathrm{t}}\boldsymbol{D}$ has an invariant vector $\boldsymbol{k}\wedge\boldsymbol{u}$, that is, an eigenvector corresponding to the eigenvalue $+1$. Hence show that the other eigenvectors are of the form $\alpha\boldsymbol{u} + \beta\boldsymbol{k}$. Find them and the corresponding eigenvalues. Hence find the symmetric positive definite factor, $\boldsymbol{A}$, of $\boldsymbol{D}$ and the orthogonal factor $\boldsymbol{R}$. Is the axis of rotation along $\boldsymbol{k}\wedge\boldsymbol{u}$?

5.(*a*) Show that, for a medium which is deformed by the propagation of a plane elastic wave of displacement $\boldsymbol{a}\cos(\boldsymbol{k}\cdot\boldsymbol{x} - \omega t)$, the change of density is given by

$$\delta\rho/\rho = (\boldsymbol{k}\cdot\boldsymbol{a})\sin(\boldsymbol{k}\cdot\boldsymbol{x} - \omega t),$$

for finite as well as infinitesimal amplitudes $\boldsymbol{a}$.

(*b*) Expand $\boldsymbol{a}\cos(\boldsymbol{k}\cdot\boldsymbol{x} - \omega t)$ about a point $\boldsymbol{x}_0$, to obtain to first order $\boldsymbol{a}\cos(\boldsymbol{k}\cdot\boldsymbol{x}_0 - \omega t) - (\boldsymbol{ak}\cdot\boldsymbol{\delta x})\sin(\boldsymbol{k}\cdot\boldsymbol{x}_0 - \omega t)$, and hence show that if the wave is transverse the local deformation is a time-dependent simple shear. What is it if the wave is longitudinal?

6. Show that the rotational part of an infinitesimal transformation described by the infinitesimal vector $\boldsymbol{u}$ is determined by the vector $\frac{1}{2}(\partial/\partial \boldsymbol{X}) \wedge \boldsymbol{u}$, whose direction gives the axis of rotation and whose magnitude gives the angle of rotation.

4

Transformations describing deformations of a material medium

Deformation theory is usually applied to continuous media. In this chapter we relate the theory to a crystal structure. Coherent transformations are discussed where neighbouring atoms or ions on a lattice remain neighbours, i.e., remain coherently related. The way in which crystal symmetry or anisotropy can be introduced into the mathematical treatment is briefly discussed. Incoherent transformations are described, the most important being those where the shape of a phase is altered by solution or crystallisation at a surface – in more general terms where the solid crystalline phase is interacting with a system such as a fluid phase whose molecules are not coherently related as to position. Another such incoherent system would be a (mobile) set of defects in a crystal.

4.1. Coherent transformations

As was emphasised in chapter 2, the transformations discussed were simply mathematical transformations of space, and now we discuss how these transformations can be used to describe the properties and behaviour of deformed substances.

The simplest model of a material medium is that of a continuum, in which we associate a point or 'particle' of the continuum with every point in the space. Most books on elastic deformations take this model. The deformations of the material medium are isomorphic to the transformations of co-ordinates described in chapter 2. That is, any deformation of the medium is described by a unique transformation $\boldsymbol{x} \to \boldsymbol{X}$. This model presupposes that the 'points' or 'particles' of the material medium remain *coherently* related in space under deformations; that is, every point, which in the undeformed state is surrounded by points on a very small closed surface such as a sphere, always remains surrounded by the same points, which will now describe a small closed surface resulting from the deformation of the original surface. The term coherent deformation will be used in a similar sense for media described in terms of atomic or ionic lattices.

Anisotropy can be introduced into the model of a continuous medium via its macroscopic properties. As an example consider the description of thermal expansion, using the continuum model to describe a crystalline substance. Consider first the simple case of an isotropic crystal. If the crystal is held uniformly at temperature T_0 and then at T_0+t, the change in shape would be described by a transformation $\boldsymbol{x}\to\boldsymbol{X}$, which would be of the form

$$\boldsymbol{X}=\boldsymbol{E}(1+t\alpha)\boldsymbol{x} \tag{4.1.1}$$

where α is the only linear coefficient required, and $\boldsymbol{E}$ is the unit matrix.

For the more general anisotropic type of thermal expansion, which is still linear in the temperature increment, we would have

$$\boldsymbol{X}=\boldsymbol{x}+t\boldsymbol{\tau}\boldsymbol{x} \tag{4.1.2}$$

where the tensor $\boldsymbol{\tau}$ describes the change per degree. Symmetry can be 'fed into' the model of the material continuum by choosing $\boldsymbol{\tau}$ to be invariant to certain symmetry operators, e.g. to the operations of a crystal point group. Similarly, anisotropy can be introduced into the treatment of elasticity by requiring that the elastic constants be invariant, not to all orthogonal transformations, but only to those corresponding to a crystal point group.

Thus, with the continuum model, anisotropy is introduced in a macroscopic way. In practice this leads to no difficulties for such phenomena as thermal expansion, pure deformations or coherent phase transitions which, neglecting crystal defect production, can be regarded as coherent processes in the sense used above.

Other phenomena which must be fitted into the general framework are the diffusion of a fluid component into a stressed (deformed) solid, and crystallisation or solution at the surface of such a solid. Why is a transformation $\boldsymbol{x}\to\boldsymbol{X}$ required and what should it be?

The approach taken here is that of a physicist who recognises that material media have microscopic structure and that the deformations of a material under chemical or other changes are related to this structure. For example, in the case of diffusion of a gas into a stressed metal or alloy, such as hydrogen into palladium, the thermodynamic treatment would be based on co-ordinates derived from coherent deformations of the metallic lattice. The lattice could be a Bravais lattice, in which the lattice vectors $\boldsymbol{a}$, $\boldsymbol{b}$, $\boldsymbol{c}$ are defined by the relative positions of certain ions (fig. 6). If hydrogen is forced into the metal which is under stress, a change of shape can be produced coherently, such that lattice metal neighbours remain neighbours. Now in a Bravais lattice the cells can be numbered by three integers l_a, l_b, l_c such that the position vector

$$\boldsymbol{x}_{\mathrm{F}}=l_a\boldsymbol{a}+l_b\boldsymbol{b}+l_c\boldsymbol{c} \tag{4.1.3}$$

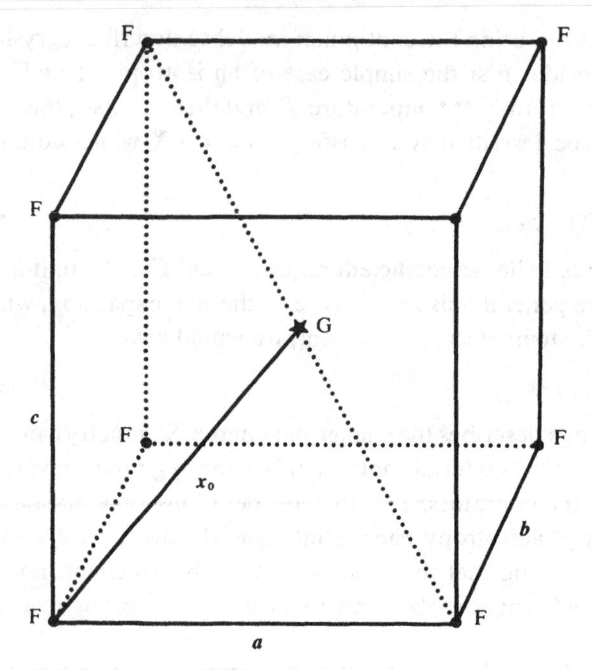

Fig. 6. A Bravais lattice unit cell.

determines the co-ordinates of that vertex of the cell which is closest to the origin. This vertex could be the position of an F-atom or ion (say); if, as in fig. 6, the crystal has a G-atom or ion at the cell centre, the co-ordinate $\boldsymbol{x}_{\mathrm{G}}$ of the G-atom or ion would be

$$\boldsymbol{x}_{\mathrm{G}} = \boldsymbol{x}_{\mathrm{F}} + \boldsymbol{x}_0. \tag{4.1.4}$$

If, in a coherent process such as diffusion or thermal expansion,

$$\boldsymbol{a} \to \boldsymbol{a}', \qquad \boldsymbol{b} \to \boldsymbol{b}', \qquad \boldsymbol{c} \to \boldsymbol{c}', \qquad \boldsymbol{x}_0 \to \boldsymbol{x}_0' \tag{4.1.5}$$

then

$$\boldsymbol{x}_{\mathrm{F}} \to \boldsymbol{X}_{\mathrm{F}}, \qquad \boldsymbol{x}_{\mathrm{G}} \to \boldsymbol{X}_{\mathrm{G}},$$

where

$$\boldsymbol{X}_{\mathrm{F}} = l_a\boldsymbol{a}' + l_b\boldsymbol{b}' + l_c\boldsymbol{c}', \tag{4.1.6}$$

and

$$\boldsymbol{X}_{\mathrm{G}} - \boldsymbol{x}_{\mathrm{G}} = \boldsymbol{X}_{\mathrm{F}} - \boldsymbol{x}_{\mathrm{F}} + (\boldsymbol{x}_0' - \boldsymbol{x}_0). \tag{4.1.7}$$

If $\boldsymbol{X}_{\mathrm{F}} - \boldsymbol{x}_{\mathrm{F}}$ represents a macroscopically measurable change of dimensions, we can neglect $\boldsymbol{x}_0' - \boldsymbol{x}_0$, which is of microscopic magnitude. Thus we can take

$$\boldsymbol{X}_G - \boldsymbol{x}_G = \boldsymbol{X}_F - \boldsymbol{x}_F. \tag{4.1.8}$$

Equation (4.1.6) may be elegantly expressed as a homogeneous transformation using the set of vectors $\boldsymbol{A}$, $\boldsymbol{B}$, $\boldsymbol{C}$, which are reciprocal to $\boldsymbol{a}$, $\boldsymbol{b}$ and $\boldsymbol{c}$, and defined by

$$\boldsymbol{A}\cdot\boldsymbol{a} = 1, \qquad \boldsymbol{A}\cdot\boldsymbol{b} = \boldsymbol{A}\cdot\boldsymbol{c} = 0, \tag{4.1.9}$$

with cyclic interchange to obtain the remaining six relations. Such reciprocal vectors are of great use in crystallography, where orthogonal axis systems are not used. If $\boldsymbol{a}$, $\boldsymbol{b}$ and $\boldsymbol{c}$ are independent vectors then any vector in three dimensions is expressible as

$$\boldsymbol{x} = m_a\boldsymbol{a} + m_b\boldsymbol{b} + m_c\boldsymbol{c}, \tag{4.1.10}$$

and, using (4.1.9), it is easy to see that

$$m_a = (\boldsymbol{x}\cdot\boldsymbol{A}); \qquad m_b = (\boldsymbol{x}\cdot\boldsymbol{B}); \qquad m_c = (\boldsymbol{x}\cdot\boldsymbol{C}), \tag{4.1.11}$$

i.e., the reciprocal vectors have been used in solving the three linear equations (4.1.10) for the unknowns m_a, m_b, m_c.

Thus

$$\boldsymbol{x} = (\boldsymbol{aA} + \boldsymbol{bB} + \boldsymbol{cC})\cdot\boldsymbol{x}, \tag{4.1.12}$$

and since $\boldsymbol{x}$ is an arbitrary vector

$$\boldsymbol{aA} + \boldsymbol{bB} + \boldsymbol{cC} = \boldsymbol{E}, \tag{4.1.13}$$

the unit matrix in three dimensions.

Hence, from (4.1.3), (4.1.6) and (4.1.8), the homogeneous transformation which describes the lattice shape change is given by

$$\boldsymbol{X} = (\boldsymbol{a}'\boldsymbol{A} + \boldsymbol{b}'\boldsymbol{B} + \boldsymbol{c}'\boldsymbol{C})\cdot\boldsymbol{x}. \tag{4.1.14}$$

For example,

$$\boldsymbol{X}_G = (\boldsymbol{a}'\boldsymbol{A} + \boldsymbol{b}'\boldsymbol{B} + \boldsymbol{c}'\boldsymbol{C})\cdot\boldsymbol{x}_G, \tag{4.1.15}$$

and similarly for the F vertices. It is easily seen that (4.1.14) describes the change of macroscopic shape of any part of the crystal. In such a description, however, the correspondence of transformation to change of shape is not one-to-one. We can choose any transformation which gives the correct displacements of the lattice atoms or ions, but the simplest choice is always adequate. For instance, in the above example, a homogeneous transformation will do all that is required.

In this treatment, we have not assumed that the shape changes were infinitesimal, and so the same description can be used for the α–β quartz transition which occurs via small displacements or rotations of the SiO_4 tetrahedra such that neighbours remain neighbours.

4.2. Incoherent transformations

The extreme case of incoherence is that of a fluid, such as a gas, where it is obviously impossible to relate average positions of individual atoms or molecules in any meaningful way to points in space. In thermodynamics, the only interest one has in the deformations of a fluid is in the change in shape or dimensions of the fluid mass, leading to such mechanical co-ordinates as volume V and surface area A.

Crystallisation or solution at the surface of a stressed solid which is in contact with a solution of the solid is also an incoherent process. The thermodynamic treatment of this was given by Gibbs (1906). In this case, the solid phase may have been coherently deformed from some standard state, and then have been altered in mass by crystallisation or solution. While the deformation of the phase in the first process may be described by a coherent transformation, the change of shape and mass of the solid phase in the second process cannot be so described. If mass is added by crystallisation at part of its surface there is an inhomogeneous change of shape (see fig. 7). If the stress throughout the solid remains homogeneous, the atomic or ionic lattice deformation remains constant. There have been changes in shape and volume, however, and these are important thermodynamically, since they determine the work done by the phase against its surroundings when the crystallisation or solution proceeds. This work contributes to the condition for chemical equilibrium, and, in

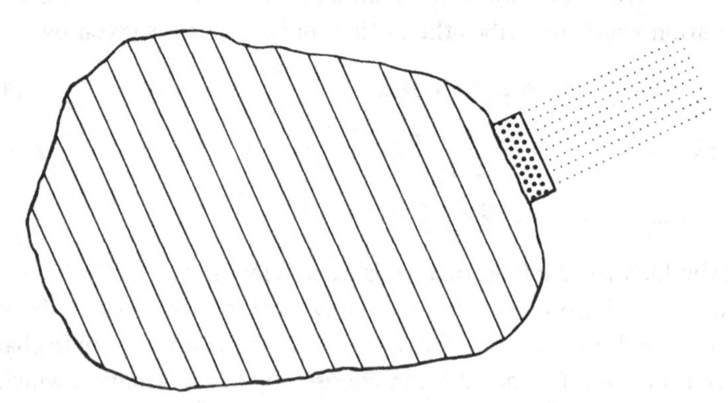

Fig. 7. An inhomogeneous change of shape arising from a localised crystallisation, from a solution in contact with the solid only over a small area.

hydrostatic cases, is the origin of the term pv in the chemical potential $\mu = u - Ts + pv$, where u, s and v are respectively the specific internal energy, specific entropy and specific volume for the phase. This change of shape may be described by many transformations and we shall show that, in thermodynamics, we must use a transformation which gives the correct work term as above, as well as the change of shape. As this transformation is in no way related to the deformation of the atomic or ionic solid lattice, but only to its external shape, we are justified in calling it an incoherent transformation and in general it will not be a homogeneous transformation. This example will be treated in more detail in chapter 21.

Production in a crystal of such defects as Schottky defects, and edge and screw dislocations may be an incoherent process. The production and distribution of such defects will depend, amongst other things, on non-hydrostatic stresses. If reversible processes can be associated with the production of defects, it should be possible to deal with their thermodynamic properties. In this book we shall be discussing the thermodynamics of solids under homogeneous conditions, and it will thus be necessary to assume the possibility of uniform production of defects in a homogeneous solid, in such a way that the solid remains homogeneous. This would be somewhat analogous to the case of diffusion into a solid when one assumes the diffusion to be homogeneous. The simplest defects are known as Schottky defects (see, e.g., Kittel, 1971). These defects are lattice vacancies, and the production of these could clearly be treated as analogous to diffusion. The production of colour centres or of Frenkel defects, in which an atom or ion is transferred from a lattice site to an interstitial site, could also be treated in a similar way.

Provided the production of these simple types of defects may be assumed or idealised to be a thermodynamically reversible process, the thermodynamic properties will depend on the energy required to produce a homogeneous distribution of such defects – and this will depend on the heat of production and the homogeneous change of shape of the solid. It may not be possible to discuss more complicated types of defects, such as edge and screw dislocations and slip, in this way. The difficulty could arise in two ways – the production might be irreversible or it might be impossible to conceive of the homogeneous production of such defects.

Summary of important equations

A. *The position vectors of the lattice points* F *and* G

$$\boldsymbol{x}_{\mathrm{F}} = l_a\boldsymbol{a} + l_b\boldsymbol{b} + l_c\boldsymbol{c}, \tag{4.1.3}$$

$$\boldsymbol{x}_{\mathrm{G}} = \boldsymbol{x}_{\mathrm{F}} + \boldsymbol{x}_0. \tag{4.1.4}$$

B. *The transformations of the lattice vectors and points*

$$\boldsymbol{a} \to \boldsymbol{a}', \qquad \boldsymbol{b} \to \boldsymbol{b}', \qquad \boldsymbol{c} \to \boldsymbol{c}', \qquad \boldsymbol{x}_0 \to \boldsymbol{x}_0', \tag{4.1.5}$$

$$\boldsymbol{X}_F = l_a\boldsymbol{a}' + l_b\boldsymbol{b}' + l_c\boldsymbol{c}', \tag{4.1.6}$$

$$\boldsymbol{X}_G - \boldsymbol{x}_G = \boldsymbol{X}_F - \boldsymbol{x}_F + (\boldsymbol{x}_0' - \boldsymbol{x}_0) \tag{4.1.7}$$

$$= \boldsymbol{X}_F - \boldsymbol{x}_F, \tag{4.1.8}$$

on neglecting $\boldsymbol{x}_0' - \boldsymbol{x}_0$.

C. *The reciprocal vectors,* $\boldsymbol{A}$, $\boldsymbol{B}$, $\boldsymbol{C}$, *of* $\boldsymbol{a}$, $\boldsymbol{b}$, $\boldsymbol{c}$

$$\boldsymbol{A} \cdot \boldsymbol{a} = 1, \qquad \boldsymbol{A} \cdot \boldsymbol{b} = \boldsymbol{A} \cdot \boldsymbol{c} = 0, \tag{4.1.9}$$

with cyclic interchange for $\boldsymbol{B}$ and $\boldsymbol{C}$.

$$\boldsymbol{aA} + \boldsymbol{bB} + \boldsymbol{cC} = \boldsymbol{E}. \tag{4.1.13}$$

The homogeneous transformation of (4.1.6) and (4.1.8) is described by

$$\boldsymbol{X} = (\boldsymbol{a}'\boldsymbol{A} + \boldsymbol{b}'\boldsymbol{B} + \boldsymbol{c}'\boldsymbol{C}) \cdot \boldsymbol{x}. \tag{4.1.14}$$

Examples

1.(*a*) Prove that the reciprocal vectors of $\boldsymbol{a}$, $\boldsymbol{b}$ and $\boldsymbol{c}$ are given by

$$\boldsymbol{A} = (\boldsymbol{b} \wedge \boldsymbol{c})/(\boldsymbol{a} \cdot \boldsymbol{b} \wedge \boldsymbol{c})$$

with cyclic interchange of $\boldsymbol{a}$, $\boldsymbol{b}$, $\boldsymbol{c}$ to obtain $\boldsymbol{B}$ and $\boldsymbol{C}$ from this expression.

(*b*) If $\boldsymbol{a}$, $\boldsymbol{b}$ and $\boldsymbol{c}$ are a set of orthogonal vectors, prove that $\boldsymbol{A}$ is of modulus a^{-1}, and is in the direction of $\boldsymbol{a}$. Note that $\boldsymbol{A}$ has dimensions reciprocal to those of $\boldsymbol{a}$ ($a = |\boldsymbol{a}|$).

(*c*) If $\boldsymbol{a}$, $\boldsymbol{b}$, $\boldsymbol{c}$ are an orthonormal set, prove that $\boldsymbol{A} = \boldsymbol{a}$ etc.

(*d*) Prove that the matrix whose rows are the vectors $\boldsymbol{A}$, $\boldsymbol{B}$, $\boldsymbol{C}$, respectively is the reciprocal of that matrix whose columns are the vectors $\boldsymbol{a}$, $\boldsymbol{b}$, $\boldsymbol{c}$, respectively.

2.(*a*) What is the reciprocal matrix of

$$\boldsymbol{a}'\boldsymbol{A} + \boldsymbol{b}'\boldsymbol{B} + \boldsymbol{c}'\boldsymbol{C},$$

where $\boldsymbol{A}$, $\boldsymbol{B}$, $\boldsymbol{C}$ are the reciprocal vectors of $\boldsymbol{a}$, $\boldsymbol{b}$, $\boldsymbol{c}$?

(*b*) Evaluate the strain tensor corresponding to the homogeneous transformation described by the above matrix.

(*c*) Prove that the determinant of the transformation is $(\boldsymbol{a}' \cdot \boldsymbol{b}' \wedge \boldsymbol{c}')/(\boldsymbol{a} \cdot \boldsymbol{b} \wedge \boldsymbol{c})$, and discuss the result.

3. Discuss the eigenvalues and eigenvectors of

$$\lambda \boldsymbol{aA} + \mu \boldsymbol{bB} + \nu \boldsymbol{cC}.$$

What sort of transformation does it describe in general? When does it describe a symmetric transformation?

4. A homogeneous transformation is described by a matrix $\boldsymbol{T}$. If this is

expressible, as in example 2, in the form $\boldsymbol{a}'\boldsymbol{A}+\boldsymbol{b}'\boldsymbol{B}+\boldsymbol{c}'\boldsymbol{C}$, show that this form of $\boldsymbol{T}$ is not unique in its choice of vectors. For example, show how to express $\boldsymbol{T}$ in the forms $\boldsymbol{a}''\boldsymbol{i}+\boldsymbol{b}''\boldsymbol{j}+\boldsymbol{c}''\boldsymbol{k}$ and $\boldsymbol{i}\boldsymbol{A}''+\boldsymbol{j}\boldsymbol{B}''+\boldsymbol{k}\boldsymbol{C}''$, where $\boldsymbol{i}, \boldsymbol{j}, \boldsymbol{k}$ are the unit vectors in the directions of the x-, y-, and z-axes respectively; discuss the meanings of the vectors $\boldsymbol{a}''$, $\boldsymbol{b}''$, $\boldsymbol{c}''$ and $\boldsymbol{A}''$, $\boldsymbol{B}''$ and $\boldsymbol{C}''$ in these two forms; and if $\boldsymbol{T}$ is symmetric show that these two sets of vectors are identical.

5. Crystalline graphite has a hexagonal structure. The thermal expansion coefficient in the direction of the c-axis, the axis of hexagonal symmetry, is large and positive and approximately constant at $\alpha_c = 28.3\times10^{-6}\,\mathrm{K}^{-1}$, whereas the thermal expansion coefficient α_a at right angles to the c-axis, i.e., in the basal graphite planes, varies from small negative to small positive values, as in the following:

Temp.	α_a (K^{-1})
0 °C	-1.3×10^{-6}
400 °C	0
800 °C	$+0.8\times10^{-6}$

(The above figures are taken from Nelson & Riley (1945).) Evaluate the volume expansion coefficients for the three temperatures given above.

6.(*a*) A homogeneous transformation is given by

$$\boldsymbol{V} = \lambda\boldsymbol{a}\boldsymbol{A} + \mu\boldsymbol{b}\boldsymbol{B} + \nu\boldsymbol{c}\boldsymbol{C},$$

where $\boldsymbol{A}, \boldsymbol{B}, \boldsymbol{C}$ are the vectors reciprocal to $\boldsymbol{a}, \boldsymbol{b}$, and $\boldsymbol{c}$ respectively, and λ, μ, ν are infinitesimally close to 1. Show that $\boldsymbol{V}$ describes an infinitesimal deformation. Find the skew-symmetric and symmetric parts of $\boldsymbol{V}$ and hence determine the polar factorisation of $\boldsymbol{V}$.

(*b*) Discuss the particular example where $\boldsymbol{c}$ is orthogonal to $\boldsymbol{a}$ and $\boldsymbol{b}$, showing for instance that the rotational part of $\boldsymbol{V}$ is a rotation about $\boldsymbol{c}$.

7.(*a*) If $\boldsymbol{i}, \boldsymbol{j}, \boldsymbol{k}$ and $\boldsymbol{i}', \boldsymbol{j}', \boldsymbol{k}'$ are two sets of orthonormal vectors show that $\boldsymbol{i}\boldsymbol{i}'+\boldsymbol{j}\boldsymbol{j}'+\boldsymbol{k}\boldsymbol{k}'$ represents an orthogonal transformation. Describe the transformation.

(*b*) Show that the transformation

$$\boldsymbol{T} = \boldsymbol{i}\boldsymbol{X} + \boldsymbol{j}\boldsymbol{Y} + \boldsymbol{k}\boldsymbol{Z}$$

where $\boldsymbol{i}, \boldsymbol{j}, \boldsymbol{k}$ are orthonormal, and $\boldsymbol{X}$, $\boldsymbol{Y}$ and $\boldsymbol{Z}$ are arbitrary vectors, may be written as

$$(\boldsymbol{i}\boldsymbol{i}'+\boldsymbol{j}\boldsymbol{j}'+\boldsymbol{k}\boldsymbol{k}')\cdot(\boldsymbol{i}'\boldsymbol{X}+\boldsymbol{j}'\boldsymbol{Y}+\boldsymbol{k}'\boldsymbol{Z}),$$

and show that if this gives the polar factorisation of $\boldsymbol{T}$ then

$$X_{y'} = Y_{x'}, \qquad Y_{z'} = Z_{y'}, \qquad Z_{x'} = X_{z'}$$

where $Y_{x'} = (\boldsymbol{i}'\cdot\boldsymbol{Y})$, $X_{y'} = (\boldsymbol{j}'\cdot\boldsymbol{X})$ etc., are the components of $\boldsymbol{X}$, $\boldsymbol{Y}$, $\boldsymbol{Z}$ referred to the axis system $\boldsymbol{i}', \boldsymbol{j}', \boldsymbol{k}'$ as x', y', z' axes respectively.

(*c*) Apply the method indicated in (*b*) to find the polar factorisation for a simple shear transformation $\boldsymbol{V}$ given by

$$\boldsymbol{V} = \boldsymbol{i}\boldsymbol{i} + \boldsymbol{j}(\boldsymbol{j}+\beta\boldsymbol{i}) + \boldsymbol{k}\boldsymbol{k} = \boldsymbol{E} + \beta\boldsymbol{j}\boldsymbol{i}.$$

Show in particular that the orthogonal factor represents a rotation α, where $\tan\alpha = \frac{1}{2}\beta$ (see fig. 4).

5

Forces

A material body is under stress if externally applied forces act on it. Such forces are known as extrinsic forces. They may be applied at the surface of the body, e.g., by a piston, and are then known as extrinsic surface forces. Long range forces, such as those of gravity, arising from an external source are known as extrinsic body forces. For a body in equilibrium under such forces, internal forces must be brought into play to keep any part of the body at equilibrium. A part V of the body which is entirely inside the body is considered, and the forces, acting across the surface of V on material inside V and produced by material in the body outside V, are assumed to be of vanishingly small range. In this case they can be described as surface forces. With this assumption the equation of motion of the material in V is derived. It is then shown that the surface forces can be described in terms of a stress tensor. The symmetry of this stress tensor – the Cauchy stress tensor – is discussed in relation to the existence of body torque. The equation of motion is discussed in terms of this stress tensor.

5.1. The stress vector

A material body is said to be under stress if forces are applied to the outside surfaces of the body, in addition to the long range forces from external sources such as the gravitational force from the earth. Such forces are known as extrinsic forces; those acting on the surface of the body from some external agent, such as a piston operated by a hydraulic press, are known as *extrinsic surface forces*; those long range forces, such as gravity, arising from an external source are known as *extrinsic body forces*, and, since each volume element of the body is subject to such a force they are sometimes termed *volume forces*. As an example, consider a rod held vertically, clamped at its top end to a rigid support, while at its bottom end is clamped a heavy load of some kind. Considering it alone as the material body mentioned above, the rod is subject to the following extrinsic surface forces:

(*a*) The force exerted over its clamped surface at the bottom, due to the weight of the load. The total force from the weight will be distributed in some way over the surface of contact between the rod and the clamping arrangement (the rod may have a hole in it through which a bolt is fastened and from which the load is suspended).

(*b*) The force exerted by the clamp at the top end of the rod. Again the total force would be distributed over the surface of contact between rod and clamp.

The rod is also subject to the extrinsic body force of gravity, which arises because the rod has weight, that is, each volume element experiences the pull of the field of the earth.

These extrinsic forces would rupture the rod if it had no strength but this tendency is counteracted by the internal forces of the rod. In fact, in this case it can be seen that the internal forces, acting on the material below a cross-section of the rod and arising from the material above the cross-section, will be such as to counterbalance exactly the extrinsic forces acting on the material of the rod below the cross-section and on the surfaces below the cross-section. In this case the total such internal force will be equal and opposite to the total weight of the rod material and load below the cross-section. If a cut were made across the rod at this cross-section, then in order to maintain the equilibrium of the part of the system below the cut an external mechanical force would have to be applied to it at the cross-section surface equal to the internal force just described. For such a picture we see that the effect of the extrinsic forces is transmitted by the internal forces. The equilibrium of the rod under the influence of gravity on load and rod and under the reaction force of the top clamp is brought about by the action of the internal forces in keeping the rod in one piece. Finally we may remark that the dimensions of a real rod would depend on the magnitude of the load: i.e., the rod is elastic.

Now consider a volume V *entirely* within such a body under the influence of extrinsic forces. Each volume element $\mathrm{d}V$ of V experiences an extrinsic body force, of the form $\boldsymbol{f}^{\mathrm{b}}\,\mathrm{d}V$, where $\boldsymbol{f}^{\mathrm{b}}$ is the force per unit volume (not per unit mass). For example, if gravity is the only such influence, then

$$\boldsymbol{f}^{\mathrm{b}} = \rho\boldsymbol{g}, \tag{5.1.1}$$

where $\boldsymbol{g}$ is the acceleration vector of the gravity field of the earth, and ρ the mass density, i.e., the mass per unit volume.

We assume the extrinsic surface forces to be effectively of zero range. As the term surface force implies, they are contact forces. Since the region V is entirely within the body and away from the external surfaces, these forces do not act on the material in V. Besides the extrinsic body force of (5.1.1), the only other forces acting on the material in V arise from forces exerted by the material surrounding V. We assume such forces also to be effectively of zero range.

Consider a surface element, $\mathrm{d}\boldsymbol{\Sigma} = \boldsymbol{N}\,\mathrm{d}\Sigma$, of the surface of V. $\boldsymbol{N}$ is the unit normal at this surface element, drawn outward from V. Since the forces are of zero range, only that matter outside V immediately adjacent to $\mathrm{d}\boldsymbol{\Sigma}$ will exert a force on material in V immediately adjacent to $\mathrm{d}\boldsymbol{\Sigma}$. This force is clearly proportional to the area of the surface element, and so we may denote it by $\boldsymbol{t}\,\mathrm{d}\Sigma$, where $\boldsymbol{t}$ is known as the surface force density or as the stress vector. As we have defined it, $\boldsymbol{t}\,\mathrm{d}\Sigma$ is the force exerted *by* material on the exterior side of $\mathrm{d}\boldsymbol{\Sigma}$ *on* the material on the interior side.

Hence, the force on material in V which arises from the surrounding material is given by

$$\int_{\Sigma} \boldsymbol{t}\,\mathrm{d}\Sigma. \tag{5.1.2}$$

For the material in the closed region V inside the body, the equation of motion is given by

$$\int_{\Sigma} \boldsymbol{t}\,\mathrm{d}\Sigma + \int_{V} \boldsymbol{f}^{\mathrm{b}}\,\mathrm{d}V = \int_{V} (\rho\,\mathrm{d}V)\frac{\mathrm{D}\boldsymbol{U}}{\mathrm{D}t}, \tag{5.1.3}$$

where the right-hand side is the rate of change of momentum of the material inside V and needs some explanation. The first integral is integrated over Σ, the surface of V; the other two throughout V. $\boldsymbol{U}$ describes the velocities of the different points of the medium and is known as the vector velocity field.

It describes the velocity in a particular way which should be carefully noted: $\boldsymbol{U}(\boldsymbol{X}, t)$ is the velocity of points or particles passing through the point $\boldsymbol{X}$ at time t. Thus $\boldsymbol{U}$ does not describe the velocity of one particular particle, but is the velocity of that particle which is at $\boldsymbol{X}$ at time t. To obtain the acceleration of a particle we must measure its change of velocity in an infinitesimal time $\mathrm{d}t$. For the particle at $\boldsymbol{X}$ at time t, its position at time $t+\mathrm{d}t$ will be $\boldsymbol{X}+\boldsymbol{U}\,\mathrm{d}t$ and hence its velocity will be $\boldsymbol{U}(\boldsymbol{X}+\boldsymbol{U}\,\mathrm{d}t, t+\mathrm{d}t)$. Its acceleration $\boldsymbol{\alpha}$ will be such that, for example,

$$\begin{aligned}\alpha_x = \lim_{\mathrm{d}t\to 0} \{&U_x(X+U_x\,\mathrm{d}t, Y+U_y\,\mathrm{d}t, Z+U_z\,\mathrm{d}t, t+\mathrm{d}t)\\ &- U_x(X, Y, Z, t)\}/\mathrm{d}t,\end{aligned} \tag{5.1.4}$$

where all the arguments of the velocity field vector component U_x have been written out. On expanding the first term in the curly bracket to first order in $\mathrm{d}t$ it is easy to see that

$$\alpha_x = \frac{\mathrm{D}U_x}{\mathrm{D}t} = \frac{\partial U_x}{\partial t} + U_x \frac{\partial U_x}{\partial X} + U_y \frac{\partial U_x}{\partial Y} + U_z \frac{\partial U_x}{\partial Z} \tag{5.1.5}$$

or in vector notation

$$\boldsymbol{\alpha} = \frac{\mathrm{D}}{\mathrm{D}t}\boldsymbol{U} = \frac{\partial \boldsymbol{U}}{\partial t} + U_\beta \frac{\partial \boldsymbol{U}}{\partial X_\beta} \tag{5.1.6}$$

where

$$\frac{\mathrm{D}}{\mathrm{D}t} = \left(\frac{\partial}{\partial t} + \boldsymbol{U} \cdot \frac{\partial}{\partial \boldsymbol{X}}\right) \tag{5.1.7}$$

is known as the hydrodynamic derivative, and can be applied to other quantities of the flow. For instance, the density field could be $\rho(\boldsymbol{X}, t)$ where ρ is the density at time t at the point $\boldsymbol{X}$. In fluid flow, we could follow the motion of a mass element of the fluid, for example, by adding a small blob of dye. In a solid medium, in which the velocity field arises from an acoustic vibration, we could similarly follow the motion of a mass element, which would be sufficiently identified by its rest position and shape in some standard state. In such cases, the density of a mass element would have meaning and its rate of change would be

$$\frac{\mathrm{D}}{\mathrm{D}t}\rho = \left(\frac{\partial}{\partial t} + \boldsymbol{U} \cdot \frac{\partial}{\partial \boldsymbol{X}}\right)\rho \tag{5.1.8}$$

where $\partial\rho/\partial t$ would mean the rate of change of density at the point $\boldsymbol{X}$. Thus the hydrodynamic derivative can be of use when we follow a particle or mass element in its actual motion.

We note, in (5.1.3), that $\rho\,\mathrm{d}V$, being the mass of a volume element, remains constant, although the volume element may be changing shape.

After this rather long aside, we return, in the next section, to the stress vector $\boldsymbol{t}$ and use (5.1.3) to show that it may be described by a tensor of second rank to be known as the Cauchy stress tensor $\boldsymbol{\Theta}$.

We conclude this section by remarking that if the internal forces were not assumed to be of infinitesimally short range it would be impossible to describe them by the stress vector $\boldsymbol{t}$, since forces of finite range could not be made up of contributions $\boldsymbol{t}\,\mathrm{d}\Sigma$. This can be seen because this assumption requires that the force, due to particles outside a closed region, acting on a particle *immediately* on the inside of $\mathrm{d}\Sigma$ arises solely from particles *immediately* outside $\mathrm{d}\Sigma$. If this assumption cannot be met, as with magnetic and electrostatic forces, the above description breaks down and a stress vector cannot, strictly speaking, be used. Reasonably satisfactory treatments in these cases can be given by making suitable assumptions as to the division of internal forces between short range forces and long range forces. A very good discussion of this is given by R. R. Birss (1967).

5.2. The stress tensor

First, we state an enormously useful mathematical theorem in its simplest and most powerful form. If ϕ is a continuous function of the position co-ordinates X_α then the theorem states that

$$\int_\Sigma \phi N_\alpha \,\mathrm{d}\Sigma = \int_V (\partial\phi/\partial X_\alpha)\,\mathrm{d}V \tag{5.2.1}$$

where the integration on the left-hand side is over the surface Σ of a closed region V, N_α is the αth component of $\boldsymbol{N}$, the unit outward drawn normal at the surface element $\mathrm{d}\Sigma$. The integration on the right-hand side is throughout the volume V. The theorem is known sometimes as Green's First Theorem.

An immediate consequence of (5.2.1), known as the Divergence Theorem, can be derived by setting ϕ equal to the component f_α of a vector $\boldsymbol{f}$. Summing over α gives

$$\int_\Sigma \boldsymbol{f}\cdot\boldsymbol{N}\,\mathrm{d}\Sigma = \int_V \left(\frac{\partial}{\partial \boldsymbol{X}}\cdot\boldsymbol{f}\right)\mathrm{d}V. \tag{5.2.2}$$

If ϕ is a scalar function, (5.2.1) becomes in vector notation

$$\int_\Sigma \phi\boldsymbol{N}\,\mathrm{d}\Sigma = \int_V \frac{\partial}{\partial \boldsymbol{X}}\phi\,\mathrm{d}V. \tag{5.2.3}$$

Equation (5.2.1) is very easy to prove for a simple region V, such as one shaped like an egg, in which any straight line which intersects the surface Σ does so at two points only. This proof may easily be generalised for more complicated regions. We shall prove it here for the simple region mentioned, and leave it to the reader to generalise. It may also be generalised for such cases as those for which ϕ has a finite discontinuity over a surface intersecting the region – such generalisations are required in electrostatics for example.

Proof of (5.2.1). Divide the region V into two parts on either side of a plane normal to the X-axis. Through the surface element $\mathrm{d}\Sigma^{(+)}$ (the superscript denotes that it is on the positive X-side of the YOZ plane) draw a cylinder whose axis is in the X-direction, as shown in fig. 8. This cylinder intersects Σ again at the surface element $\mathrm{d}\Sigma^{(-)}$ where the unit

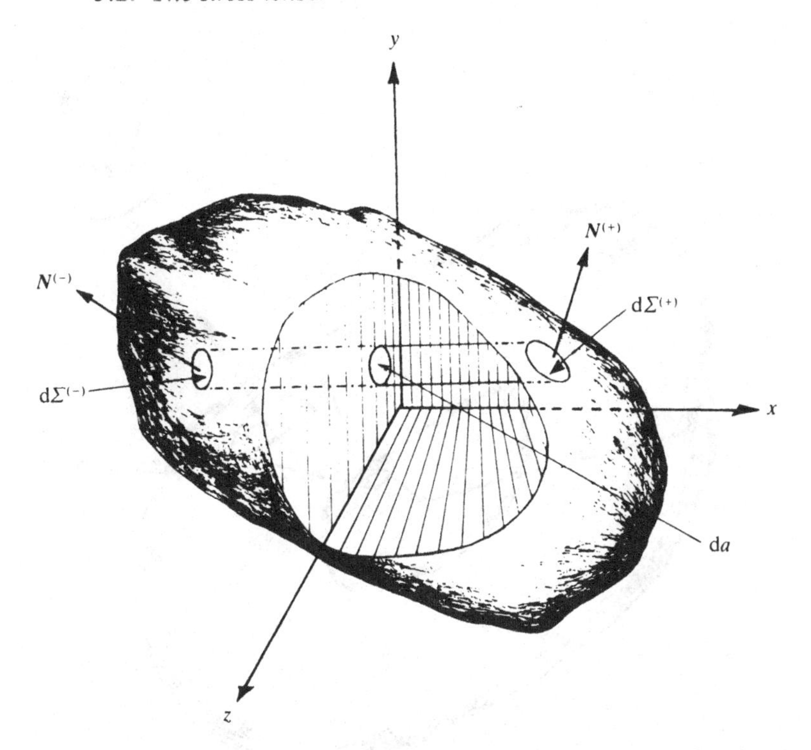

Fig. 8. This figure shows how the integration through V is related to the integration over the section with the YOZ plane.

outward normal is $\boldsymbol{N}^{(-)}$. If we integrate $(\partial\phi/\partial X)$ throughout the volume of this cylinder, it is easy to see that

$$\int_{(\text{cylinder})} (\partial\phi/\partial X)\,\mathrm{d}V = \int_{(\text{cylinder})} (\partial\phi/\partial X)\,\mathrm{d}X\,\mathrm{d}a \tag{5.2.4}$$

where $\mathrm{d}a$ = projection of $\mathrm{d}\Sigma^{(+)}$ and $\mathrm{d}\Sigma^{(-)}$ on the YOZ plane. Equation (5.2.4) becomes, on integrating the right-hand side with respect to X,

$$\{\phi^{(+)} - \phi^{(-)}\}\,\mathrm{d}a, \tag{5.2.5}$$

where, for example, $\phi^{(+)}$ is the value of ϕ at $\mathrm{d}\Sigma^{(+)}$. On completing the integration over the whole region V, we obtain

$$\int \{\phi^{(+)} - \phi^{(-)}\}\,\mathrm{d}a, \tag{5.2.6}$$

where integration is over all such cylinders, i.e., $\mathrm{d}a$ is integrated over the cross-section in plane YOZ. Now

$$\mathrm{d}a = N_x^{(+)}\,\mathrm{d}\Sigma^{(+)} = -N_x^{(-)}\,\mathrm{d}\Sigma^{(-)}, \tag{5.2.7}$$

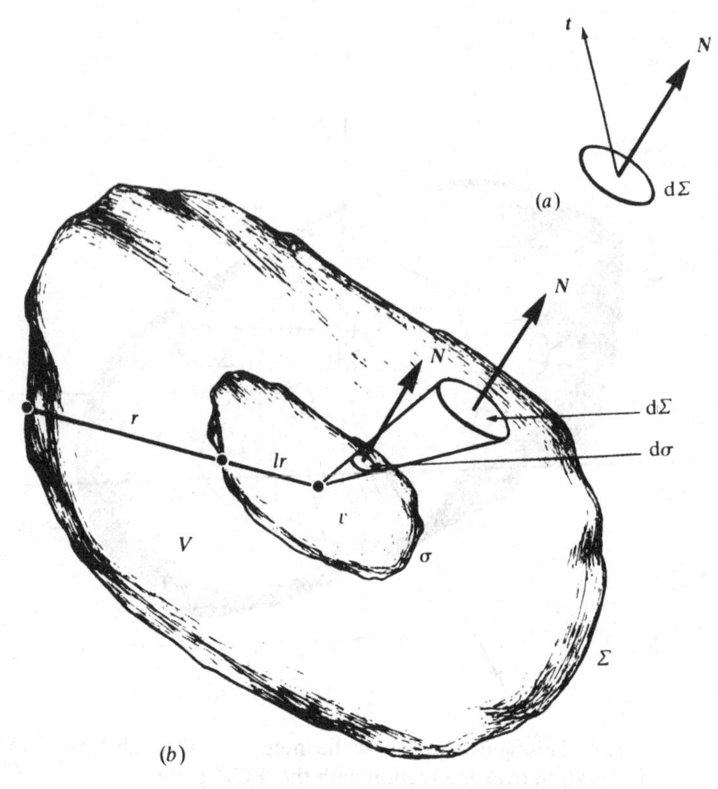

Fig. 9. (*a*) and (*b*). This figure demonstrates the shrinking process on *V*.

the minus sign arising from the 'outward normal' convention. Using (5.2.7) in (5.2.6) we see that

$$\int_V (\partial\phi/\partial X)\,\mathrm{d}V = \int_\Sigma \phi N_x\,\mathrm{d}\Sigma, \qquad (5.2.8)$$

where now $\mathrm{d}\Sigma$ is integrated over the whole surface, and we have thus proved (5.2.1).

We now have the mathematical equipment to deal with the stress tensor.

We return to the equation of motion (5.1.3), where the integrations are over the surface Σ of a closed region V. By application of a hypothetical isotropic shrinking procedure on V, we can prove that $\boldsymbol{t}$ is linearly related to a stress tensor. That is, we recognise that $\boldsymbol{t}\,\mathrm{d}\Sigma$, where $\mathrm{d}\Sigma$ is a surface element at a point, is dependent on the orientation of $\mathrm{d}\Sigma$, and so is dependent on the components of $\boldsymbol{N}$ (see fig. 9*a*). $\boldsymbol{t}$ will thus be a function of position $\boldsymbol{X}$ and orientation $\boldsymbol{N}$ and, to express this, it can be written as

$\boldsymbol{t}(X, Y, Z, N_x, N_y, N_z)$. We shall show that

$$\boldsymbol{t} = \boldsymbol{\Theta N} = \boldsymbol{\Theta}(X, Y, Z)\boldsymbol{N} \tag{5.2.9}$$

where $\boldsymbol{\Theta}$ is a (stress) tensor and is a function of position (and of time t if time-dependent processes are occurring).

For convenience, take an origin in V and consider (5.1.3) applied to the region, obtained by transforming V by the transformation

$$\boldsymbol{x} = l\boldsymbol{X}, \tag{5.2.10}$$

where eventually we shall let $l \to 0$. Under this isotropic transformation V transforms to $v = l^3 V$; the bounding surface retains the same shape; $\mathrm{d}\Sigma$ transforms to $\mathrm{d}\sigma = l^2\,\mathrm{d}\Sigma$ but the unit normal remains unchanged in direction (see fig. 9*b*). Applying (5.1.3) to volume v

$$\int_\sigma \boldsymbol{t}(\boldsymbol{x}, \boldsymbol{N})\,\mathrm{d}\sigma + \int_v \boldsymbol{f}^{\mathrm{b}}(\boldsymbol{x})\,\mathrm{d}v = \int_v \rho(\boldsymbol{x})\frac{\mathrm{D}\boldsymbol{U}}{\mathrm{D}t}\,\mathrm{d}v \tag{5.2.11}$$

where $\boldsymbol{U}$ is also a function of $\boldsymbol{x}$, the argument t not being shown.

Transforming back to V, (5.2.11) becomes

$$\int_\Sigma \boldsymbol{t}(l\boldsymbol{X}, \boldsymbol{N})l^2\,\mathrm{d}\Sigma + l^3\int_V \boldsymbol{F}^{\mathrm{b}}(l\boldsymbol{X})\,\mathrm{d}V = 0 \tag{5.2.12a}$$

where

$$\boldsymbol{F}^{\mathrm{b}} = \boldsymbol{f}^{\mathrm{b}} - \rho\frac{\mathrm{D}\boldsymbol{U}}{\mathrm{D}t}. \tag{5.2.12b}$$

On dividing by l^2 and proceeding to the limit of $l = 0$, we obtain

$$\int_\Sigma \boldsymbol{t}(O, \boldsymbol{N})\,\mathrm{d}\Sigma = 0. \tag{5.2.13}$$

That is, since the origin O is an arbitrary point in V, we have proved that if we take the stress vector $\boldsymbol{t}$ at any point O, then the integral in (5.2.13) over any arbitrary closed surface Σ enclosing the point is zero, where at any surface element $\mathrm{d}\Sigma$ the integrand is taken as if that surface element with its orientation and area were at the point O.

Since (5.2.13) is true for any surface, we take a tetrahedron $PABC$ made up of any triangle ABC, and choose the other three triangular faces by the criterion that each is a projection of ABC on three mutually perpendicular planes (see fig. 10).

PA, PB, PC are mutually perpendicular, and hence $\boldsymbol{M}_1$, $\boldsymbol{M}_2$, $\boldsymbol{M}_3$, the unit vectors parallel respectively to these three edges, form an orthonormal set and are inward drawn normals to the tetrahedron. If Δ is

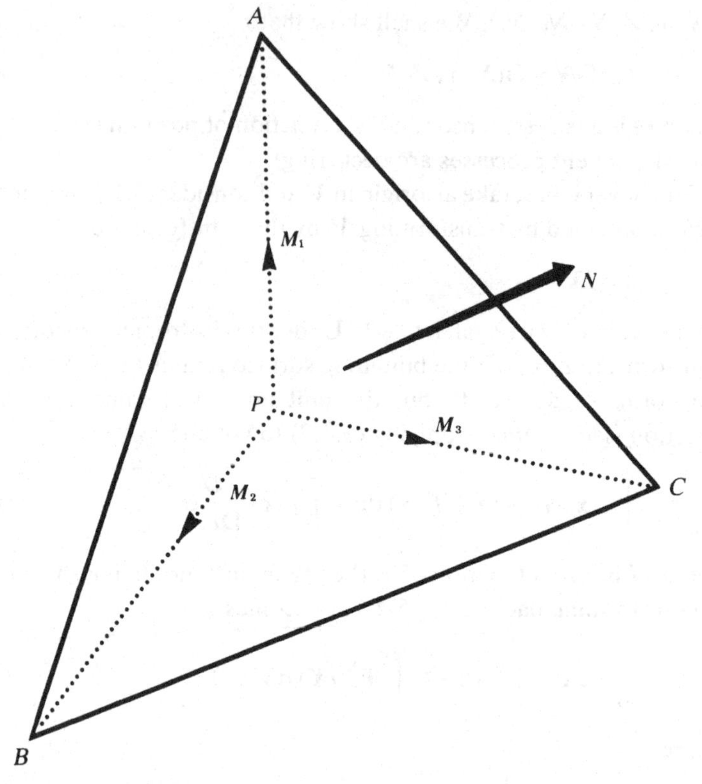

Fig. 10. The tetrahedron *PABC*, and the unit vectors $\boldsymbol{N}$, $\boldsymbol{M}_1$, $\boldsymbol{M}_2$, $\boldsymbol{M}_3$.

the area of the face *ABC* and if $\boldsymbol{N}$ is the outward unit normal to this face, then the areas of the faces *PBC*, *PCA*, *PAB* are, respectively, given by $\Delta(\boldsymbol{M}_1 \cdot \boldsymbol{N})$, $\Delta(\boldsymbol{M}_2 \cdot \boldsymbol{N})$, $\Delta(\boldsymbol{M}_3 \cdot \boldsymbol{N})$.

If O is any point inside *PABC*, if $\boldsymbol{t}$ is the stress vector at O for a surface element parallel to the face *ABC*, and if $\boldsymbol{t}_1$, $\boldsymbol{t}_2$, $\boldsymbol{t}_3$ are the stress vectors at O for surface elements parallel respectively to *PBC*, *PCA*, *PAB*, then on integrating (5.2.13) over the four faces of *PABC*, we obtain

$$\boldsymbol{t} - \boldsymbol{t}_1(\boldsymbol{M}_1 \cdot \boldsymbol{N}) - \boldsymbol{t}_2(\boldsymbol{M}_2 \cdot \boldsymbol{N}) - \boldsymbol{t}_3(\boldsymbol{M}_3 \cdot \boldsymbol{N}) = 0, \tag{5.2.14}$$

where use has been made of the expressions given above for the areas of the faces *PBC*, *PCA*, *PAB*.
Thus

$$\boldsymbol{t} = \boldsymbol{\Theta}\boldsymbol{N} \tag{5.2.15}$$

where, since $\boldsymbol{N}$ is arbitrary, $\boldsymbol{\Theta}$ is a tensor given by

$$\boldsymbol{\Theta} = \boldsymbol{t}_1\boldsymbol{M}_1 + \boldsymbol{t}_2\boldsymbol{M}_2 + \boldsymbol{t}_3\boldsymbol{M}_3; \tag{5.2.16}$$

and so $\boldsymbol{t}$ for any orientation is determined by the stress vectors and unit normals for any three mutually perpendicular surface elements at the same point.

If $\boldsymbol{M}_1, \boldsymbol{M}_2, \boldsymbol{M}_3$ are taken as defining the x_1, x_2, x_3 co-ordinate axes, then it is seen from (5.2.16) that the elements of the tensor $\boldsymbol{\Theta}$ are given by

$$\Theta_{\alpha\beta} = t_{\alpha\beta}, \tag{5.2.17}$$

where $t_{\alpha\beta}$ is the αth component of $\boldsymbol{t}_\beta$. Here the reader should pause and draw a few diagrams to make clear the meaning of $t_{\alpha\beta}$; for instance, the forces on a small cube of material could be illustrated, assuming, for simplicity, that $\boldsymbol{\Theta}$ is uniform over its surface.

$\boldsymbol{\Theta}$ is known as the Cauchy stress tensor – for hydrostatic conditions at a pressure p it is easily seen to reduce to $-p\boldsymbol{E}$, where $\boldsymbol{E}$ is the unit tensor. The minus sign arises because of the sign convention used in defining the stress vector $\boldsymbol{t}$ (see section 5.4, following (5.4.6)).

5.3. Symmetry of the stress tensor $\boldsymbol{\Theta}$

By using the 'shrinking method' again it is simple to prove that the stress tensor $\boldsymbol{\Theta}$ is symmetric, if there are no body torques acting on the material. If such a body torque exists, it will be defined by a vector function $\boldsymbol{\tau}$ such that $\boldsymbol{\tau}\,\mathrm{d}v$ is the torque acting on the volume element $\mathrm{d}v$. For a body force $\boldsymbol{f}^{\mathrm{b}}$, like that due to gravity, the torque on the material in v about an origin inside v is

$$\int_v \boldsymbol{x} \wedge \boldsymbol{f}^{\mathrm{b}}\,\mathrm{d}v. \tag{5.3.1}$$

This integral will tend to zero faster than the volume, as can be seen by using the transformation (5.2.10). For, on expressing (5.3.1) in terms of an integral over the finite volume V of the same shape as v, the torque becomes

$$l^4 \int_V \boldsymbol{X} \wedge \boldsymbol{f}^{\mathrm{b}}(l\boldsymbol{X})\,\mathrm{d}V, \tag{5.3.2}$$

and to lowest order in l, this may be taken as

$$l^4 \int_V \boldsymbol{X} \wedge \boldsymbol{f}^{\mathrm{b}}(O)\,\mathrm{d}V, \tag{5.3.3}$$

where $\boldsymbol{f}^{\mathrm{b}}(O)$ is the value of $\boldsymbol{f}^{\mathrm{b}}$ at a convenient origin O inside V. In a similar way we see that

$$\int_v \boldsymbol{x} \wedge \boldsymbol{F}^{\mathrm{b}}\,\mathrm{d}v = l^4 \int_V \boldsymbol{X} \wedge \boldsymbol{F}^{\mathrm{b}}(O)\,\mathrm{d}V, \tag{5.3.4}$$

where

$$\boldsymbol{F}^{\mathrm{b}} = \boldsymbol{f}^{\mathrm{b}} - \rho \frac{\mathrm{D}\boldsymbol{U}}{\mathrm{D}t}, \tag{5.3.5}$$

as in (5.2.12).

Consider the region v, whose bounding surface is σ. Take moments of the forces acting on the material in this region. Using the equation of motion of each volume element and integrating we obtain

$$\int_\sigma \boldsymbol{x} \wedge \boldsymbol{t}(\boldsymbol{x}, \boldsymbol{N})\, \mathrm{d}\sigma + \int_v \boldsymbol{\tau}\, \mathrm{d}v + \int_v \boldsymbol{x} \wedge \boldsymbol{F}^{\mathrm{b}}\, \mathrm{d}v = 0, \tag{5.3.6}$$

where $\boldsymbol{\tau}$ and $\boldsymbol{F}^{\mathrm{b}}$ are functions of t and $\boldsymbol{x}$, and where the origin is inside v. Using (5.2.10), (5.3.6) becomes

$$l^3 \int_\Sigma \boldsymbol{X} \wedge \boldsymbol{t}(l\boldsymbol{X}, \boldsymbol{N})\, \mathrm{d}\Sigma + l^3 \int \boldsymbol{\tau}(l\boldsymbol{X})\, \mathrm{d}V + l^4 \int \boldsymbol{X} \wedge \boldsymbol{F}^{\mathrm{b}}(l\boldsymbol{X})\, \mathrm{d}V = 0. \tag{5.3.7}$$

On dividing by l^3 and allowing $l \to 0$, (5.3.7) reduces to

$$\int_\Sigma \boldsymbol{X} \wedge \boldsymbol{\Theta}(O)\boldsymbol{N}\, \mathrm{d}\Sigma + \int_V \boldsymbol{\tau}(O)\, \mathrm{d}V = 0. \tag{5.3.8}$$

The origin O may be any point within V. Therefore $\boldsymbol{\Theta}(O)$ and $\boldsymbol{\tau}(O)$, the stress tensor and body torque, satisfy (5.3.8) at any point O within V.

We note that in the limiting process, the torque contribution from finite extrinsic body forces vanishes. Body torques, however, do arise in magnetic materials when an external magnetic field is applied, and it is to such materials that (5.3.8) may apply.

Writing (5.3.8) in suffix notation, we have

$$\delta_{\alpha\beta\gamma} \int_\Sigma X_\beta \Theta_{\gamma\mu} N_\mu\, \mathrm{d}\Sigma + \int_V \tau_\alpha\, \mathrm{d}V = 0, \tag{5.3.9}$$

dropping the argument of $\boldsymbol{\Theta}$ and $\boldsymbol{\tau}$ for simplicity. Now on applying the theorem of (5.2.1) to the first integral, we obtain

$$\{\delta_{\alpha\beta\gamma}\Theta_{\gamma\beta} + \tau_\alpha\}V = 0, \tag{5.3.10}$$

on remembering that the $\Theta_{\gamma\mu}$ and the τ_α are not, according to (5.3.8), functions of position but are the values of these elements at the origin. To obtain (5.3.10) we have used the result that $\partial X_\beta / \partial X_\mu = \delta_{\beta\mu}$.

If we write out this condition in components there results, after cancelling V,

$$\left.\begin{aligned} \Theta_{zy}-\Theta_{yz}+\tau_x&=0\\ \Theta_{xz}-\Theta_{zx}+\tau_y&=0\\ \Theta_{yx}-\Theta_{xy}+\tau_z&=0.\end{aligned}\right\} \tag{5.3.11}$$

Of course, if $\boldsymbol{\tau}$ vanishes, the stress tensor $\boldsymbol{\Theta}$ must be symmetric and we shall assume that this is so, unless it is stated *explicitly* otherwise.

The symmetric stress tensor has real eigenvalues and orthogonal eigenvectors, and its principal directions can be found such that

$$\left.\begin{aligned} \boldsymbol{\Theta N}_1&=\Theta_1\boldsymbol{N}_1\\ \boldsymbol{\Theta N}_2&=\Theta_2\boldsymbol{N}_2\\ \boldsymbol{\Theta N}_3&=\Theta_3\boldsymbol{N}_3.\end{aligned}\right\} \tag{5.3.12}$$

That is, across surface elements whose normal is $\boldsymbol{N}_1$ the stress vector $\boldsymbol{\Theta N}_1$ is normal, i.e., the force is normal and of magnitude Θ_1 per unit area.

5.4. The equation of motion

The equation of motion (5.1.3) may now be expressed in terms of volume integrals only, on using the theorem (5.2.1) and the stress tensor expression (5.2.9) for $\boldsymbol{t}$. It is easily shown to be

$$\int_V\left\{\frac{\partial\Theta_{\alpha\beta}}{\partial X_\beta}+f_\alpha^{\mathrm{b}}-\rho\frac{\mathrm{D}U_\alpha}{\mathrm{D}t}\right\}\mathrm{d}V=0. \tag{5.4.1}$$

Since the volume V is arbitrary, the integrand must vanish and so

$$\frac{\partial\Theta_{\alpha\beta}}{\partial X_\beta}+f_\alpha^{\mathrm{b}}-\rho\frac{\mathrm{D}U_\alpha}{\mathrm{D}t}=0, \tag{5.4.2}$$

while for static equilibrium

$$\frac{\partial\Theta_{\alpha\beta}}{\partial X_\beta}+f_\alpha^{\mathrm{b}}=0. \tag{5.4.3}$$

It must be remembered that, so far in this chapter, we have been considering regions completely inside the material body and thus the extrinsic surface forces have not come into consideration. For equilibrium of the whole body, the total extrinsic surface and body forces must be in equilibrium. Under the usual conditions of statics, the total forces must vanish (see example 4, chapter 6), i.e.,

$$\int_{\Sigma_T} f^{\mathrm{s}}\,\mathrm{d}\Sigma_T + \int_{V_T} f^{\mathrm{b}}\,\mathrm{d}V_T = 0. \tag{5.4.4}$$

The turning moments about any point must also vanish, that is, we must have

$$\int_{\Sigma_T} \boldsymbol{X} \wedge f^{\mathrm{s}}\,\mathrm{d}\Sigma_T + \int_{V_T} \boldsymbol{X} \wedge f^{\mathrm{b}}\,\mathrm{d}V_T = 0, \tag{5.4.5}$$

where Σ_T represents the external surface, V_T the region occupied by the whole body, and $\boldsymbol{f}^{\mathrm{s}}$, $\boldsymbol{f}^{\mathrm{b}}$ the extrinsic surface force and body force densities respectively. For instance, the consequence of (5.4.4) for a rod supported at its top end is that the force exerted by the support is equal and opposite to the weight of the rod.

To illustrate (5.4.3), let us consider the simple case of a fluid under a hydrostatic pressure. For this case, the stress tensor is given by

$$\Theta_{\alpha\beta} = -p\delta_{\alpha\beta}, \tag{5.4.6}$$

where p is the hydrostatic pressure. (The presence of the minus sign merits some explanation. The reader should use the definition of $\boldsymbol{t}$ given after (5.1.2) to show that positive Θ_{xx}, say, represents an extensive stress, that is, one tending to extend a body, whereas a positive pressure tends to compress. In fact, pressure is defined on an opposite convention to $\boldsymbol{t}$, which, in the hydrostatic case, is equal to $-p\boldsymbol{N}$.)

For example, if we have a gaseous atmosphere at equilibrium, then (5.4.3) applies and, on using (5.4.6), this gives

$$-\frac{\partial p}{\partial \boldsymbol{X}} + \boldsymbol{f} = 0, \tag{5.4.7}$$

where $\boldsymbol{f} = \rho\boldsymbol{g}$ is the force per unit volume due to gravity acting on the gas; $\boldsymbol{g}$ is the acceleration due to gravity. Hence, if the X-axis is chosen in the direction of increasing height,

$$\left.\begin{aligned} &\frac{\partial p}{\partial X} + \rho g = 0, \\ &\frac{\partial p}{\partial Y} = \frac{\partial p}{\partial Z} = 0. \end{aligned}\right\} \tag{5.4.8}$$

Therefore p is uniform in the Y- and Z-directions. To integrate the first equation of (5.4.8) we must have a relation between p, ρ and X. We also must know the dependence of g on height X. We will assume that g is a constant, and that the atmosphere is isothermal and behaves like a perfect gas, so that the required relation is, simply,

$$p=\rho kT/m, \tag{5.4.9}$$

where m is the molecular mass. Equation (5.4.8) can then be integrated, and we obtain

$$p=p_0\exp(-mgX/kT), \tag{5.4.10}$$

where p_0 is the pressure when $X=0$.

Summary of important equations

A. *Forces*

$\boldsymbol{t}\,\mathrm{d}\Sigma$ is the surface force acting on surface element $\mathrm{d}\boldsymbol{\Sigma}$,

where $\boldsymbol{t}$ is known as the surface force density or the stress vector;

$\boldsymbol{f}^{\mathrm{b}}$ is the body force per unit volume.

B. *The equation of motion*

$$\int_\Sigma \boldsymbol{t}\,\mathrm{d}\Sigma+\int_V \boldsymbol{f}^{\mathrm{b}}\,\mathrm{d}V=\int_V(\rho\,\mathrm{d}V)\frac{\mathrm{D}\boldsymbol{U}}{\mathrm{D}t}, \tag{5.1.3}$$

where the hydrodynamic derivative is given by

$$\frac{\mathrm{D}}{\mathrm{D}t}=\left(\frac{\partial}{\partial t}+\boldsymbol{U}\cdot\frac{\partial}{\partial\boldsymbol{X}}\right). \tag{5.1.7}$$

C. *Green's First Theorem*

$$\int_\Sigma \phi N_\alpha\,\mathrm{d}\Sigma=\int_V(\partial\phi/\partial X_\alpha)\,\mathrm{d}V. \tag{5.2.1}$$

D. *The Divergence Theorem*

$$\int_\Sigma \boldsymbol{f}\cdot\boldsymbol{N}\,\mathrm{d}\Sigma=\int_V\left(\frac{\partial}{\partial\boldsymbol{X}}\cdot\boldsymbol{f}\right)\mathrm{d}V. \tag{5.2.2}$$

E. *The stress vector* $\boldsymbol{t}$

$$\int_\Sigma \boldsymbol{t}(O,\boldsymbol{N})\,\mathrm{d}\Sigma=0, \tag{5.2.13}$$

$$\boldsymbol{t}=\boldsymbol{\Theta}\boldsymbol{N}, \tag{5.2.15}$$

where $\boldsymbol{\Theta}$ is the Cauchy stress tensor. The elements of $\boldsymbol{\Theta}$ satisfy

$$\Theta_{zy}-\Theta_{yz}+\tau_x=0, \tag{5.3.11}$$

with cyclic interchange yielding the remaining two equations of (5.3.11). The vector component τ_x is the x-component of the body torque.

In all considerations following (5.3.11), $\boldsymbol{\tau}$ *is assumed to be zero unless explicitly stated otherwise.*

F. *Differential equation of motion*

$$\frac{\partial \Theta_{\alpha\beta}}{\partial X_\beta} + f_\alpha^{\mathrm{b}} - \rho \frac{\mathrm{D}U_\alpha}{\mathrm{D}t} = 0. \tag{5.4.2}$$

G. *For static equilibrium*

$$\frac{\partial \Theta_{\alpha\beta}}{\partial X_\beta} + f_\alpha^{\mathrm{b}} = 0. \tag{5.4.3}$$

Examples

1.(*a*) If ϕ is the potential energy per unit mass due to gravity at the surface of the earth, show that

$$\phi(\boldsymbol{X}, t) = gX \quad \text{and} \quad \mathrm{D}\phi/\mathrm{D}t = gU_X,$$

where the X-axis is vertically upwards. Here ϕ does not contain the time explicitly.

(*b*) Such physical quantities as ϕ may also be expressed as functions of $\boldsymbol{x}$ and t, where $\boldsymbol{x}$ is a reference configuration. That is, $\boldsymbol{X}(\boldsymbol{x}, t)$, $\boldsymbol{U}(\boldsymbol{x}, t)$, $\boldsymbol{\alpha}(\boldsymbol{x}, t)$, $\phi(\boldsymbol{x}, t)$ are the position vector, velocity etc., of that particle which was at $\boldsymbol{x}$ in the reference configuration at some initial time t_0, i.e., $\boldsymbol{X}(\boldsymbol{x}, t_0) = \boldsymbol{x}$.
Show that

$$\boldsymbol{U}(\boldsymbol{x}, t) = \partial \boldsymbol{X}(\boldsymbol{x}, t)/\partial t; \qquad \boldsymbol{\alpha} = \partial \boldsymbol{U}(\boldsymbol{x}, t)/\partial t;$$

and therefore, that in this representation, generally,

$$\mathrm{D}/\mathrm{D}t = \partial/\partial t.$$

Verify that the same result is obtained as in (*a*) for $\mathrm{D}\phi/\mathrm{D}t$.

2. Show that the force per unit volume on a volume element in a material body *due to the surrounding material* is given by $(\partial/\partial \boldsymbol{X}) \cdot \boldsymbol{\Theta}$; i.e., the βth component of this force is $\partial \Theta_{\alpha\beta}/\partial X_\alpha$. (Use theorem (5.2.1).)

3.(*a*) Expand the first integral in equation (5.2.12a) to the term of order l^3 and, using $\boldsymbol{t} = \boldsymbol{\Theta N}$, show that the terms of order l^3 reduce to the equation of motion (5.4.2).

(*b*) By expanding $\boldsymbol{t}(l\boldsymbol{X}, \boldsymbol{N})$ and $\boldsymbol{\tau}(l\boldsymbol{X})$ to terms of first order in l, show similarly that the terms of order l^4 in (5.3.7) reduce to the equation of motion (5.4.2).

6

Boundary conditions and work

The boundary conditions for the Cauchy stress tensor at an interface between two different media in a body are discussed, and it is shown that the stress vector $\boldsymbol{t} = \boldsymbol{\Theta}\boldsymbol{N}$ at the interface must be continuous, where $\boldsymbol{N}$ is the normal to the interface surface. The topic of virtual work is then discussed – this is the work done in an infinitesimal displacement, which is arbitrary except that it must satisfy any constraints on the system. Such a displacement is termed virtual because, since it is arbitrary (to the above extent), it need not be a displacement in any actual motion being considered.

The work considered is the work done *on* the body *by* the extrinsic surface and body forces, i.e., it is the work done on the body by forces exerted by some external agency. This is, for example, the sense in which we use the term 'the work done on a system' in thermodynamics. Expressions for the virtual work done on the system in an infinitesimal virtual displacement are obtained for the case where the particles of the media are in motion, and for the case where the body is at equilibrium. The Principle of Virtual Work is discussed briefly. It is shown that, in general, when one wishes to express the virtual work as a sum of terms made up of a generalised force times a change in a mechanical co-ordinate, the Cauchy stress tensor cannot be used. It is possible to express the virtual work as such a sum in many ways, and in this chapter two ways are described. In the first, the mechanical co-ordinates are the nine elements $D_{\alpha\beta}$ and the corresponding force terms are the nine elements of a (usually non-symmetric) tensor $\boldsymbol{B}$, so that the virtual work *per unit reference volume* $\mathrm{d}W/v = B_{\alpha\beta}\,\mathrm{d}D_{\beta\alpha}$. $\boldsymbol{B}$ is known as the Boussinesq stress tensor. It has the property that it remains constant, under homogeneous deformations, when the total load forces on the faces are held constant. In contrast, the load on a face must increase or decrease in proportion to the area of the face, if the Cauchy stress tensor is to remain constant under homogeneous deformations. At the reference state $\boldsymbol{B} = \boldsymbol{\Theta}$, and in general $\boldsymbol{B} = J\boldsymbol{D}^{-1}\boldsymbol{\Theta}$, and thus the symmetry of $\boldsymbol{\Theta}$ requires three equations to be satisfied, namely,

$$(\boldsymbol{DB})_{ij} = (\boldsymbol{DB})_{ji}, \quad i \neq j.$$

Hence, of the nine elements of $\boldsymbol{B}$, only six are independent. This is related to the fact that strain energies depend only on the six independent elements of $\boldsymbol{\eta}$.

In the second way, the six independent elements of $\boldsymbol{\eta}$ are introduced as mechanical co-ordinates, and a symmetric stress tensor $\boldsymbol{\Lambda}$ is introduced so that

$$\mathrm{d}W/v = \Lambda_{\alpha\beta}\,\mathrm{d}\eta_{\beta\alpha}.$$

In this case $\boldsymbol{\Lambda} = J\boldsymbol{D}^{-1}\boldsymbol{\Theta}(\boldsymbol{D}^T)^{-1}$, and thus is equal to $\boldsymbol{\Theta}$ at the reference state.

Virtual work is discussed very briefly for the case when $\boldsymbol{\Theta}$ is not symmetric.

6.1. Boundary conditions

In chapter 5 we discussed the forces on a part of the material which was entirely within the body considered. The forces acting were extrinsic body forces together with the stress forces exerted by the surrounding material. Consider now an infinitesimal element of the material which has part of its enclosing surface in the external surface. The element is in the form of a cylindrical pill-box, whose height h is very small compared to the diameter of its base.

The force on the volume element due to the surrounding material is exerted on the inner base and on the curved wall of the pill-box, and may be described by a stress tensor. This stress tensor may be taken as uniform over the infinitesimal region.

The equation of motion of the material in the volume element is then given by

$$(\boldsymbol{f}^{\mathrm{s}} - \boldsymbol{\Theta}\boldsymbol{N})\,\mathrm{d}\Sigma + \left(\boldsymbol{f}^{\mathrm{b}} - \rho\frac{\mathrm{D}\boldsymbol{U}}{\mathrm{D}t}\right)h\,\mathrm{d}\Sigma = 0 \qquad (6.1.1)$$

where $\boldsymbol{f}^{\mathrm{s}}$ and $\boldsymbol{f}^{\mathrm{b}}$ are the extrinsic surface and body force densities.

Because we can make h as small as we like compared to the diameter of the base, we have neglected the forces acting on the curved wall of the cylinder (if $\boldsymbol{\Theta}$ is uniform these add up to zero in any case). For the same reason we can neglect, in the limit, the term in (6.1.1) proportional to the volume $h\,\mathrm{d}\Sigma$ of the pill-box. We then obtain the boundary condition

$$\boldsymbol{f}^{\mathrm{s}} = \boldsymbol{\Theta}\boldsymbol{N}. \qquad (6.1.2)$$

If the external surface here were regarded as the interface between two different media, a similar argument would proceed, but with the pill-box placed so that each end is in a different medium, as shown in fig. 11. That

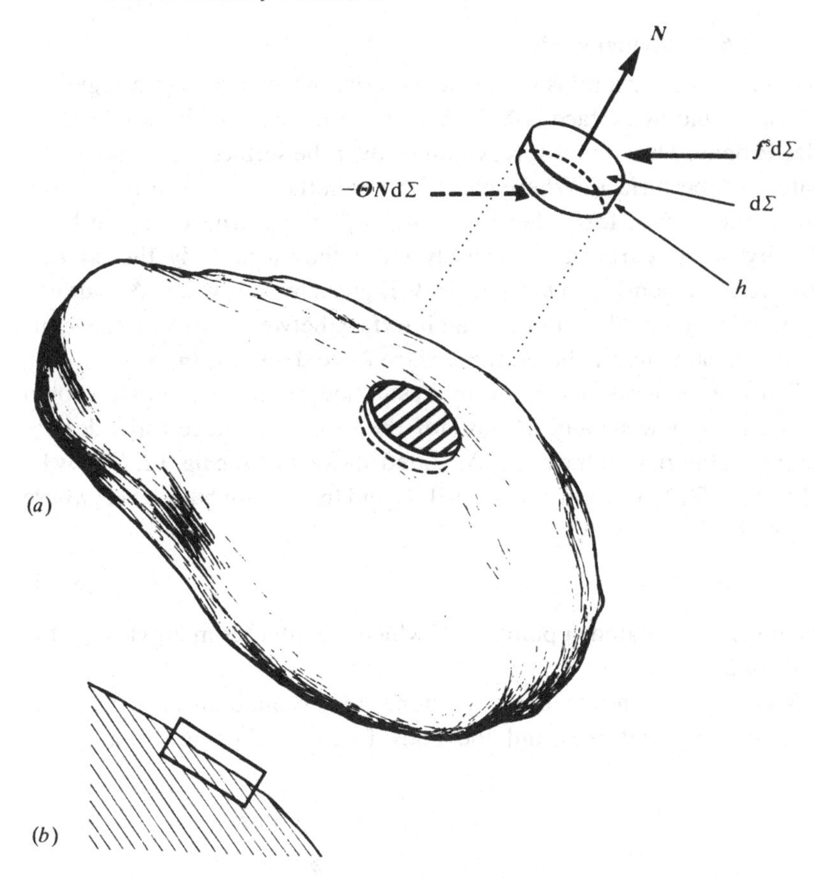

Fig. 11. The pill-box for boundary conditions.

is, the interface (regarded macroscopically as a mathematical surface of zero thickness) passes through the pill-box. It is easy to show that, instead of (6.1.2),

$$\mathbf{\Theta}_1 \boldsymbol{N} = \mathbf{\Theta}_2 \boldsymbol{N} \tag{6.1.3}$$

where $\boldsymbol{N}$ is a unit normal at a point A in the interface and $\mathbf{\Theta}_1$, $\mathbf{\Theta}_2$ are the stress tensors at points in the media, labelled 1 and 2 respectively, infinitesimally close to A. Equation (6.1.3) *states that the stress vector is continuous at an interface.* Equations (6.1.3) and (6.1.2) can mean the same thing; for example, if the external force density f^s is applied by a piston in contact with part or the whole of the surface of the body, then, at each point of the surface in contact, $\mathbf{\Theta}_1$, say, could describe the stress tensor in the body at that point, and $\mathbf{\Theta}_2$ the stress tensor in the material of the piston at the same point.

6.2. Virtual work

We now consider a mass of elastic material, which occupies a region V whose bounding surface is Σ. This mass forms part or the whole of an elastic body. The surface forces applied over the surface Σ depend on the nature of the surface. Where part of Σ lies in the external surface of the body, the surface force density acting is $\boldsymbol{f}^s$, the extrinsic surface force density; where part of Σ lies entirely within the elastic body, the external surface force density on material in V is given by $\boldsymbol{\Theta N}$, where $\boldsymbol{N}$ is drawn outward from V; if part of Σ is an interface between two homogeneous phases making up the body, the surface force density is $\boldsymbol{\Theta}_1\boldsymbol{N}$.

Thus, in general, if part of an elastic body is considered we should introduce a new symbol $\boldsymbol{h}^s$, say, for the external surface force density acting on this part of the body. As stated above, and see fig. 12, $\boldsymbol{h}^s$ may be $\boldsymbol{f}^s$, $\boldsymbol{\Theta N}$ or $\boldsymbol{\Theta}_1\boldsymbol{N}$ and, because of (6.1.2) and (6.1.3), we have, everywhere on the surface of V,

$$\boldsymbol{h}^s = \boldsymbol{\Theta N}, \tag{6.2.1}$$

where $\boldsymbol{\Theta}$ is evaluated at points in V which are infinitesimally close to the surface Σ.

We wish to consider the work done *on* the material in V by these external surface forces and the body forces when there is a virtual

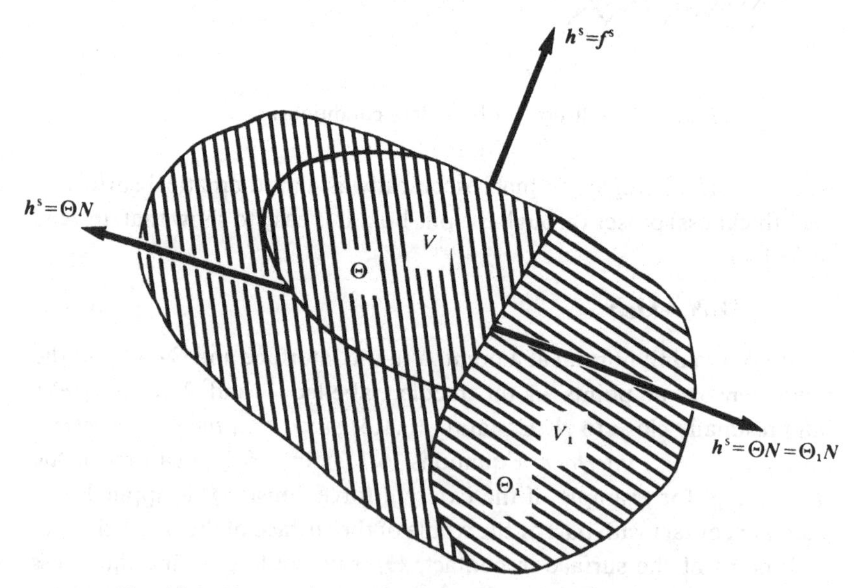

Fig. 12. V is part of an elastic body, consisting of two homogeneous phases, V lies wholly in one phase and V_1 is the other phase.

infinitesimal deformation described by $\delta \boldsymbol{u}$. That is, $\delta \boldsymbol{u}$ is an infinitesimal deformation completely arbitrary except for any constraints on the body. The expression for this work done is very important in discussions of elastic energy and the thermodynamics of deformation; it also leads to the statement of the Principle of Virtual Work.

Since work is measured by the scalar product of forces and the displacements of their points of application, we have, for the virtual work performed *on* the material in the volume V,

$$\delta W = \int_{\Sigma} \boldsymbol{h}^{\mathrm{s}} \cdot \delta \boldsymbol{u} \, \mathrm{d}\Sigma + \int_{V} \boldsymbol{f}^{\mathrm{b}} \cdot \delta \boldsymbol{u} \, \mathrm{d}V. \tag{6.2.2}$$

Using (6.2.1) for $\boldsymbol{h}^{\mathrm{s}}$,

$$\delta W = \int_{\Sigma} \delta \boldsymbol{u} \cdot \boldsymbol{\Theta N} \, \mathrm{d}\Sigma + \int_{V} \boldsymbol{f}^{\mathrm{b}} \cdot \delta \boldsymbol{u} \, \mathrm{d}V, \tag{6.2.3}$$

where the first integrand is $\delta u_\alpha \Theta_{\alpha\beta} N_\beta$.

Further, on using theorem (5.2.1) in the surface integral, we obtain

$$\delta W = \int_{V} \left(\Theta_{\alpha\beta} \frac{\partial(\delta u_\alpha)}{\partial X_\beta} + \delta u_\beta \left\{ \frac{\partial \Theta_{\beta\alpha}}{\partial X_\alpha} + f^{\mathrm{b}}_\beta \right\} \right) \mathrm{d}V, \tag{6.2.4}$$

and, on using the equation of motion (5.4.2),

$$\delta W = \int_{V} \left(\Theta_{\alpha\beta} \frac{\partial(\delta u_\alpha)}{\partial X_\beta} + \delta u_\beta \rho \frac{\mathrm{D}U_\beta}{\mathrm{D}t} \right) \mathrm{d}V, \tag{6.2.5}$$

$$= \int_{V} \left(\Theta_{\alpha\beta} \frac{\partial(\delta u_\alpha)}{\partial X_\beta} + \rho \frac{\mathrm{D}}{\mathrm{D}t} \{\tfrac{1}{2} U^2\} \delta t \right) \mathrm{d}V, \tag{6.2.6}$$

where δt is the time taken in the deformation $\delta \boldsymbol{u}$, and hence the velocity component is

$$U_\beta = \delta u_\beta / \delta t. \tag{6.2.7}$$

The external work done on the material V can therefore be interpreted as follows: the first term in (6.2.6) contains the work done in deforming elastically the material in V as well as the work done by the body forces, while the second term is the increase in the kinetic energy of this material.

When the body is at equilibrium, or when all motion is uniform, (6.2.6) reduces to a result of very great importance in thermodynamics, namely,

$$\delta W = \int_{V} \Theta_{\alpha\beta} \frac{\partial(\delta u_\alpha)}{\partial X_\beta} \mathrm{d}V. \tag{6.2.8}$$

Use of the theorem (5.2.1) and (5.4.3) takes (6.2.8) back to (6.2.3) immediately (this is left as an exercise for the reader), thus verifying the above interpretation of the first term in (6.2.6).

In thermodynamics body forces are frequently neglected. The following expression for δW may be obtained from (6.2.3) and is useful in such cases:

$$\delta W = \int_\Sigma \Theta_{\alpha\beta}\delta u_\alpha N_\beta \,\mathrm{d}\Sigma. \tag{6.2.9}$$

As an example, let us write the trace of the stress tensor

$$\Theta_{\alpha\alpha} = -3p \tag{6.2.10}$$

and define

$$\Theta'_{\alpha\beta} = \Theta_{\alpha\beta} + p\delta_{\alpha\beta}. \tag{6.2.11}$$

That is, $\Theta'_{\alpha\beta}$ has zero trace. For this example, let us assume Θ is uniform, and then (6.2.9) becomes

$$\delta W = -p\int_\Sigma \delta u_\alpha N_\alpha \,\mathrm{d}\Sigma + \int_\Sigma \Theta'_{\alpha\beta}\delta u_\alpha N_\beta \,\mathrm{d}\Sigma, \tag{6.2.12}$$

$$= -p\delta V + \int_\Sigma \Theta'_{\alpha\beta}\delta u_\alpha N_\beta \,\mathrm{d}\Sigma, \tag{6.2.13}$$

and we see the appearance of the term $-p\,\delta V$, familiar in thermodynamics, to which δW reduces if the stress is hydrostatic, i.e., if $\Theta_{\alpha\beta} = -p\delta_{\alpha\beta}$. We note again that the definition of $\boldsymbol{\Theta}$ is such that, for positive diagonal elements, the effect is tensile, whereas the pressure p is defined to be compressive when positive, thus leading to the minus sign in (6.2.10).

6.3. The principle of virtual work

The mathematical expression for this principle is obtained by equating the expressions for δW in (6.2.2) and (6.2.5) to obtain

$$\int_\Sigma \boldsymbol{h}^{\mathrm{s}}\cdot\delta\boldsymbol{u}\,\mathrm{d}\Sigma + \int_V \left(\boldsymbol{F}^{\mathrm{b}}\cdot\delta\boldsymbol{u} - \boldsymbol{\Theta}\colon\frac{\partial(\delta\boldsymbol{u})}{\partial\boldsymbol{X}}\right)\mathrm{d}V = 0, \tag{6.3.1}$$

where

$$\boldsymbol{\Theta}\colon\frac{\partial(\delta\boldsymbol{u})}{\partial\boldsymbol{X}} = \Theta_{\alpha\beta}\frac{\partial(\delta u_\beta)}{\partial X_\alpha}. \tag{6.3.2}$$

V is an arbitrary region, and $\delta\boldsymbol{u}$ is arbitrary. Equation (6.3.1) has been obtained by assuming that only surface and body forces, described by the densities $\boldsymbol{h}^{\mathrm{s}}$ and $\boldsymbol{f}^{\mathrm{b}}$, are acting. We now consider the possible introduction of constraints on the motion. There may be points, in or on the body, held rigidly fixed so that at these points $\delta\boldsymbol{u}=0$; the motion of some points may be smoothly constrained along a line or a plane. No work is done in these cases by the forces of constraint, since, at such points either $\delta\boldsymbol{u}=0$, or these forces are perpendicular to $\delta\boldsymbol{u}$ for smooth constraints. Thus (6.3.1) holds, also, for such constrained variations and the forces of constraint do not appear in it. The corresponding principle for systems of discrete particles is used to introduce the Lagrangian treatment for generalised co-ordinates (Goldstein, 1950). The principle of virtual work is not used further in this book but the interested reader may try some examples at the end of the chapter.

6.4. Alternative stress tensors

The virtual work performed on deformable material, for equilibrium or static conditions, is given by (6.2.8) in terms of the stress tensor $\boldsymbol{\Theta}$ which gives the force on a surface element as

$$\boldsymbol{\Theta N}\,\mathrm{d}\Sigma=\boldsymbol{\Theta}\,\mathrm{d}\boldsymbol{\Sigma}. \tag{6.4.1}$$

From (6.4.1) we see that if we consider the finite deformation $\boldsymbol{x}\rightarrow\boldsymbol{X}$, $\mathrm{d}\boldsymbol{\sigma}\rightarrow\mathrm{d}\boldsymbol{\Sigma}$ occurring, and if we want the stress tensor $\boldsymbol{\Theta}$ to remain constant during this deformation, the loading (total force) on the surface element must change as $\mathrm{d}\boldsymbol{\sigma}$ changes so that $\boldsymbol{\Theta}$ can remain constant. As an illustration, consider an example of a normal force $\mathrm{d}f$ acting on a surface element $\mathrm{d}\boldsymbol{\sigma}=\boldsymbol{n}\,\mathrm{d}\sigma$. Since $\mathrm{d}f$ is normal, $\boldsymbol{n}$ must be a principal direction of $\boldsymbol{\Theta}$, and

$$|\mathrm{d}f|=\mathrm{d}f=\Theta_n\,\mathrm{d}\sigma, \tag{6.4.2}$$

where Θ_n is the eigenvalue of Θ corresponding to the eigenvector $\boldsymbol{n}$. Now, suppose there is a deformation such that the unit normal $\boldsymbol{n}$ remains unchanged. If $\boldsymbol{\Theta}$ is to remain constant, the force $\mathrm{d}\boldsymbol{F}$ on the deformed surface element $\mathrm{d}\boldsymbol{\Sigma}=\boldsymbol{n}\,\mathrm{d}\Sigma$ must be parallel to $\boldsymbol{n}$ and of magnitude

$$\mathrm{d}F=\Theta_n\,\mathrm{d}\Sigma. \tag{6.4.3}$$

Thus, in this simple case, if $\boldsymbol{\Theta}$ is to be held constant in the deformation, the load on the surface element must be varied, so that

$$\mathrm{d}F/\mathrm{d}f=\mathrm{d}\Sigma/\mathrm{d}\sigma. \tag{6.4.4}$$

This property of the stress tensor $\boldsymbol{\Theta}$ leads to the result that the work done at *constant* $\boldsymbol{\Theta}$, in the deformation $x\rightarrow\boldsymbol{X}=\boldsymbol{x}+\boldsymbol{u}$, depends on the path. For example, consider a substance, in the shape of a rectangular parallelepiped of edge-lengths l_1, l_2, l_3, to be under a uniform stress $\boldsymbol{\Theta}$ whose principal axes are parallel to the edges of the solid. Under a deformation in which $\boldsymbol{\Theta}$ remains constant throughout, let the parallelepiped remain rectangular and its sides remain parallel to the original corresponding sides, while its two edge-lengths l_1, l_2 change to l_1+d_1, l_2+d_2. Let Θ_i be the principal stress (eigenvalue of $\boldsymbol{\Theta}$) corresponding to the principal direction l_i. If l_1 is changed first, the work done is $\Theta_1 d_1 l_2 l_3$, and, if l_2 is changed next, the work done is $\Theta_2 d_2(l_1+d_1)l_3$. So the work done on this path is

$$W_1 = \Theta_1 d_1 l_2 l_3 + \Theta_2 d_2 l_1 l_3 + \Theta_2 d_2 d_1 l_3. \tag{6.4.5}$$

If l_2 is changed first, and then l_1, the work done is

$$W_2 = \Theta_2 d_2 l_1 l_3 + \Theta_1 d_1 l_2 l_3 + \Theta_1 d_1 d_2 l_3, \tag{6.4.6}$$

which differs from W_1 in the final term if $\Theta_1 \neq \Theta_2$. Finally, if l_1 and l_2 are changed at the same rate, it is easy to see that the work done is

$$W_3 = \Theta_1 l_2 l_3 d_1 + \Theta_2 l_1 l_3 d_2 + \tfrac{1}{2}(\Theta_1+\Theta_2) l_3 d_1 d_2. \tag{6.4.7}$$

Again the last term is different from that in W_1 and W_2 if $\Theta_1 \neq \Theta_2$. It is not difficult to see that the expressions for W_1, W_2 and W_3 should be different as consideration of (6.4.4) shows that the loads on the outside surfaces differ, in general, for different paths.

This example illustrates the fact that when work W is done in a deformation $\boldsymbol{u}$, during which $\boldsymbol{\Theta}$ remains uniform and constant, then, in general, the value of W depends on the path taken between initial and final states. Equation (6.2.8), which can be written as

$$\delta W = \Theta_{\alpha\beta} \int_V \frac{\partial(\delta u_\alpha)}{\partial X_\beta} \, \mathrm{d}V, \tag{6.4.8}$$

cannot, in general, be integrated when $\boldsymbol{\Theta}$ remains constant during the deformation, and

$$W \neq \Theta_{\alpha\beta} \int_V (\partial u_\alpha / \partial X_\beta) \, \mathrm{d}V, \tag{6.4.9}$$

when $\boldsymbol{\Theta}$ is constant. The difficulty is that the region of integration V changes during the deformation.

Considering this, it may appear rather surprising that, when $\boldsymbol{\Theta} = -p\boldsymbol{E}$, i.e., when the stress is hydrostatic, uniform and constant, we can write

$$\delta W = -p\delta V \quad \text{and} \quad W = -p(V - V_0). \tag{6.4.10}$$

The reason is simply that, for this case, we can write (6.4.8) in the form

$$\delta W = -p\int_V \frac{\partial(\delta u_\alpha)}{\partial X_\alpha}\,\mathrm{d}V = -p\int_\Sigma \delta u_\alpha N_\beta\,\mathrm{d}\Sigma = -p\;\delta V, \qquad (6.4.11)$$

as discussed previously for (6.2.12). The general result can be expressed equivalently by the statement that δW cannot, in general, be expressed in the form $\Theta_{\alpha\beta}\delta\Omega_{\beta\alpha}$ where $\Omega_{\beta\alpha}$ is some generalised integrable co-ordinate. It is, however, possible to define two other stress tensors such that δW *is given by an integral over the fixed initial or reference region of the material.* If such a stress tensor is kept constant, δW is integrable – a fact which leads to generalised volume co-ordinates, just as (6.4.10) leads to the integrable co-ordinate V (or $V - V_0$).

The Boussinesq tensor. The first such tensor is known as the *Boussinesq* stress tensor $\boldsymbol{B}$, and this is introduced by considering (6.2.8) and transforming the region of integration to that of the reference region v.

We thus have, for an infinitesimal variation,

$$\delta W = \int_v \Theta_{\alpha\mu}\frac{\partial(\delta u_\alpha)}{\partial X_\mu} J\,\mathrm{d}v, \qquad (6.4.12)$$

using the notation of chapter 2. The integration is now over a fixed region v. If we write

$$\frac{\partial(\delta u_\alpha)}{\partial X_\mu} = \frac{\partial(\delta u_\alpha)}{\partial x_\beta}\frac{\partial x_\beta}{\partial X_\mu}, \qquad (6.4.13)$$

(6.4.12) may be expressed as

$$\delta W = \int_v J\frac{\partial x_\beta}{\partial X_\mu}\Theta_{\mu\alpha}\frac{\partial(\delta u_\alpha)}{\partial x_\beta}\,\mathrm{d}v = \int_v B_{\beta\alpha}\frac{\partial(\delta u_\alpha)}{\partial x_\beta}\,\mathrm{d}v, \qquad (6.4.14)$$

on defining

$$B_{\beta\alpha} = J(\partial x_\beta/\partial X_\mu)\Theta_{\mu\alpha}, \qquad (6.4.15\text{a})$$

that is,

$$\boldsymbol{B} = J\boldsymbol{D}^{-1}\boldsymbol{\Theta}, \qquad (6.4.15\text{b})$$

where $\boldsymbol{D}$ is the tensor defined in (2.3.8). Thus $\boldsymbol{B}$ and $\boldsymbol{\Theta}$ are equal at the reference state.

First we notice that, if $\boldsymbol{B}$ is kept uniform in the deformation,

$$\delta W = B_{\beta\alpha}\int_v \delta u_\alpha\,\mathrm{d}\sigma_\beta = B_{\beta\alpha}\int_v \delta u_\alpha n_\beta\,\mathrm{d}\sigma \qquad (6.4.16)$$

and, if $\boldsymbol{B}$ is also held constant, this integrates to

$$W = B_{\beta\alpha}\int_\sigma u_\alpha \,\mathrm{d}\sigma_\beta = B_{\beta\alpha}\int_v (\partial u_\alpha/\partial x_\beta)\,\mathrm{d}v. \tag{6.4.17}$$

Thus, a uniform constant $\boldsymbol{B}$ yields an integrable co-ordinate, which, since $\delta u_\alpha = \delta X_\alpha$, may conveniently be written

$$\Omega_{\alpha\beta} = \int_\sigma X_\alpha \,\mathrm{d}\sigma_\beta = \int_v \frac{\partial X_\alpha}{\partial x_\beta}\,\mathrm{d}v, \tag{6.4.18}$$

and

$$\delta W = B_{\beta\alpha}\,\delta\Omega_{\alpha\beta}. \tag{6.4.19}$$

The physical difference between $\boldsymbol{B}$ and $\boldsymbol{\Theta}$ is seen by considering the force on a surface element

$$\mathrm{d}\boldsymbol{F} = \boldsymbol{\Theta}\,\mathrm{d}\boldsymbol{\Sigma} = \boldsymbol{\Theta}J(\boldsymbol{D}^{\mathrm{t}})^{-1}\,\mathrm{d}\boldsymbol{\sigma} = J\{\boldsymbol{D}^{-1}\boldsymbol{\Theta}\}^{\mathrm{t}}\,\mathrm{d}\boldsymbol{\sigma} \tag{6.4.20}$$

(see (2.3.10)). Comparing (6.4.20) with (6.4.15b), the defining equation for $\boldsymbol{B}$, we see that

$$\mathrm{d}\boldsymbol{F} = \boldsymbol{\Theta}\,\mathrm{d}\boldsymbol{\Sigma} = \boldsymbol{B}^{\mathrm{t}}\,\mathrm{d}\boldsymbol{\sigma}. \tag{6.4.21}$$

That is, the force on the surface element $\mathrm{d}\boldsymbol{\Sigma}$ *is given by* $\boldsymbol{B}^{\mathrm{t}}\,\mathrm{d}\boldsymbol{\sigma}$, *where* $\mathrm{d}\boldsymbol{\sigma}$ *is the reference state of* $\mathrm{d}\boldsymbol{\Sigma}$. It is clear, from (6.4.21), that if the load $\mathrm{d}\boldsymbol{F}$ on the surface element is to remain constant under any deformation, then it is necessary and sufficient that $\boldsymbol{B}$ remain constant. To illustrate (6.4.21) further, let us consider a simple case where a surface element deforms in such a way that its unit normal $\boldsymbol{n}$ does not change. We suppose that the force acting on this surface element remains normal to the surface element. $\boldsymbol{n}$ is thus a principal direction of both $\boldsymbol{B}$ and $\boldsymbol{\Theta}$, and, using the notation of (6.4.2) and (6.4.3),

$$\mathrm{d}F = \Theta_n\,\mathrm{d}\Sigma = B_n\,\mathrm{d}\sigma, \tag{6.4.22}$$

and

$$\mathrm{d}f = \Theta_n^{(0)}\,\mathrm{d}\sigma, \tag{6.4.23}$$

where B_n is the eigenvalue corresponding to the eigenvector $\boldsymbol{n}$, and $\Theta_n^{(0)}$ is the corresponding eigenvalue of $\boldsymbol{\Theta}^{(0)}$, the Cauchy stress tensor at the reference state. If the load is to remain constant, i.e., $\mathrm{d}F = \mathrm{d}f$, then B_n must equal $\Theta_n^{(0)}$, and it is left to the reader to show that this is consistent with the fact that $\boldsymbol{B}$ must remain constant.

The strain tensor as a co-ordinate. First, we write (6.2.8) as

$$\delta W = \frac{1}{2}\int \Theta_{\gamma\omega}\left\{\frac{\partial(\delta u_\gamma)}{\partial X_\omega} + \frac{\partial(\delta u_\omega)}{\partial X_\gamma}\right\} \mathrm{d}V, \tag{6.4.24}$$

which follows from the symmetry of $\boldsymbol{\Theta}$ (see example 12, chapter 2). Now (2.1.12) may be written

$$\delta\eta_{\mu\nu} = \frac{1}{2}\left\{\frac{\partial(\delta u_\gamma)}{\partial X_\omega} + \frac{\partial(\delta u_\omega)}{\partial X_\gamma}\right\}\frac{\partial X_\gamma}{\partial x_\mu}\frac{\partial X_\omega}{\partial x_\nu}, \tag{6.4.25}$$

and, on multiplying both sides of this equation by

$$\frac{\partial x_\mu}{\partial X_\alpha}\frac{\partial x_\nu}{\partial X_\beta},$$

then, since

$$\frac{\partial x_\mu}{\partial X_\alpha}\frac{\partial X_\gamma}{\partial x_\mu} = \delta_{\alpha\gamma}, \tag{6.4.26}$$

we have

$$\frac{1}{2}\left\{\frac{\partial(\delta u_\alpha)}{\partial X_\beta} + \frac{\partial(\delta u_\beta)}{\partial X_\alpha}\right\} = \delta\eta_{\mu\nu}\frac{\partial x_\mu}{\partial X_\alpha}\frac{\partial x_\nu}{\partial X_\beta}. \tag{6.4.27}$$

Substitution of (6.4.27) in (6.4.24) then results in

$$\delta W = \int_V \left(\Theta_{\alpha\beta}\frac{\partial x_\mu}{\partial X_\alpha}\frac{\partial x_\nu}{\partial X_\beta}\right)\delta\eta_{\mu\nu}\,\mathrm{d}V. \tag{6.4.28}$$

If we wish to transform the integral to one over the region of the reference state, we have, using (2.2.6),

$$\delta W = \int_v \left(J\Theta_{\alpha\beta}\frac{\partial x_\mu}{\partial X_\alpha}\frac{\partial x_\nu}{\partial X_\beta}\right)\delta\eta_{\mu\nu}\,\mathrm{d}v. \tag{6.4.29}$$

On defining a new symmetric stress tensor,

$$\Lambda_{\mu\nu} = J\frac{\partial x_\mu}{\partial X_\alpha}\frac{\partial x_\nu}{\partial X_\beta}\Theta_{\alpha\beta}, \tag{6.4.30a}$$

i.e.,

$$\boldsymbol{\Lambda} = J\boldsymbol{D}^{-1}\boldsymbol{\Theta}(\boldsymbol{D}^{\mathrm{t}})^{-1} = \boldsymbol{B}(\boldsymbol{D}^{\mathrm{t}})^{-1}, \tag{6.4.30b}$$

we have, finally,

$$\delta W = \int_v \Lambda_{\mu\nu}\delta\eta_{\mu\nu}\,\mathrm{d}v = \int_v \mathrm{tr}(\boldsymbol{\Lambda}\delta\boldsymbol{\eta})\,\mathrm{d}v. \tag{6.4.31}$$

We note that, in infinitesimal strain theory,

$$\boldsymbol{B} = \boldsymbol{\Theta} = \boldsymbol{\Lambda}. \tag{6.4.32}$$

We see that deformations at constant $\boldsymbol{\Lambda}$ provide an integrable co-ordinate, which for a homogeneous material may be given as $v\boldsymbol{\eta}$ where v is the reference volume, such that the work per unit reference volume is

$$(\delta W/v) = \Lambda_{\alpha\beta}\delta\eta_{\beta\alpha} = \mathrm{tr}(\boldsymbol{\Lambda}\delta\boldsymbol{\eta}). \tag{6.4.33}$$

It should be noted that $\boldsymbol{B}$ and $\boldsymbol{\Omega}$ are not necessarily symmetric tensors, as can be seen from (6.4.15) and (6.4.18), whereas $\boldsymbol{\Lambda}$ and $\boldsymbol{\eta}$ are both symmetric.

In chapter 7, a new unique factorisation of $\boldsymbol{D}$ into an orthogonal matrix $\boldsymbol{V}$ and a triangular matrix $\boldsymbol{H}$ will be described, the triangular matrix $\boldsymbol{H}$ being such that $H_{\alpha\beta} = 0$ for $\alpha > \beta$. The six elements of $\boldsymbol{H}$ will provide six independent mechanical co-ordinates and it will be shown in chapter 8 how the work done per *unit reference volume* $\delta W/v = L_{\alpha\beta}\delta H_{\beta\alpha}$, where there are six non-zero $L_{\alpha\beta}$, the stress elements conjugate to the $H_{\beta\alpha}$.

In general, in the application of homogeneous finite strain theory to anisotropic crystals the tensor co-ordinates $\eta_{\alpha\beta}$ are the most useful, especially in the application of symmetry considerations where the *linear* transformations of the tensor $\boldsymbol{\eta}$ under transformations of co-ordinate axis systems are the best to use. However, the use of the six elements of $\boldsymbol{H}$ as co-ordinates is of unique importance in considering the thermodynamic stability of solid phases, and hence in phase inversions as will be discussed later. This importance follows from the fact that we can geometrically restrict the rigid rotations of a crystal so that all homogeneous deformations $\boldsymbol{D}$ are triangular, and just as the $\boldsymbol{V}$ form a group, so do the triangular matrices. There are other advantages in the use of triangular matrices as strain co-ordinates, arising from their group properties. They transform non-linearly, however, under orthogonal transformations and thus are not so useful as tensorial strain co-ordinates, which transform linearly, when the symmetry properties of crystalline material are being studied (see chapter 18).

6.5. Infinitesimal strain theory and stress tensors

In infinitesimal strain theory, the quantities u_α and their derivatives are treated as infinitesimal quantities, and thus the transition from finite to infinitesimal theory is effected by expressing every quantity and equation to the lowest possible order in these quantities. For example, the neglect of higher orders in (6.4.15b) and in (6.4.30b) results in (6.4.32), while

(6.4.14) and (6.4.31) both reduce, for uniform $\boldsymbol{\Theta}$, to

$$\delta W = \Theta_{\alpha\beta}\,\delta\mathcal{V}_{\alpha\beta} \tag{6.5.1}$$

where

$$\delta\mathcal{V}_{\alpha\beta} = \frac{1}{2}\int_v \left(\frac{\partial(\delta u_\alpha)}{\partial x_\beta} + \frac{\partial(\delta u_\beta)}{\partial x_\alpha}\right) \mathrm{d}v = \frac{1}{2}\int_\sigma \{\delta u_\alpha n_\beta + \delta u_\beta n_\alpha\}\,\mathrm{d}\sigma. \tag{6.5.2}$$

This may be integrated to the symmetric tensor

$$\mathcal{V}_{\alpha\beta} = \tfrac{1}{3}v\delta_{\alpha\beta} + \frac{1}{2}\int_\sigma (u_\alpha n_\beta + u_\beta n_\alpha)\,\mathrm{d}\sigma, \tag{6.5.3}$$

where the first term on the right is the constant of integration, chosen so that the trace of $\mathcal{V}_{\alpha\beta}$ is equal to V. (Taking the trace of (6.5.3) gives

$$\mathcal{V}_{\alpha\alpha} = v + (V - v) = V.) \tag{6.5.4}$$

Again from (6.4.31), and since $\boldsymbol{\Lambda} = \boldsymbol{\Theta}$,

$$\delta W = \Theta_{\alpha\beta}\int_v \delta\eta_{\alpha\beta}\,\mathrm{d}v, \tag{6.5.5}$$

for uniform $\boldsymbol{\Theta}$. On replacing $\boldsymbol{\eta}$ by the infinitesimal strain tensor $\boldsymbol{\varepsilon}$, we thus see that

$$\delta\mathcal{V}_{\alpha\beta} = v\delta\varepsilon_{\alpha\beta},$$

if the change $\delta\varepsilon_{\alpha\beta}$ *is homogeneous*; *and*

$$\mathcal{V}_{\alpha\beta} = v\{\tfrac{1}{3}\delta_{\alpha\beta} + \varepsilon_{\alpha\beta}\} \tag{6.5.6}$$

if the strain $\varepsilon_{\alpha\beta}$ *is homogeneous.* In chemical changes especially we may be considering infinitesimal inhomogeneous changes of shape, and thus it must be emphasised that (6.5.3) is the more general definition of the generalised volume co-ordinates – reduction to (6.5.6) being a particular case.

In this and the preceding section, two sets of extensive mechanical co-ordinates have been introduced: the $\Omega_{\alpha\beta}$ for finite strains, and the $\mathcal{V}_{\alpha\beta}$ for infinitesimal. The $\Omega_{\alpha\beta}$ or $\mathcal{V}_{\alpha\beta}$ can only be used when $\boldsymbol{B}$ or $\boldsymbol{\Theta}$, respectively, is uniform – restrictions which severely limit their use. We shall usually express extensive quantities in terms of densities; for example the internal energy U will be expressed as vu, where u is the internal energy per unit reference volume. The work done per unit reference volume will be given by

$$\delta w = B_{\alpha\beta}\delta D_{\beta\alpha} = \Lambda_{\alpha\beta}\delta\eta_{\beta\alpha}, \tag{6.5.7}$$

where, for homogeneous conditions, δw is equal to $\delta W/v$ where δW the total work done on the material under study. It will be found that the use of densities of extensive thermodynamic quantities, per unit reference volume, is economical of symbols as well as allowing the treatment of materials under inhomogeneous conditions.

6.6. Virtual work when $\boldsymbol{\Theta}$ is not symmetric

If we assume a body torque density $\boldsymbol{\tau}$, the Cauchy stress tensor $\boldsymbol{\Theta}$ will not be symmetric, the asymmetry being determined by the body torque as given in (5.3.11). If we make no further qualifications, but assume that the forces on a system are communicated by body force, body torque and surface force densities, the work done on a part or the whole of an elastic body is given by

$$\delta W = \int_\Sigma \boldsymbol{h}^{\mathrm{s}} \cdot \delta \boldsymbol{u} \, \mathrm{d}\Sigma + \int_V \boldsymbol{f}^{\mathrm{b}} \cdot \delta \boldsymbol{u} \, \mathrm{d}V + \int_V \boldsymbol{\tau} \cdot \delta \boldsymbol{\omega} \, \mathrm{d}V, \tag{6.6.1}$$

where

$$\delta \boldsymbol{\omega} = \frac{1}{2} \frac{\partial}{\partial \boldsymbol{X}} \wedge \delta \boldsymbol{u} \tag{6.6.2}$$

is the rotational part of the transformation $\delta \boldsymbol{u}$, (see example 6, chapter 3).

Proceeding as with (6.2.2), using (6.2.1) and the equation of motion (5.4.2), it is easy to obtain

$$\delta W = \int_V \left\{ \Theta_{\alpha\beta} \frac{\partial(\delta u_\alpha)}{\partial X_\beta} + \boldsymbol{\tau} \cdot \delta \boldsymbol{\omega} + \rho \frac{\mathrm{D}}{\mathrm{D}t} (\tfrac{1}{2} U^2) \, \delta t \right\} \mathrm{d}V. \tag{6.6.3}$$

It is left to the reader as a simple exercise to use (5.3.11) and (6.6.2) and show that this reduces to

$$\delta W = \int_V \left\{ \Theta^{\mathrm{s}}_{\alpha\beta} \frac{\partial(\delta u_\alpha)}{\partial X_\beta} + \rho \frac{\mathrm{d}}{\mathrm{D}t} (\tfrac{1}{2} U^2) \, \delta t \right\} \mathrm{d}V, \tag{6.6.4}$$

where $\boldsymbol{\Theta}^{\mathrm{s}}$ is the symmetric part of $\boldsymbol{\Theta}$; i.e.

$$\boldsymbol{\Theta}^{\mathrm{s}} = \tfrac{1}{2}(\boldsymbol{\Theta} + \boldsymbol{\Theta}^{\mathrm{t}}). \tag{6.6.5}$$

When the Cauchy stress tensor is symmetric, (6.6.4) reduces to the expression (6.2.6) obtained for the virtual work.

Summary of important equations

A. *Forces*

$\boldsymbol{f}^{\mathrm{b}}$ = extrinsic body force density

Summary of important equations

$$\boldsymbol{F}^{\mathrm{b}} = \boldsymbol{f}^{\mathrm{b}} - \rho \frac{\mathrm{D}\boldsymbol{U}}{\mathrm{D}t}$$

$\boldsymbol{f}^{\mathrm{s}}$ = extrinsic surface force density

$\boldsymbol{h}^{\mathrm{s}}$ = external surface force density acting on any part of a body (see fig. 12).

B. *Boundary conditions*

for an external surface $\boldsymbol{f}^{\mathrm{s}} = \boldsymbol{\Theta N}$, (6.1.2)

for an interface $\boldsymbol{\Theta}_1\boldsymbol{N} = \boldsymbol{\Theta}_2\boldsymbol{N}$, (6.1.3)

in general $\boldsymbol{h}^{\mathrm{s}} = \boldsymbol{\Theta N}$. (6.2.1)

C. *Virtual work on any part of a body*

$$\delta W = \int_V \left\{ \Theta_{\alpha\beta} \frac{\partial(\delta u_\alpha)}{\partial X_\beta} + \rho \frac{\mathrm{D}}{\mathrm{D}t}(\tfrac{1}{2}U^2)\,\delta t \right\} \mathrm{d}V. \qquad (6.2.6)$$

If the body is at equilibrium, the virtual work may be expressed in various forms, as follows,

$$\delta W = \int_V \Theta_{\alpha\beta} \frac{\partial(\delta u_\alpha)}{\partial X_\beta} \mathrm{d}V = \int_\Sigma \Theta_{\alpha\beta}\, \delta u_\alpha N_\beta\, \mathrm{d}\Sigma \qquad \begin{cases}(6.2.8)\\(6.2.9)\end{cases}$$

$$= \int_v B_{\beta\alpha}\{\partial(\delta u_\alpha)/\partial x_\beta\}\,\mathrm{d}v = \int_v B_{\beta\alpha}\,\delta D_{\alpha\beta}\,\mathrm{d}v \qquad (6.4.14)$$

$$= \int_v \Lambda_{\alpha\beta}\,\delta\eta_{\alpha\beta}\,\mathrm{d}v, \qquad (6.4.31)$$

where

$$\boldsymbol{B} = J\boldsymbol{D}^{-1}\boldsymbol{\Theta}, \qquad (6.4.15\mathrm{b})$$

and

$$\boldsymbol{\Lambda} = J\boldsymbol{D}^{-1}\boldsymbol{\Theta}(\boldsymbol{D}^{\mathrm{t}})^{-1}. \qquad (6.4.30\mathrm{b})$$

D. *Generalised volume co-ordinate*

$$\Omega_{\alpha\beta} = \int_v (\partial X_\alpha/\partial x_\beta)\,\mathrm{d}v = \int_\sigma X_\alpha\,\mathrm{d}\sigma_\beta. \qquad (6.4.18)$$

$$\delta W = B_{\beta\alpha}\,\delta\Omega_{\alpha\beta}. \qquad (6.4.19)$$

Generalised volume co-ordinate in infinitesimal strain theory,

$$\mathscr{V}_{\alpha\beta} = \tfrac{1}{3}v\delta_{\alpha\beta} + \frac{1}{2}\int_v \{(\partial u_\alpha/\partial x_\beta) + (\partial u_\beta/\partial x_\alpha)\}\,\mathrm{d}v$$

$$= \tfrac{1}{3}v\delta_{\alpha\beta} + \frac{1}{2}\int_\sigma (u_\alpha n_\beta + u_\beta n_\alpha)\,\mathrm{d}\sigma. \tag{6.5.3}$$

$$\delta W = \Theta_{\alpha\beta}\,\delta\mathscr{V}_{\alpha\beta}. \tag{6.5.1}$$

E. *Principle of virtual work*

$$\int_\Sigma \boldsymbol{h}^{\mathrm{s}}\cdot\delta\boldsymbol{u}\,\mathrm{d}\Sigma + \int_V \left(\boldsymbol{F}^{\mathrm{b}}\cdot\delta\boldsymbol{u} - \boldsymbol{\Theta}:\frac{\partial(\delta\boldsymbol{u})}{\partial\boldsymbol{X}}\right)\mathrm{d}V = 0, \tag{6.3.1}$$

where V is any part or all of an elastic body, and Σ the surface of V; $\delta\boldsymbol{u}$ is arbitrary except for any constraints which may apply.

Examples

1. Show that the change in volume, in an infinitesimal deformation, is given by

$$\delta V = \int_V \left(\frac{\partial}{\partial\boldsymbol{X}}\cdot\delta\boldsymbol{u}\right)\mathrm{d}V.$$

2. Prove that

$$\int_V \rho\frac{\mathrm{D}\boldsymbol{U}}{\mathrm{D}t}\,\mathrm{d}V = M\frac{\mathrm{d}\boldsymbol{U}_{\mathrm{C.M.}}}{\mathrm{d}t},$$

where M is the total mass of the material in V, and $\boldsymbol{U}_{\mathrm{C.M.}}$ is the velocity of its centre of mass.

3.(*a*) Show that (6.3.1) reduces to

$$\int_V \left\{\frac{\partial\Theta_{\alpha\beta}}{\partial X_\beta} + F^{\mathrm{b}}_\alpha\right\}\delta u_\alpha\,\mathrm{d}V = 0, \tag{1}$$

and that, since $\delta\boldsymbol{u}$ is arbitrary except at constraining points, the equation of motion may be derived from this, i.e.,

$$\frac{\partial}{\partial\boldsymbol{X}}\cdot\boldsymbol{\Theta} + \boldsymbol{F}^{\mathrm{b}} = 0. \tag{2}$$

(*b*) Prove $\boldsymbol{U} = \partial\boldsymbol{u}/\partial t$ for coherent deformations, and that

$$\frac{\partial}{\partial\boldsymbol{X}}\cdot\boldsymbol{U} = \frac{\partial}{\partial t}\left(\frac{\partial}{\partial\boldsymbol{X}}\cdot\boldsymbol{u}\right) = -\frac{\partial}{\partial t}(\rho/\rho_0),$$

where ρ_0 = density at the standard state, and the deformations are infinitesimal. Hence, neglecting body forces, derive from (2), the following equation which is exact to the first order only in $\rho - \rho_0$.

$$\frac{\partial}{\partial\boldsymbol{X}}\cdot\left(\frac{\partial}{\partial\boldsymbol{X}}\cdot\boldsymbol{\Theta}\right) = -\rho\frac{\partial^2}{\partial t^2}(\rho/\rho_0) = -\frac{\partial^2\rho}{\partial t^2}. \tag{3}$$

(*c*) In (*b*), if $\boldsymbol{\Theta} = -p\boldsymbol{E}$, show that (3) becomes

$$\nabla^2 p = \frac{\partial^2 \rho}{\partial t^2}, \tag{4}$$

where

$$\nabla^2 = \frac{\partial^2}{\partial X^2} + \frac{\partial^2}{\partial Y^2} + \frac{\partial^2}{\partial Z^2}.$$

Show that we can deduce, from (4), the wave equation

$$\nabla^2 p = \left(\frac{\rho_0}{\gamma p_0}\right)\frac{\partial^2 p}{\partial t^2}, \tag{5}$$

for the propagation of sound waves in a perfect gas; we must assume that the adiabatic equation of state, $p = k\rho^\gamma$, relates p and ρ in the motion; p_0 and ρ_0 are the undisturbed values of p and ρ.

4. In (6.3.1), take $\delta \boldsymbol{u}$ as appropriate to, first, an arbitrary translational displacement and, second, an arbitrary rigid body rotation, and hence (*a*) prove (5.4.4) and (5.4.5) for static equilibrium, and (*b*) discuss the equations of motion of an elastic body, comparing them with those for a rigid body.
5. Consider the boundary condition (6.1.3). Show that, if $\boldsymbol{N}$ is in the direction of the x_1-axis, the first column and row of $\boldsymbol{\Theta}_1$ must equal the first column and row of $\boldsymbol{\Theta}_2$.
6. Prove that, at an interface between a solid and liquid, a principal direction of the Cauchy stress tensor in the solid at a point on the interface is normal to the surface.

7.(*a*) Prove that

$$J\frac{\partial \Theta_{\beta\alpha}}{\partial X_\alpha} = \frac{\partial B_{\alpha\beta}}{\partial x_\alpha}, \quad \text{i.e.} \quad J\frac{\partial}{\partial \boldsymbol{X}} \cdot \boldsymbol{\Theta} = \frac{\partial}{\partial \boldsymbol{x}} \cdot \boldsymbol{B}^{\mathrm{t}},$$

and hence that

$$\frac{\partial \Theta_{\beta\alpha}}{\partial X_\alpha}\,\mathrm{d}V = \frac{\partial B_{\alpha\beta}}{\partial x_\alpha}\,\mathrm{d}v.$$

Hint: $\displaystyle\int_v \frac{\partial}{\partial \boldsymbol{x}} \cdot \{J(\boldsymbol{D}^{\mathrm{t}})^{-1}\}\,\mathrm{d}v = \int_\sigma J(\boldsymbol{D}^{\mathrm{t}})^{-1} \cdot \mathrm{d}\boldsymbol{\sigma} = \int_\Sigma \mathrm{d}\boldsymbol{\Sigma} = 0,$

by (5.2.1) and (2.3.9), where σ is the surface bounding the volume v in the reference state, and Σ is the surface bounding the corresponding deformed volume V. Since v is arbitrary, the first integrand must then be zero. There is a more direct but less elegant way of proving this without the divergence theorem, by using the fact that $JD^{-1}_{\alpha\mu}$ is the cofactor of $D_{\mu\alpha} = \partial X_\mu/\partial x_\alpha$ in the determinant J.

(*b*) Using (*a*), or otherwise, show that the β-component of the force on a volume element $\mathrm{d}V$ due to the surrounding material is given by

$$\frac{\partial B_{\alpha\beta}}{\partial x_\alpha}\,\mathrm{d}v.$$

(*c*) Hence prove that the equation of motion is given by

$$\frac{\partial B_{\alpha\beta}}{\partial x_\alpha} + \psi_\beta = \rho_0\left(\frac{\mathrm{D}U_\beta}{\mathrm{D}t}\right),$$

where $\psi_\beta = Jf^{\mathrm{b}}_\beta$, and is the body force per unit of *reference* volume, and $\rho_0 = J\rho$ is the density in the reference state. Use the result in (*a*) to reconcile this form with the equation of motion (5.4.2). (See chapter 5, example 1).

8. Consider the boundary conditions for the Boussinesq tensor, and show how the continuity of this tensor across an interface depends on the continuity of the strain quantities $D_{\alpha\beta}$.

9.(*a*) Show that for a homogeneous phase under hydrostatic conditions

$$B_{\alpha\beta}\Omega_{\beta\alpha} = -3pV; \quad B_{\alpha\beta}\,\mathrm{d}\Omega_{\beta\alpha} = -p\,\mathrm{d}V$$

and that for infinitesimal deformations

$$\Theta_{\alpha\beta}\mathcal{V}_{\alpha\beta} = -pV; \quad \Theta_{\alpha\beta}\,\mathrm{d}\mathcal{V}_{\beta\alpha} = -p\,\mathrm{d}V.$$

(*b*) Prove that for homogeneous conditions, hydrostatic or non-hydrostatic, $B_{\alpha\beta}\Omega_{\beta\alpha} = -3PV$ where $-3P$ is the trace of Θ.

10.(*a*) If the force on $\mathrm{d}\boldsymbol{\Sigma}$ is parallel to $\mathrm{d}\boldsymbol{\sigma}$, is the direction of $\mathrm{d}\boldsymbol{\sigma}$ a principal direction of $\boldsymbol{B}$ or of $\boldsymbol{B}^{\mathrm{t}}$?

(*b*) If the stress is hydrostatic show that $\boldsymbol{B}$ is symmetric when the deformation is a symmetric strain.

11. $\boldsymbol{B}$ is not necessarily symmetric although $\boldsymbol{\Theta}$ is – discuss the restrictions on the elements $B_{\alpha\beta}$ that this statement implies. Show that $\boldsymbol{B}$ is symmetric, if $\boldsymbol{D}$ is symmetric and commutes with $\boldsymbol{\Theta}$. Prove, in this case, that $\boldsymbol{\Theta}$, $\boldsymbol{D}$ and $\boldsymbol{B}$ all have the same principal axes.

12. A sample of homogeneous material is in the shape of a cube. It is maintained under stress by being confined by a system of walls and pistons; the stress tensor in the material is uniform but not symmetric. Using the boundary conditions at the surface of the sample, discuss whether the stress tensor can be uniform throughout the pistons and walls – if it is assumed that the material of these is such that in them the stress tensor is symmetric.

The cube may for instance consist of magnetic material to which a uniform magnetic field is applied, while the walls and pistons are made of non-magnetic material.

7

Another unique factorisation of $\boldsymbol{D}$

A unique factorisation is given of a real non-singular matrix $\boldsymbol{D}$ into $\boldsymbol{VH}$, where $\boldsymbol{V}$ is an orthogonal matrix and $\boldsymbol{H}$ is a triangular matrix with positive diagonal elements. The six independent elements of $\boldsymbol{H}$ may be used as strain co-ordinates. In general $\boldsymbol{\eta}$, being a tensor, will be more generally useful in discussing anisotropic materials, but such triangular matrices as $\boldsymbol{H}$ have group properties which enable a general strain $\boldsymbol{D}$ to be factorised into a product of three simple shears at right angles to each other multiplied by a diagonal matrix and then by an orthogonal matrix. In chapter 22 we shall show that the six elements of $\boldsymbol{H}$ have an important application in the treatment of the thermodynamic stability of solid phases.

The infinitesimal case of the factorisation is discussed and shown to correspond to the unique decomposition of an infinitesimal matrix into the sum of a skew-symmetric matrix and a triangular matrix.

7.1. The unique factorisation of $\boldsymbol{D}$ into an orthogonal matrix $\boldsymbol{V}$ and a triangular matrix $\boldsymbol{H}$ with positive diagonal elements

In this section we shall discuss the above factorisation, using a geometrical description, while in the next section we shall show how to obtain the factorisation analytically.

Let us express the factorisation in the form

$$\boldsymbol{V}^{\mathrm{t}}\boldsymbol{D} = \boldsymbol{H}, \tag{7.1.1}$$

where $\boldsymbol{H}$ is a triangular matrix of the form

$$\boldsymbol{H} = \begin{bmatrix} H_{11} & H_{12} & H_{13} \\ 0 & H_{22} & H_{23} \\ 0 & 0 & H_{33} \end{bmatrix} = \begin{bmatrix} H_1 & H_6 & H_5 \\ 0 & H_2 & H_4 \\ 0 & 0 & H_3 \end{bmatrix}. \tag{7.1.2}$$

The last form indicates the abbreviated notation to be used for triangular matrices. In the unique factorisation of $\boldsymbol{D}$ the diagonal elements of the triangular factor $\boldsymbol{H}$ are positive.

It can be seen from (7.1.2) that the transformation $\boldsymbol{H}$ has the following properties. First, the image of any point on the x_1-axis also lies on that axis, in other words, under the transformation the direction, if not the sense, of the x_1-axis will remain invariant. Second, the image of any point in the x_1Ox_2 plane also lies in this plane. Hence, under the transformation $\boldsymbol{H}$, the x_1Ox_2 plane and a straight line, the x_1-axis, are left invariant in orientation and direction, respectively.

Having noted this property of $\boldsymbol{H}$, we shall show that the transformation $\boldsymbol{D}$ may be followed by an orthogonal transformation $\boldsymbol{V}^{\mathrm{t}}$, suitably chosen so that under the resultant transformation $\boldsymbol{V}^{\mathrm{t}}\boldsymbol{D}$, an arbitrarily chosen plane containing the origin, and any straight line through the origin in that plane, are left invariant as to orientation and direction.

Consider a plane, containing the origin O, whose unit normal is $\boldsymbol{n}$. For brevity we call this the plane $\boldsymbol{n}$. Let Oa be any straight line through O and lying in the plane $\boldsymbol{n}$. Under the transformation $\boldsymbol{D}$, as shown in fig. 13, let the image of the plane $\boldsymbol{n}$ be the plane $\boldsymbol{N}$ and that of Oa be OA.

We now describe the required orthogonal transformation in the following two steps.

(*a*) We bring the plane $\boldsymbol{N}$ back to coincidence with the plane $\boldsymbol{n}$ by the rotation 1, about OB, the line of intersection of planes $\boldsymbol{N}$ and $\boldsymbol{n}$. In this rotation, OA moves to OA_1.

(*b*) By a rotation 2, about $\boldsymbol{n}$ as axis, we bring the line OA_1 to OA_2, that is, to collinearity with Oa. We note that OA_2 (and OA_1) need not necessarily be equal in length to Oa.

The product, $\boldsymbol{V}^{\mathrm{t}}$, of these two rotations is such that $\boldsymbol{V}^{\mathrm{t}}\boldsymbol{D}$ leaves invariant the orientation of the plane $\boldsymbol{n}$ and the direction of the line Oa. It may be noticed that fig. 13 has been drawn so that OA_2 and Oa are in the same sense. They could be oppositely orientated, although this would not occur in deformations of a material medium. If this were the case, we could follow the above rotations by an inversion operator $\bar{\boldsymbol{E}}$: $\boldsymbol{x} \to -\boldsymbol{x}$, so that finally OA_2 and Oa would be in the same sense. $\boldsymbol{V}^{\mathrm{t}}$, the product of rotations 1, 2 and $\bar{\boldsymbol{E}}$, would be orthogonal. Its determinant would be -1.

If we choose the line Oa as the x_1-axis, and the plane $\boldsymbol{n}$ as the x_1Ox_2 plane, we see that the transformation $\boldsymbol{V}^{\mathrm{t}}\boldsymbol{D}$ must be triangular.

From (7.1.1), since $\boldsymbol{V}^{\mathrm{t}}$ is the reciprocal of $\boldsymbol{V}$, we have obtained, geometrically, the factorisation

$$\boldsymbol{D} = \boldsymbol{V}\boldsymbol{H}, \tag{7.1.3}$$

where $\boldsymbol{V}$ is orthogonal, and where

$$J = \det(\boldsymbol{D}) = \det(\boldsymbol{V})\det(\boldsymbol{H}) = \pm H_1H_2H_3. \tag{7.1.4}$$

If, say, H_2 is negative, we may multiply, in (7.1.3), both $\boldsymbol{V}$ and $\boldsymbol{H}$ by the diagonal orthogonal matrix whose diagonal elements are $(1, -1, 1)$.

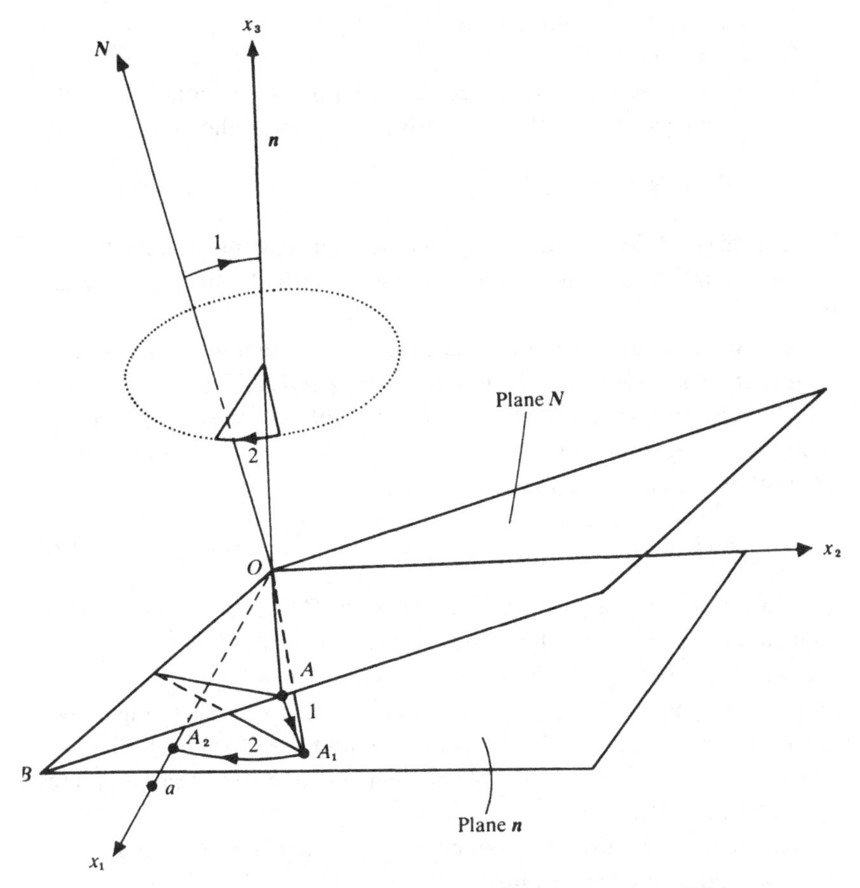

Fig. 13. How by a rotation after a deformation $\boldsymbol{D}$, a line and a plane can be returned to their original orientations. Rotation 1 about OB brings the plane $\boldsymbol{N}$ back to its original orientation. Rotation 2 about $\boldsymbol{n}$ brings the line Ox_1 finally back to its original direction Ox.

We thus obtain a factorisation of the same form as (7.1.3), but where H_2 is now positive. Since we may similarly make any other diagonal element of $\boldsymbol{H}$ positive, we have proved that $\boldsymbol{D}$ may be factorised as the product of an orthogonal matrix and a triangular matrix with *positive* diagonal elements.

It remains only to prove that such a factorisation is unique, and to do this we must first show that the triangular matrices of the form (7.1.2), with positive diagonal elements, form a group. This set of non-singular matrices is a group if the following four conditions are satisfied:

(1) the product of any two of these matrices,

(2) the reciprocal of any such matrix,
(3) the unit matrix

all belong to the set, i.e., are triangular with positive diagonal elements;
(4) the matrices satisfy the associative law of multiplication,

$$(\boldsymbol{H}_a\boldsymbol{H}_b)\boldsymbol{H}_c = \boldsymbol{H}_a(\boldsymbol{H}_b\boldsymbol{H}_c). \tag{7.1.5}$$

The fourth condition is satisfied by all finite matrices and is a result of the law of multiplication. The reader may easily verify the first three conditions.

Now we need the simple lemma that *the only triangular matrix with positive diagonal elements which is also orthogonal is the unit matrix.* The proof is elementary and direct. If $\boldsymbol{H}$ is an orthogonal matrix then, for example, its second and third rows, as given in (7.1.2), are orthogonal to each other, and thus we obtain

$$H_3H_4 = 0. \tag{7.1.6}$$

Here, since H_3 is positive, H_4 must be zero. Repeating the argument, applied to the first and third rows, we prove that H_5 must be zero. Likewise we prove that H_6 must be zero. Thus, our orthogonal triangular matrix with positive diagonal elements must be diagonal. Being orthogonal, its reciprocal must be its transpose, and therefore each diagonal element must equal its reciprocal and so must equal $+1$. This completes the proof of the lemma.

We can now establish the uniqueness of the factorisation. For suppose it is not unique and that, in fact,

$$\boldsymbol{D} = \boldsymbol{V}_1\boldsymbol{H}_1 = \boldsymbol{V}_2\boldsymbol{H}_2 \tag{7.1.7}$$

describes two possible factorisations. Then, since $\boldsymbol{H}_1$ and $\boldsymbol{H}_2$, having positive diagonal elements, are non-singular and thus have reciprocals, (7.1.7) leads to

$$\boldsymbol{V}_2^t\boldsymbol{V}_1 = \boldsymbol{H}_2\boldsymbol{H}_1^{-1}. \tag{7.1.8}$$

Because of the group properties of both types of matrix the left-hand side is an orthogonal matrix and the right-hand side is a triangular matrix with positive diagonal elements. Then, by the above lemma, both sides must equal the unit matrix, and hence the factorisation is unique.

Thus, in general, *a real non-singular matrix* $\boldsymbol{D}$ *can be uniquely factorised into the product of a triangular matrix with positive diagonal elements and an orthogonal matrix.*

7.2. How to factorise *D* analytically

In section 1 we gave a geometrical method of factorising $\boldsymbol{D}$. Now we show how it can be done analytically. We can write

$$\boldsymbol{D} = \begin{bmatrix} l_1 & m_1 & n_1 \\ l_2 & m_2 & n_2 \\ l_3 & m_3 & n_3 \end{bmatrix} \begin{bmatrix} D_1 & 0 & 0 \\ 0 & D_2 & 0 \\ 0 & 0 & D_3 \end{bmatrix}, \tag{7.2.1}$$

$$D_i = +(D_{1i}^2 + D_{2i}^2 + D_{3i}^2)^{\frac{1}{2}}, \quad i = 1, 2, 3; \tag{7.2.2}$$

that is, D_1 is the positive square root of the modulus of the first column of the matrix $\boldsymbol{D}$, etc. The first matrix, $\boldsymbol{D}'$ say, has been chosen such that

$$\begin{aligned} l_1^2 + l_2^2 + l_3^2 &= 1, \\ m_1^2 + m_2^2 + m_3^2 &= 1, \\ n_1^2 + n_2^2 + n_3^2 &= 1, \end{aligned} \tag{7.2.3}$$

i.e., the columns represent unit vectors.

If we can factorise $\boldsymbol{D}'$ in the unique manner required for $\boldsymbol{D}$, so that

$$\boldsymbol{D}' = \boldsymbol{V}\boldsymbol{H}', \tag{7.2.4}$$

then

$$\boldsymbol{D} = \boldsymbol{V}\boldsymbol{H} \tag{7.2.5}$$

is the unique factorisation required of $\boldsymbol{D}$, where $\boldsymbol{H}$ is $\boldsymbol{H}'$ multiplied on the right by the positive diagonal matrix of (7.2.1).

The problem now is to find a $\boldsymbol{V}$ such that $\boldsymbol{V}^{\mathrm{t}}\boldsymbol{D}'$ is triangular, and this can be done by choosing an orthogonal matrix,

$$\boldsymbol{V}^{\mathrm{t}} = \begin{bmatrix} l_1 & l_2 & l_3 \\ v_1 & v_2 & v_3 \\ w_1 & w_2 & w_3 \end{bmatrix} = \begin{bmatrix} \boldsymbol{l}^{\mathrm{t}} \\ \boldsymbol{v}^{\mathrm{t}} \\ \boldsymbol{w}^{\mathrm{t}} \end{bmatrix}, \tag{7.2.6}$$

where $\boldsymbol{l}^{\mathrm{t}}$ is the row vector which is the transpose of column vector $\boldsymbol{l}$, and, similarly, for $\boldsymbol{v}^{\mathrm{t}}$ and $\boldsymbol{w}^{\mathrm{t}}$.

It is clear that $\boldsymbol{V}^{\mathrm{t}}$ *is* orthogonal provided that $\boldsymbol{l}$, $\boldsymbol{v}$ and $\boldsymbol{w}$ are all unit vectors and mutually orthogonal, i.e. that $(\boldsymbol{l} \cdot \boldsymbol{v}) = (\boldsymbol{v} \cdot \boldsymbol{w}) = (\boldsymbol{w} \cdot \boldsymbol{l}) = 0$. Now it can be verified that the following choices of $\boldsymbol{v}$ and $\boldsymbol{w}$ satisfy this condition. (We note, also, that the choice of $\boldsymbol{w}$ below satisfies the condition $(\boldsymbol{w} \cdot \boldsymbol{m}) = 0$.)

$$\boldsymbol{v} = (1/s_{12})\{\boldsymbol{m} - (\boldsymbol{l} \cdot \boldsymbol{m})\boldsymbol{l}\}, \tag{7.2.7a}$$

$$\boldsymbol{w} = (1/s_{12})\boldsymbol{l} \wedge \boldsymbol{m}, \tag{7.2.7b}$$

$$s_{12} = |\boldsymbol{l} \wedge \boldsymbol{m}|. \tag{7.2.7c}$$

Because of these properties, we can immediately see that

$$\boldsymbol{V}^{\mathrm{t}}\boldsymbol{D}'=\begin{bmatrix}1 & (\boldsymbol{l}\cdot\boldsymbol{m}) & (\boldsymbol{l}\cdot\boldsymbol{n})\\ 0 & (\boldsymbol{v}\cdot\boldsymbol{m}) & (\boldsymbol{v}\cdot\boldsymbol{n})\\ 0 & 0 & (\boldsymbol{w}\cdot\boldsymbol{n})\end{bmatrix}. \tag{7.2.8}$$

If $(\boldsymbol{v}\cdot\boldsymbol{m})$ or $(\boldsymbol{w}\cdot\boldsymbol{n})$ is negative, then we can multiply $\boldsymbol{V}^{\mathrm{t}}$ by the appropriate orthogonal diagonal matrix with diagonal elements ± 1, to obtain a triangular matrix with positive diagonal elements.

So, finally, after some algebra,

$$\boldsymbol{H}=\boldsymbol{V}^{\mathrm{t}}\boldsymbol{D}=\begin{bmatrix}D_1 & D_2c_{12} & D_3c_{13}\\ 0 & D_2s_{12} & D_3(c_{23}-c_{13}c_{12})/s_{12}\\ 0 & 0 & J/D_1D_2s_{12}\end{bmatrix}. \tag{7.2.9}$$

If J is negative, we must multiply $\boldsymbol{V}$ by an appropriate diagonal matrix as described previously.

$$c_{12}=(\boldsymbol{l}\cdot\boldsymbol{m}),\qquad c_{23}=(\boldsymbol{m}\cdot\boldsymbol{n}),\qquad c_{31}=(\boldsymbol{l}\cdot\boldsymbol{n}),$$
$$s_{23}=|\boldsymbol{m}\wedge\boldsymbol{n}|;\qquad s_{31}=|\boldsymbol{l}\wedge\boldsymbol{n}|. \tag{7.2.10}$$

We have thus factorised $\boldsymbol{D}$ as required, and the uniqueness can be proved exactly as in section 1.

Now from (2.1.7) we can express the finite strain tensor as

$$\boldsymbol{\eta}=\tfrac{1}{2}(\boldsymbol{D}^{\mathrm{t}}\boldsymbol{D}-\boldsymbol{E}), \tag{7.2.11}$$

and the D_i, c_{ij}, s_{ij} of the right-hand side of (7.2.9) can be uniquely expressed in terms of the elements of $\boldsymbol{\eta}$, *provided* that J is positive (or at least its sign known). In fact

$$D_i=(2\eta_{ii}+1)^{\frac{1}{2}},\qquad c_{ij}=2\eta_{ij}/\{(2\eta_{ii}+1)(2\eta_{jj}+1)\}^{\frac{1}{2}},\quad i\neq j,$$
$$s_{ij}=(1-c_{ij}^2)^{\frac{1}{2}},\qquad J^2=\det(2\boldsymbol{\eta}+\boldsymbol{E}). \tag{7.2.12}$$

Thus, if $\boldsymbol{\eta}$ is given, $\boldsymbol{H}$ is determined. We will now show the converse.

7.3. $\boldsymbol{\eta}$ and $\boldsymbol{H}$, and the infinitesimal case

In the previous section we have shown how to determine the elements of $\boldsymbol{H}$ if those of $\boldsymbol{\eta}$ are given. The converse is a very simple matter. Since

$$\boldsymbol{D}^{\mathrm{t}}\boldsymbol{D}=\boldsymbol{H}^{\mathrm{t}}\boldsymbol{V}^{\mathrm{t}}\boldsymbol{V}\boldsymbol{H}=\boldsymbol{H}^{\mathrm{t}}\boldsymbol{H}, \tag{7.3.1}$$

it follows, from (7.2.11), that

$$2\boldsymbol{\eta}=\boldsymbol{H}^{\mathrm{t}}\boldsymbol{H}-\boldsymbol{E}. \tag{7.3.2}$$

Thus, the elements of $\boldsymbol{\eta}$ are determined, if those of $\boldsymbol{H}$ are given.

$\boldsymbol{\eta}$ depends only on the factor $\boldsymbol{H}$ and not on $\boldsymbol{V}$. This is significant in that, $\boldsymbol{\eta}$ being symmetric and $\boldsymbol{H}$ being triangular, both have six independent elements. Knowledge of one of these matrices determines, uniquely, the other. Just as the elements of $\boldsymbol{\eta}$ provide satisfactory finite strain elements, so may those of $\boldsymbol{H}$. However $\boldsymbol{\eta} \to 0$ at the reference state, whereas $\boldsymbol{H} \to \boldsymbol{E}$, hence we shall often use the elements of

$$\boldsymbol{h} = \boldsymbol{H} - \boldsymbol{E} \tag{7.3.3}$$

as finite strain elements which vanish at the reference state.

Now, if $\boldsymbol{D}$ represents an infinitesimal transformation, we can prove as follows that both $\boldsymbol{V}$ and $\boldsymbol{H}$ must be infinitesimally close to the unit matrix $\boldsymbol{E}$. First, since

$$\boldsymbol{V}^{\mathrm{t}}\boldsymbol{D} = \boldsymbol{H}, \tag{7.3.4}$$

$\boldsymbol{V}^{\mathrm{t}}$ tends to $\boldsymbol{H}$ as $\boldsymbol{D}$ tends to $\boldsymbol{E}$. Therefore, since $\boldsymbol{E}$ is the only orthogonal triangular matrix with positive diagonal elements, $\boldsymbol{V}^{\mathrm{t}}$ and $\boldsymbol{H}$ must both tend to $\boldsymbol{E}$, as $\boldsymbol{D}$ does.

Hence, in the infinitesimal case, to first order,

$$\boldsymbol{d} = \boldsymbol{l} + \boldsymbol{h}, \tag{7.3.5a}$$

where

$$\boldsymbol{d} = \boldsymbol{D} - \boldsymbol{E}, \quad \boldsymbol{l} = \boldsymbol{V} - \boldsymbol{E}. \tag{7.3.5b}$$

$\boldsymbol{l}$ is a skew-symmetric matrix (see (3.1.4)). Equation (7.3.5a) is the infinitesimal form of the unique factorisation of $\boldsymbol{D}$ into the product $\boldsymbol{VH}$. The decomposition $\boldsymbol{d}$ into the sum of a skew-symmetric and a triangular matrix is unique. This is easily demonstrated in the two dimensional case, for the simple decomposition

$$\begin{bmatrix} d_{11} & d_{12} \\ d_{21} & d_{22} \end{bmatrix} = \begin{bmatrix} 0 & -d_{21} \\ d_{21} & 0 \end{bmatrix} + \begin{bmatrix} d_{11} & d_{12}+d_{21} \\ 0 & d_{22} \end{bmatrix} \tag{7.3.6}$$

is clearly unique. The generalisation to higher dimensions is obvious and leads to

$$\begin{aligned} h_{\alpha\beta} &= d_{\alpha\beta} + d_{\beta\alpha}, \quad a < \beta \\ &= 0, \quad \alpha > \beta \\ &= d_{\alpha\beta}, \quad \alpha = \beta. \end{aligned} \tag{7.3.7}$$

From (7.3.7), in the infinitesimal case, we see that

$$\boldsymbol{h} = \begin{bmatrix} \varepsilon_{11} & 2\varepsilon_{12} & 2\varepsilon_{13} \\ 0 & \varepsilon_{22} & 2\varepsilon_{23} \\ 0 & 0 & \varepsilon_{23} \end{bmatrix} = \begin{bmatrix} e_1 & e_6 & e_5 \\ 0 & e_2 & e_4 \\ 0 & 0 & e_3 \end{bmatrix}, \tag{7.3.8}$$

where $\boldsymbol{\varepsilon}$ is the infinitesimal strain tensor $\frac{1}{2}\{\boldsymbol{d}+\boldsymbol{d}^{t}\}$, and the e_{α}, $i = 1, 2, \ldots, 6$, are infinitesimal strain elements, commonly used in standard books on solid state and crystal physics. The factor 2 occurring in off-diagonal elements will be discussed again later. Thus, these commonly used strain elements really arise from the unique decomposition discussed here.

7.4. Group properties of *D*, and factorisation of a general transformation into a product of elementary transformations

The product of two triangular matrices is

$$\boldsymbol{Hh} = \begin{bmatrix} H_1h_1 & H_1h_6+H_6h_2 & H_1h_5+H_6h_4+H_5h_3 \\ 0 & H_2h_2 & H_2h_4+H_4h_3 \\ 0 & 0 & h_3H_3 \end{bmatrix}, \tag{7.4.1}$$

where we use the abbreviated notation of (7.1.2). The simple nature of the diagonal elements of the product matrix should be noted. The reciprocal of $\boldsymbol{H}$ is easily found from (7.4.1) to be

$$\boldsymbol{H}^{-1} = \begin{bmatrix} H_1^{-1} & -H_6/(H_1H_2) & -\{H_5/(H_1H_3)\}+\{H_4H_6/(H_1H_2H_3)\} \\ 0 & H_2^{-1} & -H_4/(H_2H_3) \\ 0 & 0 & H_3^{-1} \end{bmatrix}. \tag{7.4.2}$$

From (7.4.1) and (7.4.2) we see that triangular non-singular matrices $\boldsymbol{H}$, with either or both of H_4 and H_6 zero, form a sub-group. So do matrices of the form

$$\boldsymbol{\Omega} = \begin{bmatrix} 1 & \Omega_6 & \Omega_5 \\ 0 & 1 & \Omega_4 \\ 0 & 0 & 1 \end{bmatrix}, \tag{7.4.3}$$

i.e., triangular matrices with unit diagonal elements. Further, if we define (these represent simple shears)

$$\boldsymbol{\Omega}_4 = \begin{bmatrix} 1 & 0 & 0 \\ 0 & 1 & \Omega_4 \\ 0 & 0 & 1 \end{bmatrix}; \qquad \boldsymbol{\Omega}_5 = \begin{bmatrix} 1 & 0 & \Omega_5 \\ 0 & 1 & 0 \\ 0 & 0 & 1 \end{bmatrix}, \tag{7.4.4}$$

with a similar definition for $\boldsymbol{\Omega}_6$, then we see that

$$\boldsymbol{\Omega}_4\boldsymbol{\Omega}_6 = \begin{bmatrix} 1 & \Omega_6 & 0 \\ 0 & 1 & \Omega_4 \\ 0 & 0 & 1 \end{bmatrix}, \tag{7.4.5a}$$

$$\boldsymbol{\Omega}_4\boldsymbol{\Omega}_5 = \boldsymbol{\Omega}_5\boldsymbol{\Omega}_4 = \begin{bmatrix} 1 & 0 & \Omega_5 \\ 0 & 1 & \Omega_4 \\ 0 & 0 & 1 \end{bmatrix}, \tag{7.4.5b}$$

$$\boldsymbol{\Omega}_6\boldsymbol{\Omega}_5 = \boldsymbol{\Omega}_5\boldsymbol{\Omega}_6 = \begin{bmatrix} 1 & \Omega_6 & \Omega_5 \\ 0 & 1 & 0 \\ 0 & 0 & 1 \end{bmatrix}, \tag{7.4.5c}$$

but

$$\boldsymbol{\Omega}_6\boldsymbol{\Omega}_4 = \begin{bmatrix} 1 & \Omega_6 & \Omega_4\Omega_6 \\ 0 & 1 & \Omega_4 \\ 0 & 0 & 1 \end{bmatrix} \neq \boldsymbol{\Omega}_4\boldsymbol{\Omega}_6. \tag{7.4.5d}$$

Thus, if we take care to keep $\boldsymbol{\Omega}_4$ to the left of $\boldsymbol{\Omega}_6$, we can factorise

$$\boldsymbol{H} = \begin{bmatrix} H_1 & 0 & 0 \\ 0 & H_2 & 0 \\ 0 & 0 & H_3 \end{bmatrix} \begin{bmatrix} 1 & \Omega_6 & \Omega_5 \\ 0 & 1 & \Omega_4 \\ 0 & 1 & 1 \end{bmatrix}$$

$$= \tilde{\boldsymbol{H}}\boldsymbol{\Omega}_4\boldsymbol{\Omega}_5\boldsymbol{\Omega}_6, \tag{7.4.6}$$

and

$$\boldsymbol{D} = \boldsymbol{V}\tilde{\boldsymbol{H}}\boldsymbol{\Omega}_4\boldsymbol{\Omega}_5\boldsymbol{\Omega}_6, \tag{7.4.7}$$

where $\Omega_4 = H_4/H_2$; $\Omega_5 = H_5/H_1$; $\Omega_6 = H_6/H_1$, and $\tilde{\boldsymbol{H}}$ is the diagonal matrix whose diagonal elements are H_1, H_2, H_3. Thus, we have decomposed a general deformation $\boldsymbol{D}$ into the product of three simple shears at right angles to each other, in the order $\boldsymbol{\Omega}_6$, $\boldsymbol{\Omega}_5$, $\boldsymbol{\Omega}_4$, and followed by an anisotropic dilatation $\tilde{\boldsymbol{H}}$, followed by a rotation $\boldsymbol{V}$.

Summary of important equations

A. *The unique factorisation*
of any real non-singular matrix $\boldsymbol{D}$ is given by

$$\boldsymbol{D} = \boldsymbol{V}\boldsymbol{H}, \tag{7.1.3}$$

where $\boldsymbol{V}$ is orthogonal and $\boldsymbol{H}$ is triangular with positive diagonal elements.

$$\boldsymbol{H} = \begin{bmatrix} H_1 & H_6 & H_5 \\ 0 & H_2 & H_4 \\ 0 & 0 & H_3 \end{bmatrix}, \tag{7.1.2}$$

$$\boldsymbol{D}^t\boldsymbol{D} = \boldsymbol{H}^t\boldsymbol{H}, \tag{7.3.1}$$

Another unique factorisation of $\boldsymbol{D}$

$$2\boldsymbol{\eta} = \boldsymbol{D}^{t}\boldsymbol{D} - \boldsymbol{E} = \boldsymbol{H}^{t}\boldsymbol{H} - \boldsymbol{E}, \tag{7.3.2}$$

$$\boldsymbol{h} = \boldsymbol{H} - \boldsymbol{E}. \tag{7.3.3}$$

B. *The infinitesimal case*

$$\boldsymbol{d} = \boldsymbol{l} + \boldsymbol{h}, \tag{7.3.5a}$$

where

$$\boldsymbol{d} = \boldsymbol{D} - \boldsymbol{E}, \quad \boldsymbol{l} = \boldsymbol{V} - \boldsymbol{E}. \tag{7.3.5b}$$

$$\boldsymbol{h} = \begin{bmatrix} e_1 & e_6 & e_5 \\ 0 & e_2 & e_4 \\ 0 & 0 & e_3 \end{bmatrix} = \begin{bmatrix} \varepsilon_{11} & 2\varepsilon_{12} & 2\varepsilon_{13} \\ 0 & \varepsilon_{22} & 2\varepsilon_{23} \\ 0 & 0 & \varepsilon_{33} \end{bmatrix}. \tag{7.3.8}$$

C. *The decomposition of a general deformation matrix* $\boldsymbol{D}$

$$\boldsymbol{D} = \boldsymbol{V}\tilde{\boldsymbol{H}}\boldsymbol{\Omega}_4\boldsymbol{\Omega}_5\boldsymbol{\Omega}_6, \tag{7.4.7}$$

$$\tilde{\boldsymbol{H}} = \text{Diagonal}(H_1, H_2, H_3),$$

$$\boldsymbol{\Omega}_4 = \begin{bmatrix} 1 & 0 & 0 \\ 0 & 1 & \Omega_4 \\ 0 & 0 & 1 \end{bmatrix}, \qquad \boldsymbol{\Omega}_5 = \begin{bmatrix} 1 & 0 & \Omega_5 \\ 0 & 1 & 0 \\ 0 & 0 & 1 \end{bmatrix}, \tag{7.4.4}$$

with a similar definition for $\boldsymbol{\Omega}_6$.

Examples

1. Prove that the modulus of each column of $\boldsymbol{D}$ is equal to the modulus of the corresponding column of $\boldsymbol{H}$, the triangular factor of $\boldsymbol{D}$.

2.(*a*) Prove that, in general, $\boldsymbol{H}$ has at least one eigenvector, and show what it is. Also show that $\boldsymbol{H}^{t}$ has at least one eigenvector and find what it is.

(*b*) What are the eigenvalues of $\boldsymbol{H}$?

3. Prove that, under orthogonal transformations of axes, det($\boldsymbol{H}$) is invariant.

4. Prove that, for a strain described by a triangular matrix $\boldsymbol{H}$, the orientation of the xy plane remains unchanged, by using (2.3.10), the transformation formula for a surface element.

8

Virtual work

In chapter 6 we showed that the virtual work can be expressed in terms of the infinitesimal virtual variations $\mathrm{d}\boldsymbol{D}$ or of $\mathrm{d}\boldsymbol{\eta}$. In this chapter we shall consider these variations to be homogeneous, and discuss the work done per unit reference volume. Further, it has been shown in chapter 7 that $\boldsymbol{D}$ may be factorised in terms of an orthogonal matrix times a triangular matrix $\boldsymbol{H}$ with positive diagonal elements. In considering the work done, it would be expected that the work done on the body would not be contributed to by the orthogonal matrix and it will be shown that this is so. The virtual work will be expressed in terms of $\mathrm{d}\boldsymbol{H}$, involving six independent elements. The corresponding stress elements will be discussed.

8.1. Virtual work

If the infinitesimal variations of strain described by $\mathrm{d}\boldsymbol{D}$ are homogeneous, as well as $\boldsymbol{D}$, we see from (6.4.14) and (6.4.31) that we can express the virtual work as

$$\mathrm{d}'w = B_{\alpha\beta}\,\mathrm{d}D_{\beta\alpha} = \mathrm{trace}(\boldsymbol{B}\,\mathrm{d}\boldsymbol{D}) \tag{8.1.1a}$$

$$= \Lambda_{\alpha\beta}\,\mathrm{d}\eta_{\beta\alpha} = \mathrm{trace}(\boldsymbol{\Lambda}\,\mathrm{d}\boldsymbol{\eta}), \tag{8.1.1b}$$

where

$$\boldsymbol{B} = J\boldsymbol{D}^{-1}\boldsymbol{\Theta}; \qquad \boldsymbol{\Lambda} = J\boldsymbol{D}^{-1}\boldsymbol{\Theta}(\boldsymbol{D}^{\mathrm{t}})^{-1}, \tag{8.1.2}$$

where $\boldsymbol{\Theta}$ is the symmetric Cauchy stress tensor, and $\boldsymbol{B}$ and $\boldsymbol{\Lambda}$ are both tensors; $\boldsymbol{B}$, the Boussinesq stress tensor, is not necessarily symmetric, whereas $\boldsymbol{\Lambda}$ is. $\mathrm{d}'w = \mathrm{d}'W/V_0$ is the work per unit reference volume, V_0 being the reference volume of the body. $\mathrm{d}'w$ is distinguished by the prime to show that it is not a perfect differential like $\mathrm{d}u$ and $\mathrm{d}f$, i.e., that it is not a differential of a function of the thermodynamic state variables. As is well known in thermodynamics, the work done in a process depends on the actual path and not merely on the initial and final states, and so it cannot be expressed as an increment of a state function.

In (8.1.1a), $\mathrm{d}'w$ is expressed in terms of the nine independent elements of $\mathrm{d}\boldsymbol{D}$, whereas in (8.1.1b) it is expressed in terms of the six independent elements of $\boldsymbol{\eta}$. This is because, when $\mathrm{d}\boldsymbol{D}$ represents a rotation, defined by three parameters such as the Euler angles, the work done is zero. This can be demonstrated as follows.

Because the real non-singular matrices $\boldsymbol{D}$ form a group, then $\boldsymbol{D}+\mathrm{d}\boldsymbol{D}$ must be the product of $\boldsymbol{D}$ and a transformation matrix infinitesimally close to $\boldsymbol{E}$, the unit matrix. That is,

$$\boldsymbol{D}+\mathrm{d}\boldsymbol{D}=(\boldsymbol{E}+\mathrm{d}\boldsymbol{K})\boldsymbol{D}, \tag{8.1.3}$$

and so

$$\mathrm{d}\boldsymbol{D}=\mathrm{d}\boldsymbol{K}\boldsymbol{D}; \qquad \mathrm{d}\boldsymbol{K}=\mathrm{d}\boldsymbol{D}\boldsymbol{D}^{-1}. \tag{8.1.4}$$

In general $\mathrm{d}\boldsymbol{K}$ is known as the infinitesimal operator of the group and is the basis of the treatment of continuous Lie groups. If $\mathrm{d}\boldsymbol{D}$ describes a rotation, $\mathrm{d}\boldsymbol{K}$ must be skew-symmetric, say $\mathrm{d}\boldsymbol{I}$. Thus from (8.1.1) and (8.1.2)

$$\begin{aligned} \mathrm{d}'w &= J\,\mathrm{trace}(\boldsymbol{D}^{-1}\boldsymbol{\Theta}\,\mathrm{d}\boldsymbol{D}) \\ &= J\,\mathrm{trace}(\boldsymbol{D}^{-1}\boldsymbol{\Theta}\,\mathrm{d}\boldsymbol{I}\boldsymbol{D}). \end{aligned} \tag{8.1.5}$$

Now a commonly used property of traces is

$$\mathrm{trace}(\boldsymbol{AB}\cdots\boldsymbol{CD})=\mathrm{trace}(\boldsymbol{DAB}\cdots\boldsymbol{C}), \tag{8.1.6}$$

where $\boldsymbol{A}, \boldsymbol{B}, \ldots, \boldsymbol{C}, \boldsymbol{D}$ are any matrices. This is left to the reader to prove (start with just two matrices, $\boldsymbol{A}$ and $\boldsymbol{B}$).

Hence

$$\begin{aligned} \mathrm{d}'w = J\,\mathrm{trace}(\boldsymbol{\Theta}\,\mathrm{d}\boldsymbol{I}) &= J\Theta_{\alpha\beta}\,\mathrm{d}I_{\beta\alpha} \\ &= 0, \end{aligned} \tag{8.1.7}$$

where this follows since $\boldsymbol{\Theta}$ and $\mathrm{d}\boldsymbol{I}$ are symmetric and skew-symmetric respectively (see chapter 2, example 12).

In the next sections, we shall show how (8.1.1a) may be transformed into an expression involving only the six independent elements of $\boldsymbol{H}$ or $\boldsymbol{A}$.

8.2. Expression of $\mathrm{d}'w$ in terms of $\mathrm{d}\boldsymbol{H}$

Suppose that we have a virtual variation, $\mathrm{d}\boldsymbol{D}$, then we can factorise as follows,

$$\boldsymbol{D}=\boldsymbol{V}\boldsymbol{H}, \tag{8.2.1}$$

$$\boldsymbol{D}+\mathrm{d}\boldsymbol{D}=(\boldsymbol{V}+\mathrm{d}\boldsymbol{V})(\boldsymbol{H}+\mathrm{d}\boldsymbol{H}), \tag{8.2.2}$$

where $\boldsymbol{V}+\mathrm{d}\boldsymbol{V}$, $\boldsymbol{H}+\mathrm{d}\boldsymbol{H}$ are the unique factors of $\boldsymbol{D}+\mathrm{d}\boldsymbol{D}$, just as $\boldsymbol{VH}=\boldsymbol{D}$, as discussed in chapter 7. Thus,

$$\mathrm{d}\boldsymbol{D} = \mathrm{d}\boldsymbol{V}\boldsymbol{H} + \boldsymbol{V}\,\mathrm{d}\boldsymbol{H}. \tag{8.2.3}$$

Now, by the argument of section 1, we can take

$$\boldsymbol{V} + \mathrm{d}\boldsymbol{V} = \boldsymbol{V}(\boldsymbol{E} + \mathrm{d}\boldsymbol{I}), \tag{8.2.4}$$

because orthogonal matrices form a group, and so

$$\mathrm{d}\boldsymbol{V} = \boldsymbol{V}\,\mathrm{d}\boldsymbol{I}. \tag{8.2.5}$$

Hence, using (8.1.1a), (8.1.2) and (8.2.5), we have

$$\begin{aligned} \mathrm{d}'w &= J\ \mathrm{trace}\{\boldsymbol{H}^{-1}\boldsymbol{\Theta}_v\,\mathrm{d}\boldsymbol{H} + \boldsymbol{H}^{-1}\boldsymbol{\Theta}_v\,\mathrm{d}\boldsymbol{I}\boldsymbol{H}\} \\ &= J\ \mathrm{trace}\{\boldsymbol{H}^{-1}\boldsymbol{\Theta}_v\,\mathrm{d}\boldsymbol{H} + \boldsymbol{\Theta}_v\,\mathrm{d}\boldsymbol{I}\} \\ &= \mathrm{trace}\{J\boldsymbol{H}^{-1}\boldsymbol{\Theta}_v\,\mathrm{d}\boldsymbol{H}\} = \mathrm{trace}(\boldsymbol{B}_v\,\mathrm{d}\boldsymbol{H}), \end{aligned} \tag{8.2.6}$$

where

$$\boldsymbol{\Theta}_v = \boldsymbol{V}^{\mathrm{t}}\boldsymbol{\Theta}\boldsymbol{V} = \boldsymbol{\Theta}_v^{\mathrm{t}}; \qquad \boldsymbol{B}_v = J\boldsymbol{H}^{-1}\boldsymbol{\Theta}_v. \tag{8.2.7}$$

In the second term, within the brackets of the first line of (8.2.6), we have used (8.1.6) to cancel the two matrix factors $\boldsymbol{H}^{-1}$ and $\boldsymbol{H}$. This second term then disappears from the equation since, after the above cancellation, it has reduced to the zero-trace product of a symmetric and a skew-symmetric matrix (chapter 2, example 12). $\boldsymbol{\Theta}_v$ is symmetric (i.e. equals its transpose), since (i) $\boldsymbol{\Theta}$ is symmetric and (ii) because of the well known matrix property:

$$(\boldsymbol{A}\boldsymbol{B}\cdots\boldsymbol{C}\boldsymbol{D})^{\mathrm{t}} = \boldsymbol{D}^{\mathrm{t}}\boldsymbol{C}^{\mathrm{t}}\cdots\boldsymbol{B}^{\mathrm{t}}\boldsymbol{A}^{\mathrm{t}}, \tag{8.2.8}$$

the proof of which is left as an exercise (to see the method, prove it for two matrices first, using suffix notation, $(\boldsymbol{AB})_{\alpha\beta} = A_{\alpha\mu}B_{\mu\beta}$). Now in (8.2.6), $\boldsymbol{B}_v$ is not in general triangular or symmetric, but its elements above the diagonal do not appear – because d$\boldsymbol{H}$ is triangular and thus has zero elements below the diagonal. That is, written in full

$$\begin{aligned} \mathrm{d}'w = {} & B_{v11}\,\mathrm{d}H_{11} + B_{v22}\,\mathrm{d}H_{22} + B_{v33}\,\mathrm{d}H_{33} + B_{v21}\,\mathrm{d}H_{12} \\ & + B_{v31}\,\mathrm{d}H_{13} + B_{v32}\,\mathrm{d}H_{23}. \end{aligned} \tag{8.2.9}$$

Thus it is convenient to define a triangular matrix $\boldsymbol{L}$ (with elements zero below the diagonal, like $\boldsymbol{H}$) so that

$$\begin{aligned} L_{\alpha\beta} &= B_{v\beta\alpha}, & \alpha &\leqslant \beta \\ &= 0, & \alpha &> \beta. \end{aligned} \tag{8.2.10}$$

We can thus use an abbreviated notation agreeing with that for $\boldsymbol{H}$,

$$H_\mu = H_{(\alpha\beta)_\mu}, \tag{8.2.11}$$

according to the scheme

$$\left.\begin{array}{lcccccc} \mu & 1 & 2 & 3 & 4 & 5 & 6 \\ (\alpha\beta)_\mu & 11 & 22 & 33 & 23 & 13 & 12. \end{array}\right\} \tag{8.2.12}$$

We similarly define

$$L_{(\alpha\beta)_\mu} = L_\mu, \quad \mu = 1, 2, \ldots, 6 \tag{8.2.13}$$

and then

$$d'w = L_\mu \, dH_\mu. \tag{8.2.14}$$

It may be worth mentioning that, since $\boldsymbol{B}_v = J\boldsymbol{H}^{-1}\boldsymbol{\Theta}_v$, $\boldsymbol{HB}_v$ must be symmetric. This restriction means that if the six $L_{\alpha\beta}$ are known, then from (8.2.10) the other three elements $B_{v12}, B_{v13}, B_{v23}$ can be determined (if $\boldsymbol{H}$ is also known) from the three equations resulting from this restriction. In other words, the six elements L_μ are independent of this restriction.

The meaning of $\boldsymbol{\Theta}_v$ can be understood as follows.

Consider an elastic body which may be homogeneously deformed from some reference state. The stress required for such deformations arises from forces produced in apparatus surrounding the body. The forces may be measured on suitable gauges. At setting A of the gauges, suppose that the deformation matrix $\boldsymbol{D}'$ is equal to $\boldsymbol{H}$, a triangular matrix, then the factor $\boldsymbol{V}'$ of $\boldsymbol{D}'$ is the unit matrix. The corresponding Cauchy stress tensor is $\boldsymbol{\Theta}'$. Let the forces be altered infinitesimally so that the gauge setting is B. The deformation matrix is now $\boldsymbol{D}' + d\boldsymbol{D}'$, where

$$d\boldsymbol{D}' = d\boldsymbol{V}'\boldsymbol{H} + d\boldsymbol{H}', \tag{8.2.15}$$

$d\boldsymbol{V}'$ being skew-symmetric. The work done in this variation is given by (8.2.6) and is

$$d'w = \operatorname{trace}(J\boldsymbol{H}^{-1}\boldsymbol{\Theta}' \, d\boldsymbol{H}'), \tag{8.2.16}$$

where $J = \det(\boldsymbol{H})$.

Now, suppose that we repeat this infinitesimal change of the gauge settings from A to B, after applying a finite rotation $\boldsymbol{V}$ to both the body and all of the apparatus producing the applied forces. It is clear that, at settings A and B respectively, the deformation matrices will be

$$\boldsymbol{D} = \boldsymbol{VH}, \tag{8.2.17}$$

$$\boldsymbol{D} + d\boldsymbol{D} = \boldsymbol{V}(\boldsymbol{H} + d\boldsymbol{D}'). \tag{8.2.18}$$

Therefore

$$\mathrm{d}\boldsymbol{D} = \boldsymbol{V}\,\mathrm{d}\boldsymbol{D}', \tag{8.2.19}$$

and, from (8.2.15) and (8.2.3),

$$\mathrm{d}\boldsymbol{D} = \boldsymbol{V}\,\mathrm{d}\boldsymbol{V}'\boldsymbol{H} + \boldsymbol{V}\,\mathrm{d}\boldsymbol{H}', \tag{8.2.20}$$

$$= \mathrm{d}\boldsymbol{V}\boldsymbol{H} + \boldsymbol{V}\,\mathrm{d}\boldsymbol{H}. \tag{8.2.21}$$

Since the factorisation (8.2.17) is unique, comparison of (8.2.20) and (8.2.21) yields

$$\mathrm{d}\boldsymbol{V} = \boldsymbol{V}\,\mathrm{d}\boldsymbol{V}'; \qquad \mathrm{d}\boldsymbol{H} = \mathrm{d}\boldsymbol{H}'. \tag{8.2.22}$$

The work done in the variation from setting A to B is, from (8.2.6),

$$\mathrm{d}'w = \mathrm{trace}(J\boldsymbol{H}^{-1}\boldsymbol{\Theta}_v\,\mathrm{d}\boldsymbol{H}), \tag{8.2.23}$$

where $\det(\boldsymbol{D}) = \det(\boldsymbol{H}) = J$. Hence, because $\mathrm{d}\boldsymbol{H} = \mathrm{d}\boldsymbol{H}'$, and because no work was done in the rotation $\boldsymbol{V}$, we can equate (8.2.16) and (8.2.23). Since $\mathrm{d}\boldsymbol{H}$ may be arbitrarily chosen, we must have (see example 6)

$$\boldsymbol{\Theta}_v = \boldsymbol{\Theta}', \tag{8.2.24}$$

and thus we see that $\boldsymbol{\Theta}_v$ is the Cauchy stress tensor which will produce the triangular deformation $\boldsymbol{H}$.

$$\boldsymbol{\Theta} = \boldsymbol{V}\boldsymbol{\Theta}_v\boldsymbol{V}^{\mathrm{t}} \tag{8.2.25}$$

is the Cauchy stress tensor which will produce the deformation $\boldsymbol{D} = \boldsymbol{V}\boldsymbol{H}$. Hence, if we know the Cauchy stress tensor required to produce a triangular deformation $\boldsymbol{H}$, (8.2.25) determines the stress tensor required to produce any deformation $\boldsymbol{D}$, which has $\boldsymbol{H}$ as its unique triangular factor.

In the next chapter we shall show that (8.2.25) is in the form of the well-known transformation of a second rank Cartesian tensor, under a transformation, $\boldsymbol{V}$, of the co-ordinate axis system.

8.3. Other strain co-ordinates

The unique factorisation $\boldsymbol{D} = \boldsymbol{R}\boldsymbol{A}$, where $\boldsymbol{R}$ and $\boldsymbol{A}$ are respectively an orthogonal and symmetric positive-definite real matrix, enables us, in a similar way to that used in section 2, to express the virtual work in terms of the six independent elements of $\boldsymbol{A}$. If interested the reader could carry out this treatment and obtain the corresponding stress elements conjugate to the elements of $\boldsymbol{A}$. That is, the virtual work may be obtained in the form

$$d'w = \chi_\alpha \, dA_\alpha \tag{8.3.1}$$

where $A_i = A_{ii}$, $i = 1, 2, 3$ and $A_4 = 2A_{23}$, etc. χ_α, the stress element conjugate to A_α, can be shown to be an element of a symmetric stress tensor. ***D*** and ***A***, the symmetric factor, transform as second rank tensors – obviously ***D*** does, and the proof of this, as well as of the tensor nature of ***A***, is an interesting exercise.

Other strain co-ordinates can be introduced and are useful in special cases. For example, in the infinitesimal case, one can use (8.1.1a) or (8.1.1b) or (8.2.14) to show that

$$\begin{aligned} d'w = {} & -P\,dJ + (\Theta_2 - \Theta_1)\,da_2 + (\Theta_3 - \Theta_1)\,da_3 \\ & + \Theta_4\,da_4 + \Theta_5\,da_5 + \Theta_6\,da_6 \end{aligned} \tag{8.3.2}$$

where

$$P = -\tfrac{1}{3}(\Theta_1 + \Theta_2 + \Theta_3). \tag{8.3.3}$$

$$\Theta_4 = \Theta_{23}, \qquad \Theta_5 = \Theta_{31}, \qquad \Theta_6 = \Theta_{12}; \tag{8.3.4}$$

$$a_i = e_i - \tfrac{1}{3}(e_1 + e_2 + e_3), \quad i = 1, 2, 3;$$

$$a_i = e_i, \quad i = 4, 5, 6. \tag{8.3.5}$$

P is known as the average pressure.

Thus, (8.3.2) reduces to $d'W = -P\,dV$ in the hydrostatic case, since $dV = V_0\,dJ$.

We shall generally use $\boldsymbol{\eta}$ for finite deformations, and $\boldsymbol{\varepsilon}$ or its triangular component ***e*** in the infinitesimal case, while ***H*** will be important in stability theory. Unfortunately ***H*** does not transform linearly as a tensor. In fact, the transformation of ***H***, the triangular factor of the tensor ***D***, is non-linear and can be shown (see chapter 12, section 3) to be of the form

$$\boldsymbol{H}' = \boldsymbol{UHS}^{t}, \tag{8.3.6}$$

under an orthogonal transformation of co-ordinates given by

$$\boldsymbol{x}' = \boldsymbol{Sx} \tag{8.3.7}$$

(see chapter 9). Here ***U*** is orthogonal and must be determined from (8.3.6) by the three equations requiring ***H***′ to be triangular, i.e., $H'_{21} = H'_{31} = H'_{32} = 0$. Hence ***U*** will depend not only on ***S*** but on ***H***, and the transformation is non-linear.

Since they transform non-linearly, the elements of ***H*** are not particularly useful in the study of anisotropic crystalline material. However, the fact that such matrices form a group of transformations, subject to a fixed geometrical restriction, makes them useful in stability theory.

Summary of important equations

A. *Virtual work in finite strain theory*

$$d'w = B_{\alpha\beta}\, dD_{\beta\alpha} = \text{trace}(\boldsymbol{B}\, d\boldsymbol{D}), \quad (8.1.1a)$$

$$= \Lambda_{\alpha\beta}\, d\eta_{\beta\alpha} = \text{trace}(\boldsymbol{\Lambda}\, d\boldsymbol{\eta}), \quad (8.1.1b)$$

$$= L_\mu\, dH_\mu = L_{\alpha\beta}\, dH_{\alpha\beta},\ \alpha \leqslant \beta. \quad (8.2.14)$$

where

$$\boldsymbol{B} = J\boldsymbol{D}^{-1}\boldsymbol{\Theta}, \qquad \boldsymbol{\Lambda} = J\boldsymbol{D}^{-1}\boldsymbol{\Theta}(\boldsymbol{D}^{t})^{-1}, \quad (8.1.2)$$

$$d'w = d'W/V_0 = \text{work per unit reference volume,}$$

$$V_0 = \text{reference volume.}$$

$$L_\mu = L_{(\alpha\beta)_\mu}, \qquad H_\mu = H_{(\alpha\beta)_\mu}, \quad \begin{matrix}(8.2.13)\\(8.2.11)\end{matrix}$$

where $\boldsymbol{L}$ is defined by

$$L_{\alpha\beta} = B_{v\beta\alpha}, \quad \alpha \leqslant \beta$$

$$= 0, \quad \alpha > \beta, \quad (8.2.10)$$

and

$$\boldsymbol{B}_v = J\boldsymbol{H}^{-1}\boldsymbol{\Theta}_v,\ \boldsymbol{\Theta}_v = \boldsymbol{V}^t\boldsymbol{\Theta}\boldsymbol{V}. \quad (8.2.7)$$

B. *Abbreviated notation*

μ	1	2	3	4	5	6
$(\alpha\beta)_\mu$	11	22	33	23	13	12.

(8.2.12)

C. *Virtual work in infinitesimal strain theory*

$$d'w = \Theta_\alpha\, de_\alpha$$

$$= -P\, dJ + (\Theta_2 - \Theta_1)\, da_2 + (\Theta_3 - \Theta_1)\, da_3$$

$$+ \Theta_4\, da_4 + \Theta_5\, da_5 + \Theta_6\, da_6, \quad (8.3.2)$$

where

$$P = -\tfrac{1}{3}(\Theta_1 + \Theta_2 + \Theta_3), \quad (8.3.3)$$

$$\Theta_4 = \Theta_{23}, \qquad \Theta_5 = \Theta_{31}, \qquad \Theta_6 = \Theta_{12}; \quad (8.3.4)$$

$$a_i = e_i - \tfrac{1}{3}(e_1 + e_2 + e_3), \quad i = 1, 2, 3;$$

$$a_i = e_i, \quad i = 4, 5, 6. \tag{8.3.5}$$

Examples

1. Show that

$$\boldsymbol{\Lambda} = J\boldsymbol{H}^{-1}\boldsymbol{\Theta}_v(\boldsymbol{H}^t)^{-1}.$$

2. Prove that $L_\mu H_\mu = -3PJ$, where $\Theta_{v\alpha\alpha} = \Theta_{\beta\beta} = -3P$.
3. Suppose that we restrict a body, whose six external faces are initially rectangles, so that the only homogeneous deformations which can occur are of the form

$$\tilde{\boldsymbol{H}} = \begin{bmatrix} H_1 & 0 & 0 \\ 0 & H_2 & 0 \\ 0 & 0 & H_3 \end{bmatrix}.$$

Prove that

$$\begin{aligned} \mathrm{d}'w &= H_2H_3\Theta_1\,\mathrm{d}H_1 + H_1H_3\Theta_2\,\mathrm{d}H_2 + H_1H_2\Theta_3\,\mathrm{d}H_3 \\ &= F_{11}\,\mathrm{d}H_1 + F_{22}\,\mathrm{d}H_2 + F_{33}\,\mathrm{d}H_3, \end{aligned}$$

where F_{ii} is the *normal* component of the total force acting on the faces perpendicular to the x_i-axis.
4. In example 3, if $\Theta_1 = \Theta_2 = \Theta_3 = -p$, show that

$$\mathrm{d}'w = -p\,\mathrm{d}J, \quad \mathrm{d}W = -p\,\mathrm{d}V.$$

5. For the infinitesimal case, prove that (8.2.6) reduces to $\mathrm{d}'w = \Theta_\alpha\,\mathrm{d}e_\alpha$.
6. If

$$\mathrm{trace}(\boldsymbol{H}^{-1}\boldsymbol{\Theta}_v\,\mathrm{d}\boldsymbol{H}) = \mathrm{trace}(\boldsymbol{H}^{-1}\boldsymbol{\Theta}'\,\mathrm{d}\boldsymbol{H}),$$

for arbitrary d$\boldsymbol{H}$, prove that

$$\boldsymbol{\Theta}_v = \boldsymbol{\Theta}'.$$

Hints:
(1) H_1, H_2, H_3 are all non-zero, since $\boldsymbol{H}$ is non-singular.
(2) Consider matrices d$\boldsymbol{H}^i$, such that

$$\mathrm{d}H_i^i \neq 0, \quad \mathrm{d}H_j^i = 0, \quad i \neq j.$$

(3) Substitute each d$\boldsymbol{H}^i$ in the first equation, of this example, in the sequence $i = 4, 5, 6, 1, 2, 3$, and so prove, in this sequence, that $\Theta'_{vi} = \Theta'_i$, all i.

9

Transformation of Cartesian tensors

An elementary discussion of the transformations of Cartesian tensor components is given, as well as a brief discussion of symmetry properties of crystalline materials.

9.1. Cartesian tensors

When we restrict ourselves to the use of orthonormal Cartesian coordinate axis systems, a physical quantity such as the displacement, velocity, or acceleration of a particle, is represented by a vector, the components of which are an ordered triad of numbers. For example, the displacement vector $\boldsymbol{x}$ is represented by the Cartesian vector (x_1, x_2, x_3), when referred to a given axis system, defined by a set of orthonormal vectors $\boldsymbol{i}_1, \boldsymbol{i}_2, \boldsymbol{i}_3$.

If we choose (or transform to) a different axis system, represented by the orthonormal set $\boldsymbol{i}'_1, \boldsymbol{i}'_2, \boldsymbol{i}'_3$, a vector will have a certain representation when referred to this system. It will have a different set of components, and these will be related to its components referred to the system $\boldsymbol{i}_1, \boldsymbol{i}_2, \boldsymbol{i}_3$. An observer O, using the first axis system, and an observer O', using the second, must agree on plotting the vector as to its magnitude and direction. That is, the vector is an invariant quantity, and if we use the vector $\boldsymbol{x}$ to illustrate the method, this invariance may be expressed by

$$x_\alpha \boldsymbol{i}_\alpha = x'_\gamma \boldsymbol{i}'_\gamma. \tag{9.1.1}$$

Now, we can solve for the x'_γ, by taking the scalar product of both sides of (9.1.1) with $\boldsymbol{i}'_\beta$, to obtain

$$\begin{aligned} x_\alpha(\boldsymbol{i}_\alpha \cdot \boldsymbol{i}'_\beta) &= x'_\gamma(\boldsymbol{i}'_\gamma \cdot \boldsymbol{i}'_\beta) \\ &= x'_\gamma \delta_{\gamma\beta}, \end{aligned} \tag{9.1.2}$$

from the orthonormal properties of the unit vectors $\boldsymbol{i}'_\alpha$.

Thus

$$x'_\beta = x_\alpha(\boldsymbol{i}_\alpha \cdot \boldsymbol{i}'_\beta), \tag{9.1.3}$$

which may be compared with

$$\boldsymbol{i}'_\beta = (\boldsymbol{i}'_\beta \cdot \boldsymbol{i}_\alpha)\boldsymbol{i}_\alpha. \tag{9.1.4}$$

Equation (9.1.3) may be written as

$$x'_\beta = S_{\beta\alpha} x_\alpha, \tag{9.1.5}$$

and (9.1.4) as

$$\boldsymbol{i}'_\beta = S_{\beta\alpha} \boldsymbol{i}_\alpha, \tag{9.1.6}$$

where

$$S_{\beta\alpha} = (\boldsymbol{i}'_\beta \cdot \boldsymbol{i}_\alpha). \tag{9.1.7}$$

It is left as an exercise to show that $\boldsymbol{S}$ is an orthogonal matrix (it is sufficient to prove that $\boldsymbol{S}^t\boldsymbol{S} = \boldsymbol{E}$).

Thus for observers to agree on a directed physical quantity such as displacement, electric field, velocity and so on, the components of the corresponding vector, as observed by O and O', must be related as in (9.1.5). This equation defines the transformation properties of vectors.

Now, to consider the transformation properties of such second order tensors as the stress tensor $\boldsymbol{\Theta}$, we proceed as follows. Suppose observer O observes the relationship

$$\boldsymbol{t} = \boldsymbol{\Theta N} \tag{9.1.8}$$

between $\boldsymbol{t}$, the force acting, per unit area, across a surface element, and $\boldsymbol{N}$ the unit normal to the surface element. From (9.1.5), we may write (9.1.8) as

$$\boldsymbol{S}^t\boldsymbol{t}' = \boldsymbol{\Theta S}^t\boldsymbol{N}', \tag{9.1.9}$$

where $\boldsymbol{t}'$ and $\boldsymbol{N}'$ are the representations of the force and unit normal observed by O'. On multiplying (9.1.9) by $\boldsymbol{S}$, we obtain

$$\boldsymbol{t}' = (\boldsymbol{S\Theta S}^t)\boldsymbol{N}',$$

or

$$\boldsymbol{t}' = \boldsymbol{\Theta}'\boldsymbol{N}', \tag{9.1.10}$$

where

$$\boldsymbol{\Theta}' = \boldsymbol{S\Theta S}^t \tag{9.1.11}$$

determines the representation of the Cauchy stress tensor according to observer O'. Equation (9.1.11) may be written in suffix notation as

$$\Theta'_{\alpha\beta} = S_{\alpha\mu} S_{\beta\nu} \Theta_{\mu\nu}. \tag{9.1.12}$$

A similar argument would apply to any other second order tensor describing a linear relationship between vectors.

We may discuss the transformation of a third order tensor in an analogous way. For example, we may consider a linear relation between the Cauchy stress tensor $\boldsymbol{\Theta}$ and the electric polarisation vector $\boldsymbol{P}$, which would describe simple piezo-electric phenomena.

That is,

$$P_\alpha = d_{\alpha\beta\gamma}\Theta_{\beta\gamma} \tag{9.1.13}$$

would describe the representation of the linear relation as observed by O. O' would observe the representation

$$P'_\alpha = d'_{\alpha\beta\gamma}\Theta'_{\beta\gamma}, \tag{9.1.14}$$

where the elements of $\boldsymbol{d}$ and $\boldsymbol{d}'$ determine the piezo-electric coefficients of the solid material under study. On using the property $\boldsymbol{S}^{\mathrm{t}}\boldsymbol{S} = \boldsymbol{E}$, i.e., $S_{\alpha\mu}S_{\beta\mu} = \delta_{\alpha\beta}$ together with the transformation properties of $\boldsymbol{P}$ and $\boldsymbol{\Theta}$, as given by (9.1.5) and (9.1.13), it is not difficult to obtain the transformation properties of the third rank tensor $\boldsymbol{d}$:

$$d'_{\alpha\beta\gamma} = S_{\alpha\mu}S_{\beta\nu}S_{\gamma\omega}d_{\mu\nu\omega}. \tag{9.1.15}$$

When $\boldsymbol{S}$ describes the transformation of the axis system, (9.1.5), (9.1.12), and (9.1.15) are the corresponding transformations of Cartesian tensors of the first, second, and third order (a first order tensor is usually termed a vector).

9.2. Examples of transformations

(1) A simple example is that of inversion in the origin,

$$\bar{\boldsymbol{E}}: \boldsymbol{i}'_\alpha = -\boldsymbol{i}_\alpha. \tag{9.2.1}$$

Since (9.1.1) must be satisfied for a vector with arbitrary components, we see immediately that

$$x'_\alpha = -x_\alpha, \quad \alpha = 1, 2, 3. \tag{9.2.2}$$

We note from (9.1.12) that $\boldsymbol{\Theta}$, the Cauchy stress tensor, remains unchanged, i.e.,

$$\Theta'_{\alpha\beta} = \Theta_{\alpha\beta}, \quad \text{all } \alpha, \beta, \tag{9.2.3}$$

whereas for the piezo-electric coefficient $\boldsymbol{d}$ we have a change of sign, and

$$d'_{\alpha\beta\gamma} = -d_{\alpha\beta\gamma}, \quad \text{all } \alpha, \beta, \gamma. \tag{9.2.4}$$

If a crystal has a centre of symmetry, i.e. if its crystal symmetry is such that its properties are the same when referred to either of the two sets of axes, which are related by the inversion operator $\bar{\boldsymbol{E}}$, then we see from (9.2.4) that $\boldsymbol{d}$ must be zero for such crystals. That is, from its tensor properties $\boldsymbol{d}$ must change sign on inversion, but for a crystal of this symmetry it must remain invariant. Thus, it must be zero, and so crystals with a centre of symmetry cannot exhibit a first-order piezo-electric effect.

(2) A rotation of $(\pi/2)$ about the x_3-axis. With the usual right-handed set of axes, a rotation of the axis system through $\pi/2$ in a right-handed sense about the x_3-axis results in:

$$\boldsymbol{i}_1' = \boldsymbol{i}_2; \qquad \boldsymbol{i}_2' = -\boldsymbol{i}_1; \qquad \boldsymbol{i}_3' = \boldsymbol{i}_3, \tag{9.2.5}$$

and, thus,

$$x_1' = x_2; \qquad x_2' = -x_1; \qquad x_3' = x_3, \tag{9.2.6}$$

as can be verified from (9.1.1) or (9.1.5).

So we see, for example, from (9.1.12) and (9.1.15) that we obtain the transformations of $\boldsymbol{\Theta}$, $\boldsymbol{d}$ very easily, from following the three rules: that (*a*) where we have an index 1 in the 'dashed' axis system, this changes to 2; (*b*) where there is an index 2 this changes to 1 but we must multiply the tensor element by -1; (*c*) the index 3 remains unchanged. For example, we can write down

$$\boldsymbol{\Theta}' = \begin{bmatrix} \Theta_{22} & -\Theta_{21} & -\Theta_{23} \\ -\Theta_{12} & \Theta_{11} & \Theta_{13} \\ -\Theta_{32} & \Theta_{31} & \Theta_{33} \end{bmatrix}. \tag{9.2.7}$$

We observe that if $\boldsymbol{\Theta}$ is symmetric, so is $\boldsymbol{\Theta}'$. This is generally true for any orthogonal transformation and it is left as an exercise to be proved from (9.1.12).

Similarly the elements of $\boldsymbol{d}'$ may be obtained from the above rules, e.g.,

$$d_{112}' = -d_{221}; \qquad d_{123}' = -d_{213}. \tag{9.2.8}$$

(3) Another simple example is that of a rotation of π about the x_3-axis. This is left as an exercise.

(4) A rotation of $\frac{2}{3}\pi$ about the x_3-axis. Here, for a right-handed rotation, we have

$$\left.\begin{aligned} \boldsymbol{i}_1' &= \boldsymbol{i}_1 \cos\tfrac{2}{3}\pi + \boldsymbol{i}_2 \sin\tfrac{2}{3}\pi, \\ \boldsymbol{i}_2' &= -\boldsymbol{i}_1 \sin\tfrac{2}{3}\pi + \boldsymbol{i}_2 \cos\tfrac{2}{3}\pi, \\ \boldsymbol{i}_3' &= \boldsymbol{i}_3. \end{aligned}\right\} \tag{9.2.9}$$

This is a more complicated transformation with

$$\boldsymbol{S}=\begin{bmatrix}-\frac{1}{2} & \frac{1}{2}\sqrt{3} & 0\\ -\frac{1}{2}\sqrt{3} & -\frac{1}{2} & 0\\ 0 & 0 & 1\end{bmatrix}, \tag{9.2.10}$$

which 'mixes up' the indices 1 and 2. The reader is recommended to write out the resulting transformation of $\boldsymbol{\Theta}$, which is given by (9.1.11) or (9.1.12).

(5) A rotation of $\frac{2}{3}\pi$ about an axis defined by the unit normal $(1, 1, 1)/\sqrt{3}$. This axis is the diagonal of a cube with a vertex at the origin and the three edges from this vertex lying along the x_α-axes. This is the important trigonal operation which is used when considering the symmetries of cubic lattices. It can easily be verified that if the cube is rotated by this operation it comes into coincidence with itself.

For this transformation we see that a cyclic permutation of the $\boldsymbol{i}_\alpha$ occurs, i.e.,

$$\boldsymbol{i}'_1=\boldsymbol{i}_2; \qquad \boldsymbol{i}'_2=\boldsymbol{i}_3; \qquad \boldsymbol{i}'_3=\boldsymbol{i}_1, \tag{9.2.11}$$

and so

$$x'_1=x_2; \qquad x'_2=x_3; \qquad x'_3=x_1. \tag{9.2.12}$$

Thus, for example,

$$\Theta'_{11}=\Theta_{22}, \qquad \Theta'_{12}=\Theta_{23}; \tag{9.2.13}$$

$$d'_{123}=d_{231}. \tag{9.2.14}$$

9.3. Symmetry operations

The symmetry of a crystal may be described in terms of symmetry operations. Each operation may be represented by a co-ordinate axis transformation from, say, the axis system of an observer O to that of an observer O'. Suppose that observer O measures the distribution, relative to his axis system, of the atoms, ions or molecules of the crystal, together with their spin and charge, and that observer O' makes similar measurements relative to his axis system. The transformation $O\rightarrow O'$ represents a symmetry operation, if these two measured distributions, relative to the two axis systems, are indistinguishable one from the other. Thus, when a physical property is described in the two axis systems, identical relations describing the property will be obtained. We have already discussed this in example 1 of section 2.

In infinitesimal elasticity theory, this would mean that a linear relation between $\boldsymbol{\Theta}$ and $\boldsymbol{\varepsilon}$ would be, in both sets of co-ordinates, identical. That is,

$$\Theta_{\alpha\beta} = c_{\alpha\beta\gamma\delta}\,\varepsilon_{\gamma\delta}; \qquad \Theta'_{\alpha\beta} = c_{\alpha\beta\gamma\delta}\,\varepsilon'_{\gamma\delta}. \tag{9.3.1}$$

However, from the tensor nature of $\boldsymbol{\Theta}$ and $\boldsymbol{\varepsilon}$, the stiffness elements $c_{\alpha\beta\gamma\delta}$ must also transform tensorially, so that, for a general orthogonal transformation, we should have

$$\Theta'_{\alpha\beta} = c'_{\alpha\beta\gamma\delta}\,\varepsilon'_{\gamma\delta}, \tag{9.3.2}$$

where

$$c'_{\alpha\beta\gamma\delta} = S_{\alpha\mu}S_{\beta\nu}S_{\gamma\omega}S_{\delta\tau}c_{\mu\nu\omega\tau}. \tag{9.3.3}$$

If the particular symmetry transformation is described by $\boldsymbol{S}$, we must have

$$c'_{\alpha\beta\gamma\delta} = S_{\alpha\mu}S_{\beta\nu}S_{\gamma\omega}S_{\delta\tau}c_{\mu\nu\omega\tau} = c_{\alpha\beta\gamma\delta}. \tag{9.3.4}$$

Thus, the last two members of this equation give a possible linear restriction for each element of the stiffness tensor.

As a simple example, consider a crystal which has the one symmetry operation $\bar{\boldsymbol{E}}$. The linear restriction of (9.3.4) is not really a restriction in this case, because it reduces to

$$c'_{\alpha\beta\gamma\delta} = c_{\alpha\beta\gamma\delta} = c_{\alpha\beta\gamma\delta}, \tag{9.3.5}$$

which is trivially satisfied.

Let us consider, as a second example, a crystal with one symmetry operation, namely, the transformation, corresponding to a π-rotation about the z-axis, described by $x \to -x$, $y \to -y$, $z \to +z$. Equation (9.3.4) then becomes

$$c'_{\alpha\beta\gamma\delta} = (-1)^{\omega}c_{\alpha\beta\gamma\delta} = c_{\alpha\beta\gamma\delta}, \tag{9.3.6}$$

where ω is the number of times x or y occurs in the set $\alpha\beta\gamma\delta$. For example, for the set $xxyz$, $\omega = 3$. Thus, from (9.3.6), elements such as c_{xxyz} must be zero.

Now, there are 81 possible elements of $\boldsymbol{c}$, but since $\boldsymbol{\Theta}$ and $\boldsymbol{\varepsilon}$ are symmetrical the number of distinct elements drops to 36, and we can use the abbreviated notation,

$$\Theta_{\alpha} = c_{\alpha\beta}e_{\beta}, \tag{9.3.7}$$

where the e_{α} are defined in (7.3.8), while, for $\boldsymbol{\Theta}$, we have

$$\Theta_i = \Theta_{ii}, \quad i = 1, 2, 3; \qquad \Theta_4 = \Theta_{23}, \quad \Theta_5 = \Theta_{31}, \quad \Theta_6 = \Theta_{12}. \tag{9.3.8}$$

We shall find in the thermodynamic treatment that there is a further symmetry,

$$c_{\alpha\beta} = c_{\beta\alpha}, \tag{9.3.9}$$

which results in a further reduction of the number of independent stiffness elements from 36 to 21. The symmetry restriction of (9.3.6) will reduce this number still further. Transposing from 1, 2, 3, to x, y, z, we see, from (9.3.6), that, for example,

$$c_{xxyz} = c_{yyxz} = c_{xzzz} = c_{yzzz} = 0. \tag{9.3.10}$$

Considering all the elements, we find that the following 8 must be zero:

$$c_{14}, c_{56}, c_{25}, c_{46}, c_{35}, c_{34}, c_{24}, c_{15}. \tag{9.3.11}$$

A crystal which has the π-rotation about the z-axis as its only symmetry operation is said to have the crystal point group C_2. The number of independent stiffness elements is, for such a crystal, reduced from 21 to 13.

A full thermodynamic discussion of the elastic stiffness and compliance matrices for the crystal point groups will be given in chapters 16 and 17.

Summary of important equations

A. *Transformations of Cartesian tensors*

1. First-order tensors

$$x_\alpha \boldsymbol{i}_\alpha = x'_\gamma \boldsymbol{i}'_\gamma, \tag{9.1.1}$$

$$x'_\beta = S_{\beta\alpha} x_\alpha, \tag{9.1.5}$$

$$\boldsymbol{i}'_\beta = S_{\beta\alpha} \boldsymbol{i}_\alpha, \tag{9.1.6}$$

$$S_{\beta\alpha} = (\boldsymbol{i}'_\beta \cdot \boldsymbol{i}_\alpha).. \tag{9.1.7}$$

2. Second-order tensors

$$\boldsymbol{\Theta}' = \boldsymbol{S\Theta S}^{\mathrm{t}}, \tag{9.1.11}$$

that is,

$$\Theta'_{\alpha\beta} = S_{\alpha\mu} S_{\beta\nu} \Theta_{\mu\nu}. \tag{9.1.12}$$

3. Third and fourth-order tensors

$$d'_{\alpha\beta\gamma} = S_{\alpha\mu} S_{\beta\nu} S_{\gamma\omega} d_{\mu\nu\omega}, \tag{9.1.15}$$

$$c'_{\alpha\beta\gamma\delta} = S_{\alpha\mu} S_{\beta\nu} S_{\gamma\omega} S_{\delta\tau} c_{\mu\nu\omega\tau}. \tag{9.3.3}$$

B. *The linear piezo-electric relation*

$$P_\alpha = d_{\alpha\beta\gamma}\Theta_{\beta\gamma}. \tag{9.1.13}$$

Examples

1. Show that $\Theta'_{\mu\mu} = \Theta_{\alpha\alpha}$, i.e., that the trace of $\mathbf{\Theta}$ is invariant to any orthogonal transformation.
2. Show that the eigenvalues of $\mathbf{\Theta}$ are invariant to any orthogonal transformation. How do the eigenvectors transform?
3. Show that the quadratic form $x_\alpha \Theta_{\alpha\beta} x_\beta$ is invariant, as is the bi-quadratic form $x_\alpha \Theta_{\alpha\beta} y_\beta$, where $\boldsymbol{x}$ and $\boldsymbol{y}$ are Cartesian vectors.

Part 2

Non-hydrostatic thermodynamics

10

The thermodynamic basis

We discuss the fundamental assumption that must be made for classical thermodynamics – i.e., that the state of a phase can be uniquely specified by a set of independent thermodynamic variables. The co-ordinates to be used in non-hydrostatic thermodynamics are described and the internal energy, the entropy, the Helmholtz and Gibbs free energy functions, are discussed briefly.

10.1. The fundamental thermodynamic assumption

The equilibrium thermodynamic properties of solid phases, when under non-hydrostatic stresses, will be studied in this part of the book. In this discussion we shall usually assume that the system studied can be divided into phases in each of which conditions are homogeneous. For example, throughout such a phase we shall be able to take the deformation matrix $\boldsymbol{D}$ as independent of position.

A much more important assumption is required. This is the assumption that the state of a homogeneous solid phase can be uniquely specified by a set of independent thermodynamic variables or co-ordinates. That is, we are assuming that the state of the phase is independent of the previous history of processes. This is obviously true with a simple fluid, liquid or gas, for which the state can be specified uniquely by the pressure, volume, and surface area (if surface energy is important). No matter what may have been done previously to this fluid in heating it, stirring it, etc., its properties, at an equilibrium state specified by p, V, Σ (surface area), will be independent of these processes and dependent only on the values of these variables. If the fluid contains specified amounts of two chemical species A and B, which can react to form a species AB, the state will be specified by p, V and the mass of A (say) not in combination as AB. Again, if the phase is fluid, all properties of the state will depend only on the values of these co-ordinates.

However, when a solid is studied, we cannot always assume that its state can be uniquely specified by state variables. For example, consider a material such as a rubber, in which large strains can be easily observed. The state of a length of rubber could perhaps be specified by the force of extension F and the temperature T. But it is well known that, if the rubber is stretched from some initial state F_0, T_0 by increasing F, then, on returning the force and temperature to their initial values, the length of the rubber will not have returned to its initial value. Rubber exhibits the phenomenon of elastic hysteresis. Thus, strictly speaking equilibrium thermodynamics cannot be applied to rubber. It is known, however, that a sample can be conditioned, by repeated heating and stretching, until it behaves fairly accurately as an ideal elastic material, i.e., an elastic material whose properties depend only on state variables. Sometimes a material can behave ideally in this respect if sufficient time is allowed to elapse to permit all relaxation processes to occur so that the system can return to or pass to a true equilibrium state.

Another example of a system to which the assumption of state variables cannot apply is a sample of iron, when one studies its magnetic properties. The state of a piece of iron cannot be uniquely specified by the applied magnetic field and the temperature, because it exhibits magnetic hysteresis arising from irreversible processes in the magnetic domains of the sample. Thus, the magnetic state of the iron must be determined from its previous treatment. An example of such an irreversible process is the sudden release and passage of a domain wall past a defect, i.e., the incorporation of a defect into a domain. Another example is the splitting of a large domain into two oppositely magnetised domains. (Under certain circumstances, a single domain may be treated thermodynamically provided it may be regarded as an entity in which all changes may be carried out in reversible steps. This requires that any process can be carried out in infinitesimal steps so that the system can always be considered at equilibrium. In just the same way, a chemical process may be considered reversible if it can be carried out in infinitesimal steps.)

Nevertheless, there are many crystalline substances which satisfy very closely the assumption of state variables. This, alone, justifies the study and use of classical equilibrium thermodynamics. It can also be useful as a limiting treatment for less tractable substances, for which the assumption can be held to be accurate in some particular range of states – for example, at very low or high temperatures, or over a small range of variables.

It is sometimes argued that no non-hydrostatic situation is a true equilibrium state, since, if we wait long enough, flow or other transport phenomena will occur so that only a hydrostatic stress can be permanently sustained. However, it is a fact of experience that many crystalline solids can sustain non-hydrostatic stresses without measurable creep. Thus our assumption as to state variables will apply to such systems over a wide range of conditions and time. This point has already been discussed in the introduction of this book.

10.2. The thermodynamic variables for homogeneous deformations

From the first law of thermodynamics we can infer the existence of an internal energy function for an elastic phase. It is a well-known consequence of this law that the work done in an adiabatic process in going from one state to another is independent of the path. The internal energy U of a state is measured by the work done under adiabatic conditions in going from a standard state to the state in question.

The second law of thermodynamics leads to the introduction of another function of the state, i.e. the entropy, and also to the introduction of the absolute temperature scale (see for example Zemansky, 1968). The internal energy U is then expressed most usefully as a function of the entropy S and the mechanical co-ordinates. In hydrostatics we have only the one mechanical co-ordinate V, and the differential of U is

$$dU = T\,dS - p\,dV, \tag{10.2.1}$$

where the first and second term denote, respectively, the heat absorbed and the work done in an infinitesimal variation. From (10.2.1) it follows that

$$T = (\partial U/\partial S)_V, \qquad p = -(\partial U/\partial V)_S. \tag{10.2.2}$$

In non-hydrostatics it is convenient to express extensive quantities such as U and S as densities referred to the reference volume v. That is, u is defined as U/v and is the internal energy per unit reference volume, and s as S/v is the entropy density per unit reference volume. Specific quantities are often used for extensive functions in thermodynamics. For example, the specific energy $\tilde{u}$ and the specific entropy $\tilde{s}$ are defined as the energy and entropy per unit mass. Clearly such quantities are related to the above densities by

$$\left.\begin{aligned} \tilde{u} &= u(v/M), \\ \tilde{s} &= s(v/M), \end{aligned}\right\} \tag{10.2.3}$$

where M is the mass of the phase at the state to which u and $\tilde{u}$, s and $\tilde{s}$ refer.

The application of the first and second laws of thermodynamics to an elastic homogeneous phase results in

$$du = T\,ds + \tau_\alpha\,d\gamma_\alpha \tag{10.2.4}$$

or

$$dU = T\,dS + \tau_\alpha\,d(\gamma_\alpha v) \tag{10.2.5}$$

where α is summed from 1, 2, . . . , 6.

Here we have taken the γ_α as the set of six independent mechanical co-ordinates which may for instance be the A_α; $J, a_2, a_3, \ldots, a_6$; the η_α; the H_α; i.e., the various sets of mechanical co-ordinates determining the strain, which were introduced in part 1. τ_α is the stress element conjugate to the co-ordinate γ_α. Thus, we regard u as a function of seven variables, s and the six γ_α, and, from (10.2.4), we have

$$T = \partial u/\partial s; \qquad \tau_\alpha = \partial u/\partial \gamma_\alpha. \tag{10.2.6}$$

Using Legendre transformations, other functions of the state may be formed from u and s (see, e.g., Goldstein, 1950). For instance, f, the Helmholtz free energy per unit reference volume is defined as

$$f = u - Ts, \tag{10.2.7}$$

and, on taking the differential of this equation and using (10.2.4) we obtain

$$df = -s\,dT + \tau_\alpha\,d\gamma_\alpha, \tag{10.2.8}$$

and

$$s = -(\partial f/\partial T)_\gamma, \qquad \tau_\alpha = (\partial f/\partial \gamma_\alpha)_{T,\bar{\gamma}_\alpha}, \tag{10.2.9}$$

where $\bar{\gamma}_\alpha$ denotes the set of five γ's which does not include γ_α.

Equations (10.2.9) are a useful way of expressing the equations of state of an elastic material. The first member of this equation will, when f is given as a function of T and the co-ordinates γ_α, give s as a function of these variables, i.e., T and the six γ_α. The second member of (10.2.9) will give each stress element τ_α as a function of these variables. Equation (10.2.9) is thus analogous to the hydrostatic equations

$$S = -(\partial F/\partial T)_V; \qquad p = -(\partial F/\partial V)_T. \tag{10.2.10}$$

We can also define another function, which we shall call g by analogy to the Gibbs free energy function in hydrostatics, as follows

$$g = u - Ts - \tau_\alpha \gamma_\alpha, \tag{10.2.11}$$

and so

$$dg = -s\,dT - \gamma_\alpha\,d\tau_\alpha. \tag{10.2.12}$$

However, we should be warned that, in the hydrostatic limit, g as defined in (10.2.11) for generalised co-ordinates τ_α does not always reduce to the hydrostatic value $u - Ts + pJ$. For example, if we take $\gamma_i = H_i$, or A_i, $i = 1, 2, \ldots, 6$, then it can be seen that g reduces to $u - Ts + 3pJ$ (chapter 8, example 2). If we take $\gamma_1 = J$, $\gamma_i = a_i$, $i = 2, 3, \ldots, 6$, then in this case g does reduce to $u - Ts + pJ$.

Other functions can be obtained by 'mixed transformations', e.g., we could define

$$g_1 = u - Ts - \gamma_1 \tau_1, \tag{10.2.13}$$

and

$$dg_1 = -s dT - \gamma_1 \, d\tau_1 + \sum_{i=2}^{6} \tau_i \, d\gamma_i. \tag{10.2.14}$$

Summary of important equations

A. *Specific extensive thermodynamic quantities related to their densities referred to the reference state*

$$\tilde{u} = u(v/M),$$

$$\tilde{s} = s(v/M), \tag{10.2.3}$$

where $\tilde{u}$ and $\tilde{s}$ are the specific internal energy and entropy, while u and s are the internal energy and entropy densities referred to the reference state. In (10.2.3), v is the reference volume and M is the mass of the phase at the state at which u and s are measured.

B. *Fundamental thermodynamic equations in generalised co-ordinates*

$$du = T \, ds + \tau_\alpha \, d\gamma_\alpha, \quad \alpha = 1, 2, \ldots, 6. \tag{10.2.4}$$

$$dU = T \, dS + \tau_\alpha \, d(\gamma_\alpha v), \tag{10.2.5}$$

$$T = \partial u / \partial s; \quad \tau_\alpha = \partial u / \partial \gamma_\alpha. \tag{10.2.6}$$

C. *The Helmholtz free energy density referred to the reference state*

$$f = u - Ts, \tag{10.2.7}$$

$$df = -s \, dT + \tau_\alpha \, d\gamma_\alpha. \tag{10.2.8}$$

Examples

1. Two chemical species A and B react chemically according to $A + B \rightleftharpoons AB$. If the total mole number of A in the system is N_A and of B is N_B, and if the mole number of species A not in combination in AB is n_A, show that the single quantity n_A determines the quantities of B and AB.
2. Show that $\tau_\alpha \gamma_\alpha = -3PJ$, when the γ_α are the A_α or the H_α.

11

Thermodynamic relations

Thermodynamic relations are derived as consequences of the integrability conditions of the Pfaffian $\mathrm{d}u = T\,\mathrm{d}s + \tau_\alpha\,\mathrm{d}\gamma_\alpha$. The thermodynamic relations, which lead to such important results as the symmetry of the elastic stiffness elements and many relations between experimental quantities, depend on the existence of thermodynamic functions such as the internal energy, and the Helmholtz free energy. The mathematical technique of Jacobians is used to express these relations. They are discussed first of all for the relatively simple case of hydrostatics, and then are derived for non-hydrostatics.

11.1 Thermodynamic relations in hydrostatics

The fact that thermodynamic quantities such as T, s, γ_α, τ_α can be expressed as partial derivatives of a thermodynamic function, suitably chosen, leads to relations between these quantities.

For example, in hydrostatics, we have

$$\mathrm{d}U = T\,\mathrm{d}S - p\,\mathrm{d}V, \tag{11.1.1}$$

Where U, S, V are the internal energy, entropy and volume of the phase under study. Since

$$T = \frac{\partial U}{\partial S}, \qquad p = -\frac{\partial U}{\partial V}, \tag{11.1.2}$$

then

$$\frac{\partial T}{\partial V} = -\frac{\partial p}{\partial S} = \frac{\partial^2 U}{\partial S\,\partial V}. \tag{11.1.3}$$

The resulting relationship

$$(\partial T/\partial V)_S = -(\partial p/\partial S)_V \tag{11.1.4}$$

is well known as one of Maxwell's relations in thermodynamics. Similarly for the Helmholtz free energy, $F = U - TS$, since

$$S = -(\partial F/\partial T)_V; \qquad p = -(\partial F/\partial V)_T, \tag{11.1.5}$$

we can deduce that

$$(\partial S/\partial V)_T = (\partial p/\partial T)_V, \tag{11.1.6}$$

another Maxwell relation. From $G = U - TS + pV$, we can, since

$$S = -(\partial G/\partial T)_p, \qquad V = (\partial G/\partial p)_T, \tag{11.1.7}$$

deduce

$$(\partial S/\partial p)_T = -(\partial V/\partial T)_p, \tag{11.1.8}$$

which is yet another Maxwell relation. The fourth and last Maxwell relation may be similarly obtained from the enthalpy function $H = U + pV$.

Much time has been spent on schemes and mnemonics for reproducing these relations, but the method of Jacobians gives a power and flexibility in transforming from one relation to another which is unmatched by any mnemonic or chart.

Before returning to non-hydrostatic thermodynamics, the Jacobian method will, for simplicity, be illustrated for hydrostatics.

For u and v, two functions of the variables x and y, the Jacobian is a determinant which is defined as follows,

$$\frac{\partial(u, v)}{\partial(x, y)} = \begin{vmatrix} \dfrac{\partial u}{\partial x} & \dfrac{\partial u}{\partial y} \\ \dfrac{\partial v}{\partial x} & \dfrac{\partial v}{\partial y} \end{vmatrix}. \tag{11.1.9}$$

If in turn we regard x and y as functions of two other variables z and w, then it is easy to see that

$$\frac{\partial(u, v)}{\partial(z, w)} = \frac{\partial(u, v)}{\partial(x, y)} \frac{\partial(x, y)}{\partial(z, w)} \tag{11.1.10}$$

$$= \begin{vmatrix} \left(\dfrac{\partial u}{\partial x}\dfrac{\partial x}{\partial z} + \dfrac{\partial u}{\partial y}\dfrac{\partial y}{\partial z}\right) & \left(\dfrac{\partial u}{\partial x}\dfrac{\partial x}{\partial w} + \dfrac{\partial u}{\partial y}\dfrac{\partial y}{\partial w}\right) \\ \left(\dfrac{\partial v}{\partial x}\dfrac{\partial x}{\partial z} + \dfrac{\partial v}{\partial y}\dfrac{\partial y}{\partial z}\right) & \left(\dfrac{\partial v}{\partial x}\dfrac{\partial x}{\partial w} + \dfrac{\partial v}{\partial y}\dfrac{\partial y}{\partial w}\right) \end{vmatrix} \tag{11.1.11}$$

$$= \begin{vmatrix} \dfrac{\partial u}{\partial z} & \dfrac{\partial u}{\partial w} \\ \dfrac{\partial v}{\partial z} & \dfrac{\partial v}{\partial w} \end{vmatrix}. \tag{11.1.12}$$

Equation (11.1.10) gives us the cancelling rule for Jacobians, i.e., we may write

$$\frac{\partial(u, v)}{\partial(x, y)}=\frac{J(u, v)}{J(x, y)}, \tag{11.1.13}$$

where $J(u, v)$ and $J(x, y)$ are $\partial(u, v)/\partial(z, w)$ and $\partial(x, y)/\partial(z, w)$, the Jacobians of (u, v) and (x, y) with respect to the same pair of variables (z, w in this case).

If, for example, we choose $(z, w)=(u, v)$, then, as can be seen from (11.1.9), $J(u, v)=1$, and (11.1.13) yields

$$\frac{\partial(u, v)}{\partial(x, y)}=1\bigg/\left\{\frac{\partial(x, y)}{\partial(u, v)}\right\}. \tag{11.1.14}$$

The cancelling relation (11.1.13) is most useful in transformation of variables, in conjunction with the following simple relations:

$$\frac{\partial(u, v)}{\partial(x, y)}=-\frac{\partial(u, v)}{\partial(y, x)}=-\frac{\partial(v, u)}{\partial(x, y)}=+\frac{\partial(v, u)}{\partial(y, x)} \tag{11.1.15}$$

$$\frac{\partial(u, y)}{\partial(x, y)}=\left(\frac{\partial u}{\partial x}\right)_y. \tag{11.1.16}$$

These relations follow simply from the definition (11.1.9).

On using (11.1.9), (11.1.15) and (11.1.16), we can write thermodynamic relations (11.1.4), (11.1.6), and (11.1.8), respectively as

$$\left.\begin{aligned}
\frac{\partial(T, S)}{\partial(V, S)}&=\frac{\partial(p, V)}{\partial(V, S)},\\
\frac{\partial(T, S)}{\partial(T, V)}&=\frac{\partial(p, V)}{\partial(T, V)},\\
\frac{\partial(T, S)}{\partial(T, p)}&=\frac{\partial(p, V)}{\partial(T, p)}.
\end{aligned}\right\} \tag{11.1.17}$$

The fourth can obviously be written as

$$\frac{\partial(T, S)}{\partial(p, S)}=\frac{\partial(p, V)}{\partial(p, S)}, \tag{11.1.18a}$$

and on using (11.1.16), this reduces to

$$(\partial T/\partial p)_S=(\partial V/\partial S)_p. \tag{11.1.18b}$$

Thus, the Maxwell relations can be summed up as

$$J(T, S)=J(p, V), \tag{11.1.19}$$

as can be proved by applying the cancelling rule (11.1.13) to any of the four equalities of (11.1.17) or (11.1.18a). For example, according to (11.1.13) we may write the first equality of (11.1.17) as

$$\frac{J(T, S)}{J(V, S)} = \frac{J(p, V)}{J(V, S)}, \tag{11.1.20}$$

and we can obtain (11.1.19) by cancelling $J(V, S)$ from this equation, provided of course that all of the Jacobians in it are referred to the same pair of independent variables.

As another example of the use of Jacobians, let us prove the well-known relation $\kappa^{(T)} = \gamma\kappa^{(s)}$, where $\kappa^{(s)}$ and $\kappa^{(T)}$ are respectively the adiabatic and isothermal compressibility and $\gamma = c_p/c_v$ is the ratio of the heat capacities at constant pressure and constant volume respectively.

We have

$$\kappa^{(T)} = -\left(\frac{1}{V}\right)\left(\frac{\partial V}{\partial p}\right)_T = -\left(\frac{1}{V}\right)\frac{\partial(V, T)}{\partial(V, S)}\frac{\partial(V, S)}{\partial(p, S)}\frac{\partial(p, S)}{\partial(p, T)} \tag{11.1.21}$$

on using (11.1.16) and then (11.1.13). Now,

$$\frac{\partial(p, S)}{\partial(p, T)} = \left(\frac{\partial S}{\partial T}\right)_p; \quad \frac{\partial(V, T)}{\partial(V, S)} = 1\bigg/\left(\frac{\partial S}{\partial T}\right)_V, \tag{11.1.22}$$

and

$$c_v = T(\partial S/\partial T)_V; \quad c_p = T(\partial S/\partial T)_p. \tag{11.1.23}$$

Thus (11.1.21) becomes

$$\kappa^{(T)} = (c_p/c_v)(-1/V)(\partial V/\partial p)_S = \gamma\kappa^{(s)}. \tag{11.1.24}$$

We may consider other thermodynamic systems, each with only one mechanical co-ordinate, e.g., a surface film where

$$\mathrm{d}U = T\,\mathrm{d}S + \gamma\,\mathrm{d}\Sigma, \tag{11.1.25}$$

or a magnetic system where

$$\mathrm{d}U = T\,\mathrm{d}S - H\,\mathrm{d}M, \tag{11.1.26}$$

γ being the surface tension, Σ the area of the surface film, H the applied magnetic field, and M the magnetic moment of the magnetic material.

The Maxwell relations are similar and may be summed up, analogously to (11.1.19), for the two examples, respectively, as

$$J(T, S) = J(\Sigma, \gamma) \tag{11.1.27}$$

$$J(T, S) = J(H, M), \tag{11.1.28}$$

where the differing order between co-ordinate Σ and stress element γ, co-ordinate M and stress element H takes account of the negative sign in (11.1.26).

11.2. Jacobians for non-hydrostatics

The Jacobian for any number of variables is defined as

$$\frac{\partial(y)}{\partial(x)}=\frac{\partial(y_0, y_1, \ldots, y_K)}{\partial(x_0, x_1, \ldots, x_K)}=\begin{vmatrix}(\partial y_0/\partial x_0) & \cdots & (\partial y_0/\partial x_K)\\ \vdots & \cdots & \vdots \\ (\partial y_K/\partial x_0) & \cdots & (\partial y_K/\partial x_K)\end{vmatrix} \tag{11.2.1}$$

$$=\det(\partial y_\alpha/\partial x_\beta). \tag{11.2.2}$$

If we have another set of variables z_α, $\alpha=0, 1, \ldots, K$, the cancelling rule is again easy to prove as

$$\frac{\partial(y_0, y_1, \ldots, y_K)}{\partial(x_0, x_1, \ldots, x_K)}\frac{\partial(x_0, x_1, \ldots, x_K)}{\partial(z_0, z_1, \ldots, z_K)}=\det\left(\frac{\partial y_\alpha}{\partial x_\beta}\frac{\partial x_\beta}{\partial z_\gamma}\right)$$

$$=\det(\partial y_\alpha/\partial z_\gamma)$$

$$=\frac{\partial(y)}{\partial(z)}, \tag{11.2.3}$$

on using the multiplication rule for determinants, as well as the relation

$$\frac{\partial y_\alpha}{\partial z_\gamma}=\frac{\partial y_\alpha}{\partial x_\beta}\frac{\partial x_\beta}{\partial z_\gamma}. \tag{11.2.4}$$

Thus

$$\frac{\partial(y)}{\partial(z)}=\frac{J(y)}{J(z)}, \tag{11.2.5}$$

where $J(y)$ and $J(z)$ are Jacobians with respect to the same (arbitrary) set of independent variables.

The following simple property which relates a partial derivative to a Jacobian can easily be seen to be true from the defining equation (11.2.1), i.e.,

$$\left(\frac{\partial w}{\partial a}\right)_{b,c,d,e,\ldots}=\frac{\partial(w, b, c, d, e, \ldots)}{\partial(a, b, c, d, e, \ldots)}$$

$$=\frac{\partial(b, w, c, d, e, \ldots)}{\partial(b, a, c, d, e, \ldots)}$$

$$=\frac{\partial(b, c, w, d, e, \ldots)}{\partial(b, c, a, d, e, \ldots)}$$

$$=\cdots. \tag{11.2.6}$$

This property is more easily comprehended by using symbols without suffixes.

If Py is some permutation of $y_0, y_1, \ldots, y_K$, then, from the well-known properties of a determinant which result when rows or columns are permuted, we can prove

$$\frac{\partial(Py)}{\partial(x)} = \delta_P \frac{\partial(y)}{\partial(x)} = \frac{\partial(y)}{\partial(Px)}, \tag{11.2.7}$$

and

$$\frac{\partial(Py)}{\partial(Px)} = \frac{\partial(y)}{\partial(x)}, \tag{11.2.8}$$

where δ_P is the signature of the permutation, i.e., $\delta_P = +1$ if P is an even permutation, and -1 if P is odd.

11.3. Thermodynamic relations for non-hydrostatics

As discussed in chapter 10, transformations such as (10.2.13) are known as Legendre transformations. These transformations are of importance particularly in thermodynamics and classical mechanics (see e.g. Goldstein, 1950). In these two topics we may have functions expressed in terms of a set of independent variables, e.g., in hydrostatic thermodynamics, U is regarded as a function of S and V. Often T is a much more useful variable than S. The Legendre transformation which introduces the Helmholtz free energy F, as a function of T and V, is

$$F = U - S(\partial U/\partial S)_V = U - TS. \tag{11.3.1}$$

Thus

$$\begin{aligned} \mathrm{d}F &= \mathrm{d}U - T\,\mathrm{d}S - S\,\mathrm{d}T \\ &= -S\,\mathrm{d}T - p\,\mathrm{d}V, \end{aligned} \tag{11.3.2}$$

and, therefore,

$$S = -(\partial F/\partial T)_V, \qquad p = -(\partial F/\partial V)_T. \tag{11.3.3}$$

These equations are very important in thermodynamics, particularly as F is the thermodynamic function evaluated from the canonical distribution of statistical mechanics. In classical mechanics, the most important example of a Legendre transformation is that from the Lagrangian L to the Hamiltonian H.

The Gibbs function, $G = U - TS + pV$, is similarly a Legendre transformation of F or U, and we regard G as a function of T and p, while its derivatives are

$$S = -(\partial G/\partial T)_p; \qquad V = (\partial G/\partial p)_T. \tag{11.3.4}$$

In non-hydrostatics, we can have a large number of such Legendre transformations, obtained by subtracting from u any set of product pairs chosen from Ts, and $\tau_1\gamma_1, \tau_2\gamma_2, \ldots, \tau_6\gamma_6$, where the set can consist of $1, 2, \ldots, 7$ members. Thus, in general, we can obtain a thermodynamic function g', where

$$g' = u - \text{any choice of } \{Ts, \tau_1\gamma_1, \tau_2\gamma_2, \ldots, \tau_6\gamma_6\}, \tag{11.3.5}$$

$$\text{e.g.} \quad g' = u - Ts - \tau_3\gamma_3 - \tau_5\gamma_5. \tag{11.3.6}$$

The differential of g' will be given in a general notation by

$$dg' = \mu_\alpha \, d\nu_\alpha, \tag{11.3.7}$$

where, here, the summation is over the seven values, $0, 1, 2, \ldots, 6$, of α. In this notation,

$$\textit{the ordered pair } \mu_0, \nu_0 = \textit{either } T, s \textit{ or } -s, T, \tag{11.3.8}$$

$$\textit{the ordered pair } \mu_a, \nu_a = \textit{either } \tau_a, \gamma_a \textit{ or } -\gamma_a, \tau_a, \quad \textit{for} \quad a = 1, 2, \ldots, 6. \tag{11.3.9}$$

Thus, we may choose ν_0 as either T or s, and ν_a, $a \neq 0$, as either τ_a or γ_a, and g' may be regarded as a function of the seven ν_α, so chosen. Each ν_α may be chosen in two ways; so there are $2^7 = 128$ such possible functions! Since

$$\mu_a = \partial g'/\partial \nu_a, \tag{11.3.10}$$

the 21 relations ($a, b = 0, 1, 2, \ldots, 6$, $a \neq b$),

$$\partial \mu_a/\partial \nu_b = \partial \mu_b/\partial \nu_a = \partial^2 g'/\partial \nu_a \partial \nu_b, \tag{11.3.11}$$

can be obtained analogously to the hydrostatic case discussed in Section 1. On using (11.2.6), these relations can be expressed in Jacobians, and we shall do this, using an abbreviated notation which will be explained below. Thus (11.3.11) becomes

$$\frac{\partial(\mu_a, \bar{\nu}_b)}{\partial(\nu_b, \bar{\nu}_b)} = \frac{\partial(\mu_b, \bar{\nu}_a)}{\partial(\nu_a, \bar{\nu}_a)}, \tag{11.3.12}$$

where $\bar{\nu}_a$ is the ordered set given by

$$\bar{\nu}_a = \{\nu_0, \nu_1, \nu_2, \ldots, \nu_6\} - \nu_a, \tag{11.3.13}$$

that is, we strike out ν_a from the ordered set of the seven ν's, and leave the relative order of the remainder unaltered. For example,

$$\bar{\nu}_2 = \{\nu_0, \nu_1, \nu_3, \nu_4, \nu_5, \nu_6\}. \tag{11.3.14}$$

We extend this definition to an ordered set of five ν's such that

$$\begin{aligned} \bar{\nu}_{ab} &= (\bar{\nu}_a)_b = (\bar{\nu}_b)_a \\ &= \{\nu_0, \nu_1, \ldots, \nu_6\} - \nu_a - \nu_b, \end{aligned} \tag{11.3.15}$$

that is, we strike out ν_a and ν_b from the ordered set $\{\nu_\alpha\}$ and leave the relative order of the rest unaltered.

Thus, we see that $\{\nu_a, \bar{\nu}_a\}$ is an ordered set containing all seven ν's, but is a permutation of the ordered set $\{\nu_\alpha\}$. Also we see that

$$\{\nu_a, \bar{\nu}_{ab}\} = P_a^{(b)} \bar{\nu}_b, \tag{11.3.16}$$

where $P_a^{(b)}$ is the permutation required to put the ordered set of six ν's (not containing ν_b) on the right-hand side into the order of the left-hand side.

So we can write (11.3.12) as

$$\frac{\partial(\mu_a, \nu_a, \bar{\nu}_{ab})}{\partial(\nu_b, \nu_a, \bar{\nu}_{ab})} = \frac{\partial(\mu_b, \nu_b, \bar{\nu}_{ab})}{\partial(\nu_a, \nu_b, \bar{\nu}_{ab})} \tag{11.3.17}$$

where the ordered set $\bar{\nu}_b$ on the left-hand side of (11.3.12) has been re-ordered to the set $\{\nu_a, \bar{\nu}_{ab}\}$ by the same permutation for numerator and denominator. A similar operation has been carried out on the right-hand side. On applying (11.2.8), (11.3.17) then follows.

We now note in (11.3.17) that if we interchange ν_a and ν_b in the denominator on the right-hand side, we obtain the same denominator as on the left-hand side. By (11.2.7), if we do this, we introduce a factor of -1 (since a simple interchange is an odd permutation). Thus, from the cancelling rule (11.2.5), we have

$$J(\mu_a, \nu_a, \bar{\nu}_{ab}) = -J(\mu_b, \nu_b, \bar{\nu}_{ab}), \tag{11.3.18}$$

which, in general notation, is the required set of 21 thermodynamic relations when a and b take all possible sets of unequal values.

We shall state these relations in a form giving T and s a status distinct from the six pairs τ_α, γ_α. First, if in (11.3.18) we choose $\mu_a = T$, $\nu_b = s$, then we obtain six relations

$$J(T, s, \bar{\nu}_{0b}) = -J(\tau_b, \gamma_b, \bar{\nu}_{0b}), \tag{11.3.19}$$

where $\bar{\nu}_{0b}$ is a set of five variables chosen from either τ_i or γ_i, $i \neq b$. We notice that if, for μ_a, ν_a, we choose $-s, T$ instead of T, s, or $-\gamma_b, \tau_b$ instead of τ_b, γ_b for μ_b, ν_b, then, because of (11.2.7) we would not get a new relation. If we do not choose the pair T, s for either μ_a, ν_a or μ_b, ν_b, i.e., if $a \neq 0$ or $b \neq 0$, we obtain the fifteen relations,

$$J(\tau_a, \gamma_a, \bar{\nu}_{ab}) = -J(\tau_b, \gamma_b, \bar{\nu}_{ab}), \quad a \neq b, \tag{11.3.20}$$

where $\bar{\nu}_{ab}$ is a set of five variables which must include either T or s, the remaining four being chosen from either τ_i or γ_i, $i \neq a$, $i \neq b$. It should be emphasised that the independent variables in equations (11.3.18, 19 and 20) must be the same on both sides, but can be any set of seven independent thermodynamic functions or variables. For example, if there is one state and one only with given values of $u, T, s, f, \gamma_1, \gamma_2, \gamma_3$ (at least in a given region of phase space), then these could be chosen as the independent variables – although the resulting relation would not be of much use!

As the number of relations is so large for seven variables, it may be helpful to consider a case of three variables. For instance, consider a fluid where the effect of surface tension is important. Here

$$dU = T\,ds - p\,dV + \gamma\,d\Sigma, \tag{11.3.21}$$

where Σ is the surface area and γ the surface tension. Clearly the relation (11.3.18) would, in this case, reduce to

$$\left.\begin{aligned} J(T, S, \Sigma) &= J(p, V, \Sigma), \\ J(T, S, \gamma) &= J(p, V, \gamma), \\ J(T, S, V) &= -J(\gamma, \Sigma, V), \\ J(T, S, p) &= -J(\gamma, \Sigma, p), \\ J(p, V, S) &= J(\gamma, \Sigma, S), \\ J(p, V, T) &= J(\gamma, \Sigma, T). \end{aligned}\right\} \tag{11.3.22}$$

The first pair would give very similar relations for corresponding sets of independent variables, for example, if we take T, V, Σ as independent variables in the first equation, we obtain the relation

$$(\partial S/\partial V)_{T,\Sigma} = (\partial p/\partial T)_{V,\Sigma}, \tag{11.3.23}$$

while if T, V, γ are taken as independent variables in the second equation, we see that

$$(\partial S/\partial V)_{T,\gamma} = (\partial p/\partial T)_{V,\gamma}. \tag{11.3.24}$$

The last pair of equations in (11.3.22) yield interesting relations – for example, if the independent variables in the last equation are chosen as Σ, V, T, we obtain the 'symmetry relation'

$$(\partial p/\partial \Sigma)_{V,T} = -(\partial \gamma/\partial V)_{\Sigma,T}, \tag{11.3.25}$$

and, if Σ, V, S are chosen as the independent variables in the fifth equation, one obtains

$$(\partial p/\partial \Sigma)_{V,S} = -(\partial \gamma/\partial V)_{\Sigma,S}. \tag{11.3.26}$$

These two equations are examples of what are termed symmetry relations in thermodynamics. Because, however, p and γ are associated in (11.3.21) with, respectively, a negative and a positive sign, the symmetry referred to is not obvious in these equations.

To see clearly the symmetry inherent in these equations, let us consider the corresponding general result obtained when we use generalised co-ordinates in the non-hydrostatic case. In (11.3.20) let us choose $\bar{\nu}_{ab} = \bar{\gamma}_{ab}, T$ where $\bar{\gamma}_{ab}$ is the ordered set of four variables, chosen from $\boldsymbol{\gamma} = \gamma_1, \gamma_2, \ldots, \gamma_6$, which does not contain γ_a or γ_b. Let us choose $\boldsymbol{\gamma}$, T as the set of seven independent variables for the Jacobians in this equation. Then by using (11.2.6) and (11.2.7), we may reduce (11.3.20) to

$$C_{ab}^{(T)} = (\partial \tau_a/\partial \gamma_b)_{T,\bar{\gamma}_b} = (\partial \tau_b/\partial \gamma_a)_{T,\bar{\gamma}_a} = C_{ba}^{(T)}, \tag{11.3.27}$$

where we have introduced the symbol denoting $C_{ab}^{(T)}$, an element of the 6×6 isothermal stiffness matrix. Equation (11.3.27) is the thermodynamic proof of the symmetry of this matrix. The symbol $\bar{\gamma}_b$ denotes the set $\boldsymbol{\gamma}$ with γ_b omitted. By choosing s instead of T in $\bar{\nu}_{ab}$ and in the set of independent variables in (11.3.20), we obtain the corresponding result for the symmetry of the adiabatic stiffness matrix:

$$C_{ab}^{(s)} = (\partial \tau_a/\partial \gamma_b)_{s,\bar{\gamma}_b} = (\partial \tau_b/\partial \gamma_a)_{s,\bar{\gamma}_a} = C_{ba}^{(s)}. \tag{11.3.28}$$

These relations may be obtained directly, without the use of Jacobians, by applying the integrability conditions (11.3.11) to f, for the isothermal case, and to u, for the adiabatic.

As an example of the use of (11.3.19), let us choose $\bar{\nu}_{0b} = \bar{\tau}_b$. If, further, we choose T, τ_b, $\bar{\tau}_b$ as independent variables (these are the same as the set $\boldsymbol{\tau}$, T, but they have been written as an ordered set to simplify application of (11.2.6)), then (11.3.19) reduces to

$$(\partial s/\partial \tau_b)_{T,\bar{\tau}_b} = (\partial \gamma_b/\partial T)_{\boldsymbol{\tau}}, \tag{11.3.29}$$

the right-hand side being a generalised thermal expansion coefficient.

For yet another example we shall obtain a formula for the difference between the specific heat capacities C_τ and C_γ, where these are measured at constant τ and γ respectively. These heat capacities are generalisations of c_p and c_v of hydrostatics. (There are many other choices of specific heats – consider $c_{\bar{\nu}_0}$!).

If we take s as a function of T and the γ_α, then

$$T\,\mathrm{d}s = T(\partial s/\partial T)_\gamma \mathrm{d}T + T(\partial s/\partial \gamma_\alpha)\,\mathrm{d}\gamma_\alpha. \tag{11.3.30}$$

Therefore

$$C_\tau = C_\gamma + \{(T/\rho_0)(\partial s/\partial \gamma_\alpha)_{T,\bar{\gamma}_\alpha}\}(\partial \gamma_\alpha/\partial T)_\tau. \tag{11.3.31}$$

The reference state mass density ρ_0 occurs in this equation since (s/ρ_0) is the entropy per unit mass.

On using (11.3.19), with the set $\bar{\gamma}_\alpha$ chosen for the set $\bar{\nu}_{0b}$, we can transform $(\partial s/\partial \gamma_\alpha)_{T,\bar{\gamma}_\alpha}$ as follows.

$$\left(\frac{\partial s}{\partial \gamma_\alpha}\right)_{T,\bar{\gamma}_\alpha} = \frac{\partial(s, T, \bar{\gamma}_\alpha)}{\partial(\gamma_\alpha, T, \bar{\gamma}_\alpha)} = -\frac{\partial(\gamma_\alpha, \tau_\alpha, \bar{\gamma}_\alpha)}{\partial(\gamma_\alpha, T, \bar{\gamma}_\alpha)}$$

$$= -\left(\frac{\partial \tau_\alpha}{\partial T}\right)_\gamma, \tag{11.3.32}$$

where we *must not* use the summation convention for α in these expressions.

Thus,

$$C_\tau - C_\gamma = -(T/\rho_0)(\partial \tau_\alpha/\partial T)_\gamma(\partial \gamma_\alpha/\partial T)_\tau, \tag{11.3.33}$$

where the use of the summation convention is restored. It can be shown (see example 4) that

$$(\partial \tau_\alpha/\partial T)_\gamma = -C^{(T)}_{\alpha\beta} M_\beta, \tag{11.3.34}$$

where $M_\beta = (\partial \gamma_\beta/\partial T)_\tau$ is the generalised thermal expansion coefficient. Hence (11.3.33) can be expressed as

$$C_\tau - C_\gamma = (T/\rho_0) C^{(T)}_{\alpha\beta} M_\alpha M_\beta. \tag{11.3.35}$$

This is a well-known result, which is a generalisation of the following result in hydrostatics,

$$c_p - c_v = T\tilde{V}\beta^2/\kappa^{(T)}, \tag{11.3.36}$$

where β is the volume coefficient of thermal expansion, $\kappa^{(T)}$ is the isothermal compressibility $-(V^{-1})(\partial V/\partial p)_T$, and $\tilde{V}$ is the specific volume.

11.4. Multivariate thermodynamics

The thermodynamic co-ordinates used in this chapter have been based on six mechanical strain co-ordinates and a thermodynamic variable such as T. There are other thermodynamic systems, particularly chemical systems, where there would be a number of other thermodynamic variables such as the molar quantities of the chemical species. We may be studying a phase in which the amounts of two chemical species, A and B, may be varied by diffusion, for instance, from the surroundings of the phase. (Such a system, not closed with respect to chemical composition, is often termed an open system.) The basic thermodynamic equation, for hydrostatics, for this phase would be

$$dU = T\,dS - p\,dV + \mu_A\,dN_A + \mu_B\,dN_b, \tag{11.4.1}$$

where N_A and N_B are the molar quantities, and μ_A and μ_B are the corresponding chemical potentials. It is left as an exercise to the interested reader to derive, for this system, relations analogous to (11.3.19) and (11.3.20). For the use of Jacobians in such cases the book by Wilson (1957) may be consulted.

Summary of important equations

A. *Properties of Jacobians*

$$\frac{\partial(u, v)}{\partial(x, y)} = \begin{vmatrix} \dfrac{\partial u}{\partial x} & \dfrac{\partial u}{\partial y} \\ \dfrac{\partial v}{\partial x} & \dfrac{\partial v}{\partial y} \end{vmatrix}, \tag{11.1.9}$$

$$= \frac{J(u, v)}{J(x, y)}. \tag{11.1.13}$$

$$J(u, v) = \partial(u, v)/\partial(z, w).$$

$$J(x, y) = \partial(x, y)/\partial(z, w).$$

$$\partial(u, y)/\partial(x, y) = (\partial u/\partial x)_y. \tag{11.1.16}$$

$$J(T, S) = J(p, V). \tag{11.1.19}$$

$$\frac{\partial(y)}{\partial(x)} = \frac{\partial(y_0, y_1, \ldots, y_K)}{\partial(x_0, x_1, \ldots, x_K)} = \begin{vmatrix} (\partial y_0/\partial x_0) & \cdots & (\partial y_0/\partial x_K) \\ \vdots & \cdots & \vdots \\ (\partial y_K/\partial x_0) & \cdots & (\partial y_K/\partial x_K) \end{vmatrix} \tag{11.2.1}$$

$$= \det(\partial y_\alpha/\partial x_\beta). \tag{11.2.2}$$

$$\frac{\partial y_\alpha}{\partial z_\gamma} = \frac{\partial y_\alpha}{\partial x_\beta}\frac{\partial x_\beta}{\partial z_\gamma}. \tag{11.2.4}$$

$$\frac{\partial(y)}{\partial(x)} = \frac{J(y)}{J(x)}. \tag{11.2.5}$$

$$\begin{aligned}\left(\frac{\partial w}{\partial a}\right)_{b,c,d,e,\ldots} &= \frac{\partial(w, b, c, d, e, \ldots)}{\partial(a, b, c, d, e, \ldots)} \\ &= \frac{\partial(b, w, c, d, e, \ldots)}{\partial(b, a, c, d, e, \ldots)} \\ &= \frac{\partial(b, c, w, d, e, \ldots)}{\partial(b, c, a, d, e, \ldots)} \\ &= \cdots.\end{aligned} \tag{11.2.6}$$

$$\frac{\partial(Py)}{\partial(x)} = \frac{\partial(y)}{\partial(Px)} = \delta_P \frac{\partial(y)}{\partial(x)}. \tag{11.2.7}$$

$$\frac{\partial(Py)}{\partial(Px)} = \frac{\partial(y)}{\partial(x)}. \tag{11.2.8}$$

$\delta_P = +1$ for P an even permutation,

$= -1$ for P an odd permutation.

B. *The fundamental Jacobian relations for non-hydrostatics*

$$\mathrm{d}g' = \mu_\alpha \, \mathrm{d}\nu_\alpha, \tag{11.3.7}$$

The ordered pair $\mu_0, \nu_0 =$ *either* T, s *or* $-s, T$, (11.3.8)

The ordered pair $\mu_\alpha, \nu_\alpha =$ *either* $\tau_\alpha, \gamma_\alpha$ *or* $-\gamma_\alpha, \tau_\alpha$ *for* $\alpha = 1, 2, \ldots, 6.$

$$\bar{\nu}_a = \{\nu_\alpha\} - \nu_a. \tag{11.3.13}$$

$$\bar{\nu}_{ab} = (\bar{\nu}_a)_b = (\bar{\nu}_b)_a = \bar{\nu}_{ba} = \{\nu_\alpha\} - \nu_a - \nu_b. \tag{11.3.15}$$

$$\{\nu_a, \bar{\nu}_{ab}\} = P_a^{(b)} \bar{\nu}_b. \tag{11.3.16}$$

$$J(\mu_a, \nu_a, \bar{\nu}_{ab}) = -J(\mu_b, \nu_b, \bar{\nu}_{ab}), \quad (a, b = 0, 1, \ldots, 6, a \neq b). \tag{11.3.18}$$

$$J(T, s, \bar{\nu}_{0b}) = -J(\tau_b, \gamma_b, \bar{\nu}_{0b}), \quad b = 1, 2, \ldots, 6. \tag{11.3.19}$$

$$J(\tau_a, \gamma_a, \bar{\nu}_{ab}) = -J(\tau_b, \gamma_b, \bar{\nu}_{ab}), \quad (a, b = 1, 2, \ldots, 6; a \neq b). \tag{11.3.20}$$

C. *Some important thermodynamic relations derived in section 11.3*

$$C_{ab}^{(T)} = (\partial\tau_a/\partial\gamma_b)_{T,\bar{\gamma}_b} = C_{ba}^{(T)}. \tag{11.3.27}$$

$$C_{ab}^{(s)} = (\partial\tau_a/\partial\gamma_b)_{s,\bar{\gamma}_b} = C_{ba}^{(s)}. \tag{11.3.28}$$

$$(\partial s/\partial\tau_b)_{T,\bar{\tau}_b} = (\partial\gamma_b/\partial T)_{\boldsymbol{\tau}} = M_b. \tag{11.3.29}$$

$$C_{\boldsymbol{\tau}} - C_{\boldsymbol{\gamma}} = (T/\rho_0)C_{\alpha\beta}^{(T)}M_\alpha M_\beta. \tag{11.3.35}$$

ρ_0 = reference state mass density.

$C_{\boldsymbol{\tau}}$ and $C_{\boldsymbol{\gamma}}$ are the specific heats evaluated, respectively, at constant $\boldsymbol{\tau}$ and constant $\boldsymbol{\gamma}$.

Examples

1. Use (11.1.28) to prove

$$\left(\frac{\partial T}{\partial H}\right)_S = \left(\frac{T}{C_H}\right)\left(\frac{\partial M}{\partial T}\right)_H,$$

where C_H = heat capacity of the magnetic material evaluated at constant H. To what experimental technique is this relation relevant?

2. Using Jacobians, or otherwise, prove that, in hydrostatics,
   (i) $(\partial U/\partial p)_V = C_v(\partial T/\partial p)_V$;
   (ii) $(\partial U/\partial V)_p = (C_p/\beta V) - p$
   where $\beta = (1/V)(\partial V/\partial T)_p$ is the volume coefficient of thermal expansion; and C_p, C_v are heat capacities of the system
   (iii) $(\partial U/\partial V)_T = T^2(\partial\{p/T\}/\partial T)_V$.

3. Prove that, in hydrostatics,

$$c_p - c_v = T\tilde{V}\beta^2/\kappa^{(T)},$$

where $\kappa^{(T)}$ is defined in (11.1.24), β is the volume coefficient of thermal expansion, and $\tilde{V}$ is the specific volume.

4. Use $\mathrm{d}\tau_\alpha = (\partial\tau_\alpha/\partial\gamma_\beta)_{\bar{\gamma}_\beta,T}\mathrm{d}\gamma_\beta + (\partial\tau_\alpha/\partial T)_{\boldsymbol{\gamma}}\mathrm{d}T$ to show that

$$(\partial\tau_\alpha/\partial T)_{\boldsymbol{\gamma}} = -C_{\alpha\beta}^{(T)}M_\beta,$$

where $C_{\alpha\beta}^{(T)}$ is an isothermal stiffness element, and $M_\beta = (\partial\gamma_\beta/\partial T)_{\boldsymbol{\tau}}$ is a thermal expansion coefficient.

5. A principal minor of a determinant is one whose principal diagonal is part of that of the determinant. For example we may obtain a principal minor if we cut out the rth row and column, and the sth row and column.

   Prove that

$$\frac{\partial(y_0, y_1, x_2, x_3, \ldots, x_K)}{\partial(x_0, x_1, x_2, x_3, \ldots, x_K)}$$

is a principal minor of $\partial(y)/\partial(x)$.

   Discuss other examples of this property and show that (11.2.6) is such an example.

6. Prove

$$(C_\tau/C_\gamma)=\left\{\frac{\partial(\tau_1,\ldots,\tau_6)}{\partial(\gamma_1,\ldots,\gamma_6)}\right\}_s\Big/\left\{\frac{\partial(\tau_1,\ldots,\tau_6)}{\partial(\gamma_1,\ldots,\gamma_6)}\right\}_T$$

where the numerator on the right-hand side is the Jacobian of the six variables τ_α with respect to the variables γ_α at constant s, while the denominator is the same Jacobian except that it is evaluated at constant temperature. Compare this result with (11.1.24).

7. Prove

$$(\partial s/\partial\gamma_a)_{T,\bar{\gamma}_a}=-(\partial\tau_a/\partial T)_\gamma.$$

12

Thermodynamic functions, equations of state

The thermodynamic functions u and f are introduced as functions of s or T and six independent mechanical co-ordinates η_i or h_i. The conventions for the abbreviated notation for matrix elements are given again. The equations of state for non-hydrostatics are discussed. It is shown that the triangular factor $\boldsymbol{H}$ of $\boldsymbol{D}$ transforms non-linearly under orthogonal transformations of axes.

12.1. Thermodynamic functions

From (8.2.14), (8.1.2), and (10.2.8), we can write

$$\begin{aligned} df &= -s\,dT + L_\alpha\,dh_\alpha, \\ &= -s\,dT + \Lambda_{\alpha\beta}\,d\eta_{\beta\alpha}, \end{aligned} \tag{12.1.1}$$

where we use $\boldsymbol{h} = \boldsymbol{H} - \boldsymbol{E}$, as in (7.3.3). Since $\boldsymbol{H}$ and $\boldsymbol{\eta}$ do not change if a rotation follows a deformation, no change of f occurs, provided that the rotation occurs isothermally. Hence we may write

$$f(T, \boldsymbol{D}) = f(T, \boldsymbol{SD}), \tag{12.1.2}$$

and, similarly,

$$u(s, \boldsymbol{D}) = u(s, \boldsymbol{SD}), \tag{12.1.3}$$

where $\boldsymbol{S}$ is an orthogonal matrix representing a rotation. It may easily be verified that $\boldsymbol{D}$ and $\boldsymbol{SD}$ have the same strain tensor $\boldsymbol{\eta}$ and triangular factor $\boldsymbol{H}$.

We note that we *cannot* say that

$$f(T, \boldsymbol{D}) = f(T, \boldsymbol{DS}). \tag{12.1.4}$$

$\boldsymbol{DS}$ and $\boldsymbol{D}$ do not have the same factor $\boldsymbol{H}$ or the same strain tensor $\boldsymbol{\eta}$. The two deformations $\boldsymbol{SD}$ and $\boldsymbol{DS}$ are illustrated in fig. 14. In this figure the simple shear of (2.4.36) is considered, i.e.,

$$\boldsymbol{D} = \begin{bmatrix} 1 & \beta & 0 \\ 0 & 1 & 0 \\ 0 & 0 & 0 \end{bmatrix}, \tag{12.1.5}$$

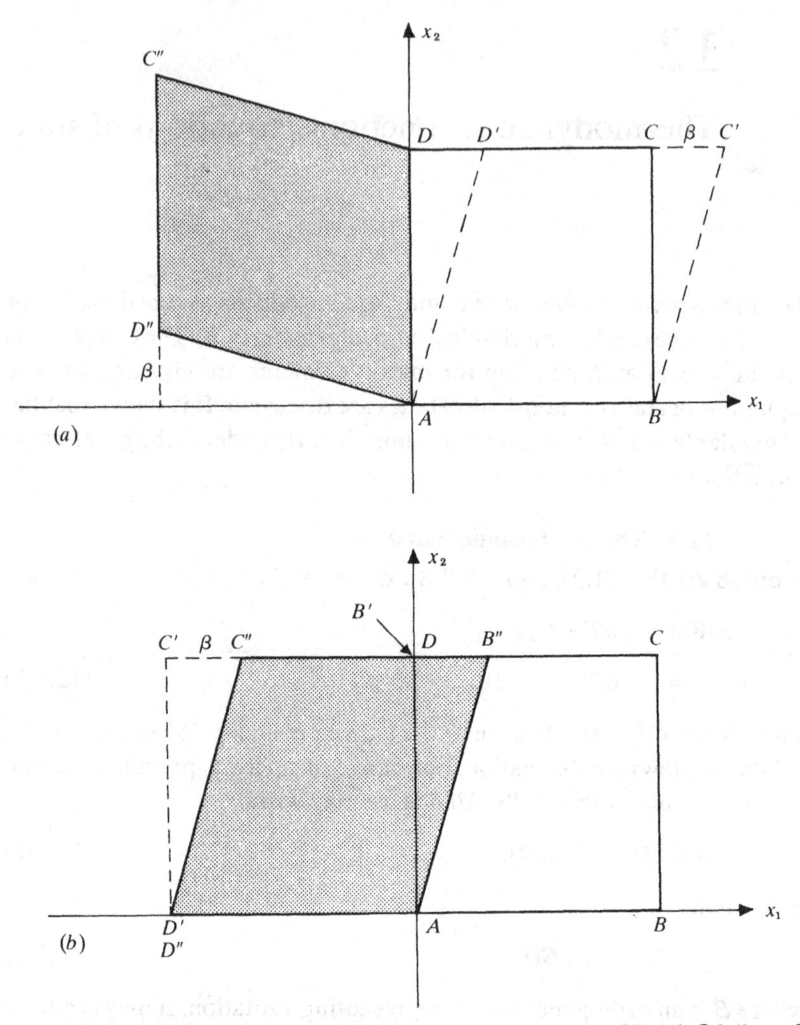

Fig. 14. (*a*) ***D*** followed by ***S***, a rotation of 90°, i.e., ***SD***. (*b*) ***S*** followed by ***D***, i.e., ***DS***.

and ***S*** is taken as a rotation of 90° about the x_3-axis, perpendicular to the plane of the paper. The deformations ***D***, ***SD***, and ***DS*** are applied to a two-dimensional material, which, in its reference state, is a unit square. In fig. 14(*a*), the deformations ***D*** and ***SD*** are successively applied, while in fig. 14(*b*), ***S*** and ***DS*** are applied. In fig. 14(*a*) we see that the diagonal AC is lengthened to $AC' = AC''$, while, in fig. 14(*b*) after ***DS***, AC is finally shortened. If the material is not isotropic, we see that while the energies of the states determined by ***D*** and ***SD*** are clearly equal, they will not necessarily be equal to the energy of the state determined by ***DS***.

Thus we see that $\boldsymbol{\eta}$ and $\boldsymbol{h}$ provide suitable mechanical co-ordinates (as would $\boldsymbol{A}$ or, better, $\boldsymbol{A}-\boldsymbol{E}$, where $\boldsymbol{A}$ was introduced in chapter 8, section 3). Because the transformations of $\boldsymbol{\eta}$ under orthogonal co-ordinate transformations (see chapter 9) are *linear* while those of $\boldsymbol{H}$ are *non-linear* (see section 3 of this chapter) it is much more simple to use $\boldsymbol{\eta}$ when discussing the effect of the symmetry properties of anisotropic crystals. Then, if we wish to use the h_i as co-ordinates, we can express $\boldsymbol{\eta}$ in terms of $\boldsymbol{h}$ as follows:

$$2\boldsymbol{\eta}=\boldsymbol{H}^{\mathrm{t}}\boldsymbol{H}-\boldsymbol{E} \tag{12.1.6}$$

$$=\boldsymbol{h}+\boldsymbol{h}^{\mathrm{t}}+\boldsymbol{h}^{\mathrm{t}}\boldsymbol{h}. \tag{12.1.7}$$

Now, according to the scheme of (8.2.12), we may write (8.2.6) and (8.1.2) as

$$L_\mu=J(\boldsymbol{H}^{-1}\boldsymbol{\Theta}_v)^{\mathrm{t}}_\mu,\quad \mu=1,2,\ldots,6, \tag{12.1.8}$$

and

$$\Lambda_\mu=J\{\boldsymbol{D}^{-1}\boldsymbol{\Theta}(\boldsymbol{D}^{\mathrm{t}})^{-1}\}_\mu. \tag{12.1.9}$$

Because of the symmetry of $\boldsymbol{\Lambda}$ and $\boldsymbol{\eta}$ we will have, in (12.1.1), 3 pairs of equal terms, namely

$$\Lambda_{ij}\,\mathrm{d}\eta_{ji}=\Lambda_{ji}\,\mathrm{d}\eta_{ij},\quad i\neq j. \tag{12.1.10}$$

Hence, for *strain tensors only*, as mentioned in chapter 8, we must introduce a factor 2 in the off-diagonal elements. If we choose as our convention for $\boldsymbol{\eta}$ (and $\boldsymbol{\varepsilon}$ and $\boldsymbol{A}$),

$$\eta_1=\eta_{11};\qquad \eta_2=\eta_{22};\qquad \eta_3=\eta_{33};\qquad \eta_4=2\eta_{23};$$
$$\eta_5=2\eta_{13};\qquad \eta_6=2\eta_{12}, \tag{12.1.11}$$

we may write

$$\mathrm{d}f=-s\,\mathrm{d}T+\Lambda_\mu\,\mathrm{d}\eta_\mu, \tag{12.1.12}$$

where $\mu=1,2,\ldots,6$.

So, finally, we may write f as a function of seven independent variables, e.g. as $f(T,\eta_i)$ or $f(T,h_i)$. Some writers retain the nine variables $\eta_{\alpha\beta}$ and perform all mathematical operations, treating these variables as independent. Only in the sequel do they take advantage of the fact that $\boldsymbol{\eta}$ is symmetric. However, this method would make the discussion of thermodynamic relations rather clumsy. In this book, we shall normally use only six independent strain variables.

12.2. Equations of state

In hydrostatics, if we are given the Helmholtz free energy F as a function of T and V, we may obtain the pressure p as a function of T and V, from

$$p = -(\partial F/\partial V)_T. \tag{12.2.1}$$

This is known as the *equation of state*. Further, we have the entropy as a function of T and V, from

$$S = -(\partial F/\partial T)_V. \tag{12.2.2}$$

This is sometimes known as the *thermal equation of state*. From this we can find the heat capacity

$$C_v = T(\partial S/\partial T)_V. \tag{12.2.3}$$

Equations (12.2.1) and (12.2.2) thus provide all the thermodynamic information for a simple chemically inert fluid phase.

In non-hydrostatics we have, from (12.1.1),

$$\Lambda_i = (\partial f/\partial \eta_i)_{T,\bar{\eta}_i}, \tag{12.2.4}$$

and

$$L_i = (\partial f/\partial h_i)_{T,\bar{h}_i}. \tag{12.2.5}$$

Thus, knowledge of f as a function of T and six strain co-ordinates gives us, analogously to p, six equations of state for the stress elements. We obtain a further thermal equation of state from

$$s = -(\partial f/\partial T)_{\boldsymbol{h}} = -(\partial f/\partial T)_{\boldsymbol{\eta}}. \tag{12.2.6}$$

These seven equations of state contain all the thermodynamic information, if f is a known function.

In the infinitesimal regime the strain elements η_i (and h_i) are approximated by e_i (*cf* equations (3.1.8) and (2.1.6b)) while the stress elements Λ_i (and L_i) are approximated by the Cauchy stress tensor elements Θ_i (*cf* equation (6.4.32)). With these approximations, the equations of state (12.2.4), (12.2.5), and (12.2.6) may be written

$$\Theta_i = (\partial f/\partial e_i)_{T,\bar{e}_i}, \tag{12.2.7}$$

$$s = -(\partial f/\partial T)_{\boldsymbol{e}}. \tag{12.2.8}$$

Similarly, from the internal energy density function u, we have approximately

$$\Theta_i = (\partial u/\partial e_i)_{s,\bar{e}_i}, \tag{12.2.9}$$

$$T=(\partial u/\partial s)_{\boldsymbol{e}}, \tag{12.2.10}$$

while e_i and T may be expressed as partial derivatives of the Gibbs free energy density function g (an exercise for the reader).

It must be emphasised, however, that these four equations (12.2.7)–(12.2.10) are only approximations, convenient in infinitesimal theory, and that Θ_i, for example, cannot in general be expressed as a partial derivative with respect to e_i of a thermodynamic function. This follows from the work of chapter 6, section 4. We return to this point in chapter 16.

12.3. Transformation of the triangular factor $\boldsymbol{H}$

Under an orthogonal transformation of the co-ordinate axis system, defined by

$$x'_\alpha=S_{\alpha\beta}x_\beta, \tag{12.3.1}$$

$\boldsymbol{D}$, a second rank tensor, will, as in (9.1.11), be transformed as follows,

$$\boldsymbol{D}'=\boldsymbol{S}\boldsymbol{D}\boldsymbol{S}^{\mathrm{t}} \tag{12.3.2}$$

We can factorise $\boldsymbol{D}'$ uniquely, so that

$$\boldsymbol{D}'=\boldsymbol{V}'\boldsymbol{H}'=\boldsymbol{S}\boldsymbol{V}\boldsymbol{H}\boldsymbol{S}^{\mathrm{t}} \tag{12.3.3}$$

where $\boldsymbol{V}$ and $\boldsymbol{V}'$ are the orthogonal factors of $\boldsymbol{D}$ and $\boldsymbol{D}'$ respectively, and $\boldsymbol{H}$ and $\boldsymbol{H}'$ are the corresponding triangular factors, with positive diagonal elements.

Thus

$$\boldsymbol{H}'=\boldsymbol{U}\boldsymbol{H}\boldsymbol{S}^{\mathrm{t}} \tag{12.3.4}$$

where $\boldsymbol{U}=\boldsymbol{V}'^{\mathrm{t}}\boldsymbol{S}\boldsymbol{V}$ is an orthogonal matrix. The matrices $\boldsymbol{H}'$ and $\boldsymbol{H}$ have the same determinant, since $\boldsymbol{D}'$ and $\boldsymbol{D}$ do. Thus $\boldsymbol{U}$ and $\boldsymbol{S}$ from (12.3.4) have the same determinant. Now, an orthogonal matrix of known determinant requires three independent parameters to determine it, e.g., the Euler angles θ, ϕ, ψ. In (12.3.4) we know $\boldsymbol{S}^{\mathrm{t}}$ and $\boldsymbol{H}$ and so we can determine $\boldsymbol{U}$ completely by the three requirements that

$$H'_{21}=H'_{31}=H'_{32}=0. \tag{12.3.5}$$

If we solve these equations for the above three parameters, we can determine the elements of $\boldsymbol{U}$, in terms of those of $\boldsymbol{S}$ and $\boldsymbol{H}$. Thus,

$$\boldsymbol{H}'=\boldsymbol{U}(\boldsymbol{S},\boldsymbol{H})\boldsymbol{H}\boldsymbol{S}^{\mathrm{t}}. \tag{12.3.6}$$

This is a non-linear transformation, with interesting properties, but we shall not study it in this book.

Summary of important equations

A. *Thermodynamic functions*

$$\begin{aligned} \mathrm{d}f &= -s\,\mathrm{d}T + L_\alpha\,\mathrm{d}h_\alpha, \\ &= -s\,\mathrm{d}T + \Lambda_{\alpha\beta}\,\mathrm{d}\eta_{\beta\alpha}, && (12.1.1) \\ &= -s\,\mathrm{d}T + \Lambda_\mu\,\mathrm{d}\eta_\mu. && (12.1.12) \end{aligned}$$

$$L_\mu = J(\boldsymbol{H}^{-1}\boldsymbol{\Theta}_v)^{\mathrm{t}}_\mu. \qquad (12.1.8)$$

$$\Lambda_\mu = J\{\boldsymbol{D}^{-1}\boldsymbol{\Theta}(\boldsymbol{D}^{\mathrm{t}})^{-1}\}_\mu. \qquad (12.1.9)$$

B. *Abbreviated notation*

$$\eta_1 = \eta_{11}; \quad \eta_2 = \eta_{22}; \quad \eta_3 = \eta_{33}; \quad \eta_4 = 2\eta_{23};$$
$$\eta_5 = 2\eta_{13}; \quad \eta_6 = 2\eta_{12}. \qquad (12.1.11)$$

C. *The relation between* $\boldsymbol{\eta}$ *and* $\boldsymbol{H}$

$$2\boldsymbol{\eta} = \boldsymbol{H}^{\mathrm{t}}\boldsymbol{H} - \boldsymbol{E}, \qquad (12.1.6)$$
$$= \boldsymbol{h} + \boldsymbol{h}^{\mathrm{t}} + \boldsymbol{h}^{\mathrm{t}}\boldsymbol{h}. \qquad (12.1.7)$$

D. *Equations of state*

$$\Lambda_i = (\partial f/\partial \eta_i)_{T,\bar{\eta}_i}. \qquad (12.2.4)$$

$$L_i = (\partial f/\partial h_i)_{T,\bar{h}_i}. \qquad (12.2.5)$$

$$\Theta_i = (\partial f/\partial e_i)_{T,\bar{e}_i} \quad \text{(infinitesimal case).} \qquad (12.2.7)$$

E. *Thermal equations of state*

$$s = -(\partial f/\partial T)_{\boldsymbol{h}} = -(\partial f/\partial T)_{\boldsymbol{\eta}} \qquad (12.2.6)$$
$$= -(\partial f/\partial T)_{\boldsymbol{e}} \quad \text{(infinitesimal case).} \qquad (12.2.8)$$

F. *The transformation of the triangular factor* $\boldsymbol{H}$

$$\boldsymbol{D}' = \boldsymbol{S}\boldsymbol{D}\boldsymbol{S}^{\mathrm{t}} \qquad (12.3.2)$$

$$\boldsymbol{D}' = \boldsymbol{V}'\boldsymbol{H}'; \; \boldsymbol{D} = \boldsymbol{V}\boldsymbol{H}.$$

$$\boldsymbol{H}' = (\boldsymbol{V}'^{\mathrm{t}}\boldsymbol{S}\boldsymbol{V})\boldsymbol{H}\boldsymbol{S}^{\mathrm{t}}$$
$$= \boldsymbol{U}(\boldsymbol{S}, \boldsymbol{H})\boldsymbol{H}\boldsymbol{S}^{\mathrm{t}} \qquad (12.3.6)$$

where U is an orthogonal matrix, such that

$$\det(\boldsymbol{U}) = \det(\boldsymbol{S}),$$

and is determined from

$$H'_{21} = H'_{31} = H'_{32} = 0. \qquad (12.3.5)$$

Examples

1. The free energy of a crystal, which has cubic symmetry in its reference state, is, neglecting higher powers,

$$f = f^*(T) + a(\eta_1 + \eta_2 + \eta_3)$$
$$+\tfrac{1}{2}\{\alpha(\eta_1+\eta_2+\eta_3)^2 + \beta(\eta_1^2+\eta_2^2+\eta_3^2) + \gamma(\eta_4^2+\eta_5^2+\eta_6^2)\},$$

where, once the reference state has been chosen, $f^*(T)$, a, α, β, γ are functions of temperature only.

(*a*) Show that the Cauchy stress tensor $\boldsymbol{\Theta}^*$ at the reference state is $a\boldsymbol{E}$, i.e., that a crystal with cubic symmetry must be under hydrostatic conditions.

(*b*) For a certain state, η_1 is the only non-zero strain element. Show that Λ_1, Λ_2, Λ_3 are the only non-zero elements of $\boldsymbol{\Lambda}$ and evaluate these. Can you find $\boldsymbol{\Theta}$ for this state without further information?

(*c*) In (*b*) find $\boldsymbol{\Theta}$ for the state described, assuming that $\boldsymbol{D}$ is triangular. Hints: Show that $J = \sqrt{(1+\eta_1)}$; $h_1 = \sqrt{(1+\eta_1)} - 1$, $h_i = 0$, $i \neq 1$. Answer:

$$\boldsymbol{\Theta} = \begin{bmatrix} \sqrt{(1+\eta_1)}\Lambda_1 & 0 & 0 \\ 0 & \Lambda_2 & 0 \\ 0 & 0 & \Lambda_2 \end{bmatrix},$$

$$\boldsymbol{\Lambda} = \begin{bmatrix} a+(\alpha+\beta)\eta_1 & 0 & 0 \\ 0 & a+\alpha\eta_1 & 0 \\ 0 & 0 & a+\alpha\eta_1 \end{bmatrix}.$$

(*d*) Repeat (*c*) for the case $\eta_i = 0$, $i \neq 3$.

2. Using the free energy of example 1, show that, if we define the isothermal stiffness element as

$$\mathring{c}^{(T)}_{\mu\nu} = (\partial\Lambda_\mu/\partial\eta_\nu)^* = (\partial\Lambda_\nu/\partial\eta_\mu)^*,$$

where $(\ )^*$ signifies evaluation at the reference state, then

$$\mathring{c}^{(T)}_{12} = \mathring{c}^{(T)}_{23} = \mathring{c}^{(T)}_{31} = \alpha,$$

$$\mathring{c}^{(T)}_{11} = \mathring{c}^{(T)}_{22} = \mathring{c}^{(T)}_{33} = \alpha + \beta,$$

and

$$\mathring{c}^{(T)}_{44} = \mathring{c}^{(T)}_{55} = \mathring{c}^{(T)}_{66} = \gamma,$$

with all other elements zero. With this free energy the stiffness matrix $\mathring{\boldsymbol{c}}^{(T)}$ is independent of $\boldsymbol{\eta}$.

3. Express f of example 1, in terms of the elements of $\boldsymbol{h}$, to second order only. Show that the stiffness elements for the reference state, $\hat{c}^{(T)}_{\alpha\beta} = \partial L_\alpha/\partial h_\beta$, are given by

$$\hat{c}^{(T)}_{\mu\nu} = \mathring{c}^{(T)}_{\mu\nu} - p, \quad \mu = \nu,$$

$$\hat{c}^{(T)}_{\mu\nu} = \mathring{c}^{(T)}_{\mu\nu}, \quad \mu \neq \nu,$$

and that the number of independent $\hat{c}^{(T)}_{\mu\nu}$ is the same as that of the $\mathring{c}^{(T)}_{\mu\nu}$, where the $\mathring{c}^{(T)}_{\mu\nu}$ are those of example 2. $p = -a$ is the hydrostatic pressure.

4. Derive the general relation

$$\hat{c}^{(T)}_{\alpha\beta} = \mathring{c}^{(T)}_{\mu\nu}(\partial\eta_\mu/\partial h_\alpha)(\partial\eta_\nu/\partial h_\beta) + (\partial f/\partial\eta_\mu)(\partial^2\eta_\mu/\partial h_\alpha \partial h_\beta)$$

and relate this to example 3.

5. If, in example 1, $\beta = 0$ for some reference state, show that

$$f - f^* = 0, \quad \text{for a strain such that}$$

$$\eta_1 + \eta_2 + \eta_3 = 0, \quad \eta_4 = \eta_5 = \eta_6 = 0.$$

If $\boldsymbol{\eta}$ is infinitesimal, what would the volume change be?

6.(*a*) For a crystal of the cubic class (in its reference state), the Gibbs free energy per unit reference volume is given, to terms of second power, by

$$g = g^*(T) - a'(\Lambda_1 + \Lambda_2 + \Lambda_3) - \tfrac{1}{2}\{\alpha'(\Lambda_1 + \Lambda_2 + \Lambda_3)^2,$$
$$+ \beta'(\Lambda_1^2 + \Lambda_2^2 + \Lambda_3^2) + \gamma'(\Lambda_4^2 + \Lambda_5^2 + \Lambda_6^2)\},$$

(*cf* example 1), where a', α', β', γ' are functions of temperature only. Show that, at the reference temperature, $a' = 0$. Hint: $\boldsymbol{\eta} = 0$ at the reference state.

(*b*) Prove that, at the reference state, the thermal expansion coefficients are given by $(\partial a'/\partial T)^* = \mathring{m}_i$, $i = 1, 2, 3$, and that $\mathring{m}_4 = \mathring{m}_5 = \mathring{m}_6 = 0$, where $\mathring{m}_i = (\partial\eta_i/\partial T)^*_\Lambda$, and $(\;\;)^*$ denotes evaluation at the reference state. The significance of the notation $\mathring{m}_i$ will be explained in chapter 13.

(*c*) Discuss, carefully, the meaning of $g^*(T)$.

7.(*a*) If $f = f^*(T) + a_\mu\eta_\mu + \tfrac{1}{2}\mathring{c}_{\alpha\beta}\eta_\alpha\eta_\beta$ is the quadratic approximation for f, show that $\mathring{c}_{\alpha\beta}$ is the isothermal stiffness element, and a_μ is Θ^*_μ (at the reference state).

Prove that

$$(\partial a_\mu/\partial T)^* = \mathring{c}_{\mu\nu}\mathring{m}_\nu,$$

where $\mathring{m}_\nu = (\partial\eta_\nu/\partial T)^*$. Hint: $(\partial a_\mu/\partial T)^* = (\partial\Lambda_\mu/\partial T)^* = (\partial\Lambda_\mu/\partial\eta_\nu)^*\mathring{m}_\nu$, where $(\;\;)^*$ signifies evaluation at the reference state.

(*b*) Correspondingly, show that the quadratic approximation for the Gibbs free energy is:

$$g = g^*(T) - d_\mu\Lambda_\mu - \tfrac{1}{2}\mathring{s}_{\alpha\beta}\Lambda_\alpha\Lambda_\beta,$$

where $\mathring{\boldsymbol{s}} = \mathring{\boldsymbol{c}}^{-1}$ is the isothermal compliance matrix, and $(\partial d_\mu/\partial T)^* = \mathring{m}_\mu$.

Show that d_μ, which is a function of T, must vanish at the reference state.

8.(*a*) Use the lemma that the only orthogonal matrix which is triangular with positive diagonal elements is the unit matrix (chapter 7, section 1) to show that, as $\boldsymbol{H} \to \boldsymbol{E}$, then $\boldsymbol{U} \to \boldsymbol{S}$ in (12.3.4), and thus that $\boldsymbol{H}' \to \boldsymbol{E}$.

(*b*) In the infinitesimal case, since $\boldsymbol{U} = (\boldsymbol{E} + \boldsymbol{I})\boldsymbol{S}$, where $\boldsymbol{I}$ is an infinitesimal skew-symmetric matrix, show that $\boldsymbol{h} = \boldsymbol{H}' - \boldsymbol{E}$ is the triangular component in the unique decomposition of $\boldsymbol{ShS}'$ into a sum of a triangular matrix and a skew-symmetric matrix (chapter 7, section 3).

13

Thermodynamic quantities, definitions, and geometrical situation

The geometrical restriction $\boldsymbol{V}=\boldsymbol{E}$ is normally applied. In this case, the strain matrix $\boldsymbol{h}$ is equal to the strain matrix $\boldsymbol{e}$. Notations are introduced for thermodynamic quantities, such as thermal expansion coefficients, stiffness and compliance elements, specific heats, which are evaluated at the reference state.

13.1. Definitions and notation

In previous chapters we have introduced in a general way thermodynamic quantities such as thermal expansion coefficients and stiffness elements. Before going on to study the properties of such quantities in detail, we must be more precise in the statement of their respective definitions. For example, we may define a thermal expansion coefficient as $(\partial \eta_\alpha/\partial T)^*_{\boldsymbol{\Lambda}}$ or $(\partial h_\alpha/\partial T)^*_{\boldsymbol{L}}$, or, in the infinitesimal regime, as $(\partial e_\alpha/\partial T)^*_{\boldsymbol{\Theta}}$. The asterisk indicates that these quantities are evaluated at the reference state. These three alternative expressions are related to each other, as discussed in chapter 14.

Similarly the stiffness elements may be defined as $(\partial \Lambda_\alpha/\partial \eta_\beta)^*$, $(\partial L_\alpha/\partial h_\beta)^*$, or $(\partial \Theta_\alpha/\partial e_\beta)^*$. Again these are evaluated at the reference state and are related to each other (see chapter 16). They may be measured under isothermal or adiabatic conditions.

The heat capacity or the specific heat of a solid may be measured under different conditions. In hydrostatics, for example, c_p is the specific heat measured at constant pressure, while c_v is measured at constant volume. In non-hydrostatic thermodynamics, we may, for example, measure the specific heat at constant $\boldsymbol{\eta}$, $\boldsymbol{h}$, $\boldsymbol{e}$, $\boldsymbol{\Lambda}$, $\boldsymbol{L}$ or $\boldsymbol{\Theta}$. Under the first three of these conditions, the measurement will be under the same condition, as constant $\boldsymbol{\eta}$ or constant $\boldsymbol{h}$ or, for the infinitesimal regime, constant $\boldsymbol{e}$, all mean that the shape and size of the specimen are held fixed. There are thus at least four possible specific heats, apart from c_p and c_v. The relations between them will be discussed in chapter 15.

Wherever possible in dealing with a homogeneous phase we shall consider that the geometrical restriction corresponding to $\boldsymbol{V}=\boldsymbol{E}$ is applied, where $\boldsymbol{V}$ is, of course, the orthogonal factor in the unique factorisation $\boldsymbol{D}=\boldsymbol{VH}$. In this case, all deformation matrices will be of the type $\boldsymbol{H}$, that is, triangular with positive diagonal elements. As discussed previously, this restriction can be achieved by fixed constraints (see also chapter 22, section 5). We then have for the stress matrices $\boldsymbol{\Lambda}$ and $\boldsymbol{L}$ the following relations with $\boldsymbol{\Theta}$. First, from (8.1.2) with $\boldsymbol{V}=\boldsymbol{E}$, we have

$$\boldsymbol{\Lambda}=J\boldsymbol{H}^{-1}\boldsymbol{\Theta}(\boldsymbol{H}^{\mathrm{t}})^{-1}, \tag{13.1.1}$$

and, from (8.2.10) with $\boldsymbol{V}=\boldsymbol{E}$,

$$L_{\alpha\beta}=(J\boldsymbol{H}^{-1}\boldsymbol{\Theta})_{\beta\alpha}=J\{\boldsymbol{\Theta}(\boldsymbol{H}^{\mathrm{t}})^{-1}\}_{\alpha\beta}, \quad \alpha\leqslant\beta. \tag{13.1.2}$$

Thus, with this geometrical restriction, the six elements $L_{\alpha\beta}$ are equal to the corresponding six elements of $\boldsymbol{B}^{\mathrm{t}}$, where $\boldsymbol{B}$ is the Boussinesq stress tensor appropriate to the restriction. Hence $\boldsymbol{\Theta}$ may be found if $\boldsymbol{H}$, or $\boldsymbol{h}$, together with either $\boldsymbol{\Lambda}$ or $\boldsymbol{L}$ are given. However, except in the approximation of the infinitesimal regime, $\boldsymbol{\Theta}$ cannot be expressed, as can $\boldsymbol{\Lambda}$ and $\boldsymbol{L}$, as a derivative of f or u.

When $\boldsymbol{V}=\boldsymbol{E}$ for all deformations, we have $\boldsymbol{h}=\boldsymbol{e}$ for both the finite and the infinitesimal cases. This follows since $\boldsymbol{D}$ is triangular and therefore, for example,

$$\varepsilon_{11}=e_1=\partial u_1/\partial x_1=h_1, \tag{13.1.3}$$

$$2\varepsilon_{23}=e_4=\partial u_2/\partial x_3+\partial u_3/\partial x_2$$

$$=\partial u_2/\partial x_3=h_4, \tag{13.1.4}$$

since $\partial u_3/\partial x_2$ is zero when $\boldsymbol{D}$ is triangular – as are $\partial u_2/\partial x_1$ and $\partial u_3/\partial x_1$.

Thus, when the restriction $\boldsymbol{V}=\boldsymbol{E}$ is applied, we may write $\boldsymbol{e}$ instead of $\boldsymbol{h}$ for homogeneous deformations, and we shall commonly do this.

As for notation, we shall usually be interested in evaluating such thermodynamic quantities as those mentioned above at the reference state. We shall use lower-case symbols for such reference state quantities. When the strain and stress matrices are $\boldsymbol{\eta}$ and $\boldsymbol{\Lambda}$ a $^\circ$ will be placed above the symbol standing for the quantity, while $\hat{}$ will be placed above the symbol when the strain and stress matrices are $\boldsymbol{h}=\boldsymbol{e}$ and $\boldsymbol{L}$, under the condition $\boldsymbol{V}=\boldsymbol{E}$. For the infinitesimal regime nothing will be placed above the symbol. Thus, we define the thermal expansion coefficients

$$\mathring{m}_\alpha=(\partial\eta_\alpha/\partial T)^*_{\boldsymbol{\Lambda}}, \qquad \hat{m}_\alpha=(\partial e_\alpha/\partial T)^*_{\boldsymbol{L}}, \qquad m_\alpha=(\partial e_\alpha/\partial T)^*_{\boldsymbol{\Theta}}, \tag{13.1.5}$$

where ( )* denotes evaluation at the reference state. When we use the general strain and stress matrices $\boldsymbol{\gamma}$ and $\boldsymbol{\tau}$, upper case letters will be employed as symbols. Thus, the generalised thermal expansion coefficient is

$$M_\alpha = (\partial \gamma_\alpha / \partial T)^*_{\tau}. \tag{13.1.6}$$

Stiffness elements will be defined as follows

$$\left.\begin{aligned} \mathring{c}_{\alpha\beta} &= (\partial \Lambda_\alpha / \partial \eta_\beta)^*, & \hat{c}_{\alpha\beta} &= (\partial L_\alpha / \partial e_\beta)^* \\ c_{\alpha\beta} &= (\partial \Theta_\alpha / \partial e_\beta)^*, & C_{\alpha\beta} &= (\partial \tau_\alpha / \partial \gamma_\beta)^*. \end{aligned}\right\} \tag{13.1.7}$$

An exactly similar notation applies to compliance elements.

We shall need to distinguish adiabatic stiffness and compliance elements from isothermal. We shall use the superscripts s or T to denote, respectively, the adiabatic or isothermal quantities. For example,

$$\hat{c}^{(s)}_{\alpha\beta} = (\partial L_\alpha / \partial e_\beta)^*_{s,\bar{e}_\beta}, \tag{13.1.8}$$

and

$$\mathring{s}^{(T)}_{\alpha\beta} = (\partial \eta_\alpha / \partial \Lambda_\beta)^*_{T,\bar{\Lambda}_\beta}. \tag{13.1.9}$$

For specific heat, we shall use the symbols $\mathring{c}_{\boldsymbol{\eta}}$, $\mathring{c}_{\boldsymbol{\Lambda}}$, $\hat{c}_{\boldsymbol{e}}$, $\hat{c}_{\boldsymbol{L}}$, where $\mathring{c}_{\boldsymbol{\eta}}$, for example, is measured at constant $\boldsymbol{\eta}$. $\hat{c}_{\boldsymbol{e}}$ and $\hat{c}_{\boldsymbol{L}}$ are to be measured at constant $\boldsymbol{e}$ and $\boldsymbol{L}$, respectively, when the restriction $\boldsymbol{V} = \boldsymbol{E}$ is applied. To be consistent with the notation introduced in chapter 10 for specific energies, a tilde should be placed above the symbol c. However, in thermodynamics, this is not common practice for specific heats and we shall not follow it here.

For functions, such as the generalised Gibbs free energy and enthalpy densities, we adopt a similar notation, as follows:

$$\hat{g}(T, \boldsymbol{L}) = f - L_\alpha h_\alpha, \qquad \hat{h}(s, \boldsymbol{L}) = u - L_\alpha h_\alpha, \tag{13.1.10}$$

$$\hat{g}(T, \boldsymbol{L}) = f - L_\alpha e_\alpha, \qquad \hat{h}(s, \boldsymbol{L}) = u - L_\alpha e_\alpha, \quad \text{when } \boldsymbol{V} = \boldsymbol{E}. \tag{13.1.11}$$

$$\mathring{g}(T, \boldsymbol{\Lambda}) = f - \Lambda_\alpha \eta_\alpha; \qquad \mathring{h}(s, \boldsymbol{\Lambda}) = u - \Lambda_\alpha \eta_\alpha. \tag{13.1.12}$$

$$g^{\tau}(T, \boldsymbol{\tau}) = f - \tau_\alpha \gamma_\alpha; \qquad h^{\tau}(s, \boldsymbol{\tau}) = u - \tau_\alpha \gamma_\alpha. \tag{13.1.13}$$

$$g^{\mathrm{i}}(T, \boldsymbol{\Theta}) = f - \Theta_\alpha e_\alpha; \qquad h^{\mathrm{i}}(s, \boldsymbol{\Theta}) = u - \Theta_\alpha e_\alpha, \tag{13.1.14}$$

for the infinitesimal case – the superscript i being necessary to avoid confusion with the Gibbs free energy density of hydrostatics. In chapter 28, we shall use

$$g'(T, \boldsymbol{B}) = f - B_{\alpha\beta} D_{\beta\alpha}. \tag{13.1.15}$$

Summary of important equations

A. *The geometrical restriction*

$$\boldsymbol{D}=\boldsymbol{H}, \quad \boldsymbol{V}=\boldsymbol{E}.$$

B. *Relations between stress matrices*

$$\boldsymbol{\Lambda}=J\boldsymbol{H}^{-1}\boldsymbol{\Theta}(\boldsymbol{H}^{\mathrm{t}})^{-1}, \tag{13.1.1}$$

$$L_{\alpha\beta}=(J\boldsymbol{H}^{-1}\boldsymbol{\Theta})_{\beta\alpha}=J\{\boldsymbol{\Theta}(\boldsymbol{H}^{\mathrm{t}})^{-1}\}_{\alpha\beta}, \quad \alpha\leqslant\beta. \tag{13.1.2}$$

C. *Thermal expansion coefficients*

$$\mathring{m}_\alpha=(\partial\eta_\alpha/\partial T)^*_{\boldsymbol{\Lambda}}, \qquad \hat{m}_\alpha=(\partial e_\alpha/\partial T)^*_{\boldsymbol{L}}, \qquad m_\alpha=(\partial e_\alpha/\partial T)^*_{\boldsymbol{\Theta}}. \tag{13.1.5}$$

$$M_\alpha=(\partial\gamma_\alpha/\partial T)^*_{\boldsymbol{\tau}}. \tag{13.1.6}$$

D. *Stiffness and compliance elements*

$$\left.\begin{aligned} \mathring{c}_{\alpha\beta}&=(\partial\Lambda_\alpha/\partial\eta_\beta)^*, & \hat{c}_{\alpha\beta}&=(\partial L_\alpha/\partial e_\beta)^* \\ c_{\alpha\beta}&=(\partial\Theta_\alpha/\partial e_\beta)^*, & C_{\alpha\beta}&=(\partial\tau_\alpha/\partial\gamma_\beta)^* \\ \mathring{s}_{\alpha\beta}&=(\partial\eta_\alpha/\partial\Lambda_\beta)^*, & \hat{s}_{\alpha\beta}&=(\partial e_\alpha/\partial L_\beta)^* \\ s_{\alpha\beta}&=(\partial e_\alpha/\partial\Theta_\beta)^*, & S_{\alpha\beta}&=(\partial\gamma_\alpha/\partial\tau_\beta)^*. \end{aligned}\right\} \tag{13.1.7}$$

E. *Generalised Gibbs free energies*

$$\hat{g}(T,\boldsymbol{L})=f-L_\alpha h_\alpha, \qquad \hat{h}(s,\boldsymbol{L})=u-L_\alpha h_\alpha. \tag{13.1.10}$$

$$\hat{g}(T,\boldsymbol{L})=f-L_\alpha e_\alpha, \qquad \hat{h}(s,\boldsymbol{L})=u-L_\alpha e_\alpha, \quad \text{when } \boldsymbol{V}=\boldsymbol{E}. \tag{13.1.11}$$

$$\mathring{g}(T,\boldsymbol{\Lambda})=f-\Lambda_\alpha\eta_\alpha, \qquad \mathring{h}(s,\boldsymbol{\Lambda})=u-\Lambda_\alpha\eta_\alpha. \tag{13.1.12}$$

$$g^\tau(T,\boldsymbol{\tau})=f-\tau_\alpha\gamma_\alpha, \qquad h^\tau(s,\boldsymbol{\tau})=u-\tau_\alpha\gamma_\alpha. \tag{13.1.13}$$

$$g^{\mathrm{i}}(T,\boldsymbol{\Theta})=f-\Theta_\alpha e_\alpha, \qquad h^{\mathrm{i}}(s,\boldsymbol{\Theta})=u-\Theta_\alpha e_\alpha. \tag{13.1.14}$$

$$g'(T,\boldsymbol{B})=f-B_{\alpha\beta}D_{\beta\alpha}. \tag{13.1.15}$$

14

Thermal expansion coefficients

The relations between the thermal expansion coefficients $\hat{\boldsymbol{m}}, \mathring{\boldsymbol{m}}, \boldsymbol{m}$ for the different strain co-ordinate systems are discussed. It is shown that they are all equal when the reference state Cauchy stress tensor is zero. For other cases the relations are derived. The symmetry scheme for these quantities is given.

14.1. Relation between $\hat{\boldsymbol{m}}$ and $\boldsymbol{m}$

The thermal expansion matrices $\hat{\boldsymbol{m}}, \mathring{\boldsymbol{m}}, \boldsymbol{m}$, are as defined in chapter 13. The difference between these quantities (which are all evaluated at the reference state) lies in the fact that they are evaluated under different constant stress conditions. For example $\hat{\boldsymbol{m}}$ is a measure of the thermal expansion occurring at constant $\boldsymbol{L}$, $\mathring{\boldsymbol{m}}$ at constant Λ, $\boldsymbol{m}$ at constant Cauchy stress $\boldsymbol{\Theta}$.

Thus, the difference between these quantities depends on the relations between Λ, $\boldsymbol{L}$ and $\boldsymbol{\Theta}$, as well as on the elastic properties.

For example, consider the relation between $\hat{\boldsymbol{m}}$ and $\boldsymbol{m}$. To determine $\boldsymbol{m}$, we measure the strain $\mathrm{d}\boldsymbol{e}$ induced by a change $\mathrm{d}T$ occurring at constant $\boldsymbol{\Theta}$. Because of the factor $J\boldsymbol{H}^{-1}$ in the relation (13.1.2) between $\boldsymbol{L}$ and $\boldsymbol{\Theta}$, there will be a change $\mathrm{d}\boldsymbol{L}$ in $\boldsymbol{L}$ accompanying the temperature change, as follows,

$$\mathrm{d}L_{\alpha\beta} = \{\boldsymbol{\Theta}^*(\mathrm{d}J\,\boldsymbol{E} - \mathrm{d}\boldsymbol{e}^{\mathrm{t}})\}_{\alpha\beta}, \quad \alpha \leq \beta. \tag{14.1.1}$$

This equation has been obtained from (13.1.2) where, after taking the differential, we have set $J = 1$ and $\boldsymbol{H} = \boldsymbol{E}$, since we are evaluating $\mathrm{d}\boldsymbol{L}$ at the reference state. (We note that $\mathrm{d}\boldsymbol{L} = 0$, if $\boldsymbol{\Theta}^* = 0$, which is a common situation in practice. Then the change occurs at constant $\boldsymbol{L}$ and constant $\boldsymbol{\Theta}^*$, so that $\hat{\boldsymbol{m}} = \boldsymbol{m}$.)

In abbreviated suffix notation we may write

$$\mathrm{d}L_{\mu} = \{\boldsymbol{\Theta}^*(\mathrm{d}J\,\boldsymbol{E} - \mathrm{d}\boldsymbol{e}^{\mathrm{t}})\}_{\mu}. \tag{14.1.2}$$

Now

$$\mathrm{d}e_{\alpha} = (\partial e_{\alpha}/\partial T)\,\mathrm{d}T + (\partial e_{\alpha}/\partial L_{\mu})\,\mathrm{d}L_{\mu}, \tag{14.1.3}$$

and on substituting (14.1.2) in (14.1.3) we obtain

$$m_\alpha = \hat{m}_\alpha + \hat{s}^{(T)}_{\alpha\mu}\{\Theta^*(\beta E - m^t)\}_\mu, \tag{14.1.4}$$

where

$$\beta = (\partial J/\partial T)^*_\Theta \tag{14.1.5}$$

is the volume coefficient of expansion.

Writing (14.1.4) concisely we have

$$\boldsymbol{m} = \hat{\boldsymbol{m}} + \hat{\boldsymbol{s}}^{(T)}\{\boldsymbol{\Theta}^*(\beta \boldsymbol{E} - \boldsymbol{m}^t)\}. \tag{14.1.6}$$

The (T) superscript for isothermal conditions should not be confused with that for the transpose. On noting that $\beta = m_1 + m_2 + m_3$, we may regard (14.1.6) as a set of linear equations in the m_i, $i = 1, 2, \ldots, 6$. If $\hat{\boldsymbol{m}}$ and $\hat{\boldsymbol{s}}^{(T)}$ are known, we may solve these equations and determine $\boldsymbol{m}$. If, instead of $\boldsymbol{\Theta}$, we keep $\boldsymbol{L}$ constant during the temperature change, it is not difficult to derive the relation

$$\hat{\boldsymbol{m}} = \boldsymbol{m} - \boldsymbol{s}^{(T)}\{\boldsymbol{\Theta}^*(\hat{\beta}\boldsymbol{E} - \hat{\boldsymbol{m}}^t)\}. \tag{14.1.7}$$

On comparison of (14.1.6) and (14.1.7) we obtain the equation

$$\hat{\boldsymbol{s}}^{(T)}\{\boldsymbol{\Theta}^*(\beta\boldsymbol{E} - \boldsymbol{m}^t)\} = \boldsymbol{s}^{(T)}\{\boldsymbol{\Theta}^*(\hat{\beta}\boldsymbol{E} - \hat{\boldsymbol{m}}^t)\}. \tag{14.1.8}$$

As an example of these relations, consider a crystal which, at its reference state, has cubic or isotropic symmetry. For such a state, we will show later that $\boldsymbol{\Theta}^*$, $\boldsymbol{m}$, $\hat{\boldsymbol{m}}$ must be multiples of $\boldsymbol{E}$, say, $-p^*\boldsymbol{E}$, $m\boldsymbol{E}$, $\hat{m}\boldsymbol{E}$. The state must be under a hydrostatic pressure p^*. Equation (14.1.8) then reduces to an equation independent of p^*, i.e.,

$$m\hat{s}^{(T)}_{\alpha\mu}E_\mu = \hat{m}s^{(T)}_{\alpha\mu}E_\mu, \tag{14.1.9}$$

(remember $\beta = 3m$). Equation (14.1.7) reduces to

$$\hat{m}E_\alpha = mE_\alpha + 2\hat{m}p^* s^{(T)}_{\alpha\mu}E_\mu, \tag{14.1.10}$$

and on adding over $\alpha = 1, 2, 3$, we obtain

$$3\hat{m} = 3m + 2\hat{m}p^*(E_\alpha s^{(T)}_{\alpha\mu}E_\mu), \tag{14.1.11}$$

i.e.,

$$\hat{\beta} = \beta + \tfrac{2}{3}\hat{\beta}\kappa^{(T)}p^*, \tag{14.1.12}$$

where $\kappa^{(T)} = E_\alpha s^{(T)}_{\alpha\mu}E_\mu$ is the bulk isothermal compressibility $-(1/V)(\partial V/\partial p)_T$ (see example 2 at the end of this chapter). Thus,

$$\hat{\beta} = \beta/(1 - \tfrac{2}{3}\kappa^{(T)}p^*). \tag{14.1.13}$$

Similar relations follow from (14.1.9) and (14.1.6). To reconcile these with this last equation would involve the relation between $\kappa^{(T)}$ and $\hat{\kappa}^{(T)}$, which may be obtained by summing (14.1.9) over $\alpha = 1, 2, 3$, when there results

$$\hat{\beta}/\beta = \hat{m}/m = \hat{\kappa}^{(T)}/\kappa^{(T)}. \qquad (14.1.14)$$

The quantity $\hat{\kappa}^{(T)}(=E_\alpha \hat{s}^{(T)}_{\alpha\mu} E_\mu)$ is a kind of isothermal volume compressibility, i.e., the fractional volume change resulting from an isothermal change of $\bar{L}$, where $\boldsymbol{L} = -\bar{L}\boldsymbol{E}$, analogous, but not identical, to a hydrostatic situation.

We note from (14.1.13) that if $p^* = 0$, then $\hat{\beta} = \beta$. This is a particular case of the more general result that $\hat{\beta} = \beta$ for all crystal classes if the reference state is at zero stress – as can be seen from (14.1.6) when $\boldsymbol{\Theta}^*$ is taken equal to the null tensor.

14.2. Relation between $\mathring{\boldsymbol{m}}$ and $\boldsymbol{m}$

A change $\mathrm{d}T$ in T, at constant $\boldsymbol{\Theta}$, will in general involve a change in $\boldsymbol{\Lambda}$, through the relation (13.1.1), such that

$$\mathrm{d}\boldsymbol{\Lambda} = \mathrm{d}J\boldsymbol{\Theta}^* - \mathrm{d}\boldsymbol{e}\boldsymbol{\Theta}^* - \boldsymbol{\Theta}^*\,\mathrm{d}\boldsymbol{e}^{\mathrm{t}}. \qquad (14.2.1)$$

Thus, following the method of section 1, we may derive the following relations,

$$\boldsymbol{m} = \mathring{\boldsymbol{m}} + \mathring{\boldsymbol{s}}^{(T)}(\beta\boldsymbol{\Theta}^* - \boldsymbol{m}\boldsymbol{\Theta}^* - \boldsymbol{\Theta}^*\boldsymbol{m}^{\mathrm{t}}), \qquad (14.2.2)$$

and

$$\mathring{\boldsymbol{m}} = \boldsymbol{m} - \boldsymbol{s}^{(T)}(\mathring{\beta}\boldsymbol{\Theta}^* - \mathring{\boldsymbol{m}}\boldsymbol{\Theta}^* - \boldsymbol{\Theta}^*\mathring{\boldsymbol{m}}^{\mathrm{t}}). \qquad (14.2.3)$$

On comparison of these two equations, we see that

$$\mathring{\boldsymbol{s}}^{(T)}(\beta\boldsymbol{\Theta}^* - \boldsymbol{m}\boldsymbol{\Theta}^* - \boldsymbol{\Theta}^*\boldsymbol{m}^{\mathrm{t}}) = \boldsymbol{s}^{(T)}(\mathring{\beta}\boldsymbol{\Theta}^* - \mathring{\boldsymbol{m}}\boldsymbol{\Theta}^* - \boldsymbol{\Theta}^*\mathring{\boldsymbol{m}}^{\mathrm{t}}). \qquad (14.2.4)$$

If, again, we take a cubic or isotropic crystal as an example, (14.2.4) yields (since $\mathring{\boldsymbol{m}} = \mathring{m}\boldsymbol{E}$ in this case)

$$m\mathring{s}^{(T)}_{\alpha\mu}E_\mu = \mathring{m}s^{(T)}_{\alpha\mu}E_\mu, \qquad (14.2.5)$$

giving

$$\mathring{m}/m = \mathring{\beta}/\beta = \mathring{\kappa}^{(T)}/\kappa^{(T)}, \qquad (14.2.6)$$

where $\mathring{\kappa}^{(T)}(=E_\alpha \mathring{s}^{(T)}_{\alpha\mu} E_\mu)$ is a bulk compressibility for the situation $\boldsymbol{\Lambda} = -\Lambda\boldsymbol{E}$, analogous to hydrostatics. Equation (14.2.3) yields

$$\mathring{m}E_\alpha = mE_\alpha + \mathring{m}p^* s^{(T)}_{\alpha\mu}E_\mu, \qquad (14.2.7)$$

and, on adding over $\alpha = 1, 2, 3$, we obtain

$$\mathring{\beta} = \beta + \tfrac{1}{3}\mathring{\beta}\kappa^{(T)}p^*, \tag{14.2.8}$$

or

$$\mathring{\beta} = \beta/(1 - \tfrac{1}{3}\kappa^{(T)}p^*). \tag{14.2.9}$$

So, by comparison of (14.1.13) and (14.2.9), we obtain

$$\mathring{\beta} = \hat{\beta}\{(1 - \tfrac{2}{3}\kappa^{(T)}p^*)/(1 - \tfrac{1}{3}\kappa^{(T)}p^*)\}. \tag{14.2.10}$$

The reason for the presence of two correction factors in the numerator and denominator respectively lies in the presence of the hydrostatic compressibility $\kappa^{(T)}$. Had we proceeded to obtain directly a relation between $\boldsymbol{\Lambda}$ and $\boldsymbol{L}$ we would have obtained a result involving $\hat{\kappa}^{(T)}$ or $\mathring{\kappa}^{(T)}$, each of which is a kind of bulk compressibility not in common use.

14.3. Symmetry results for $\mathring{\boldsymbol{m}}$, $\hat{\boldsymbol{m}}$, $\boldsymbol{m}$

By definition, the thermal expansion coefficients $\mathring{\boldsymbol{m}}$, $\hat{\boldsymbol{m}}$ and $\boldsymbol{m}$, are related to the symmetric tensors $\boldsymbol{\eta}$ and $\boldsymbol{\varepsilon}$, through such relations as

$$\left.\begin{aligned} \mathring{m}_1 &= (\partial\eta_{11}/\partial T)^*_{\Lambda}, & \mathring{m}_4 &= 2(\partial\eta_{23}/\partial T)^*_{\Lambda}; \\ \hat{m}_1 &= (\partial\varepsilon_{11}/\partial T)^*_{L}, & \hat{m}_4 &= 2(\partial\varepsilon_{23}/\partial T)^*_{L}; \\ m_1 &= (\partial\varepsilon_{11}/\partial T)^*_{\Theta}, & m_4 &= 2(\partial\varepsilon_{23}/\partial T)^*_{\Theta}. \end{aligned}\right\} \tag{14.3.1}$$

As discussed in chapter 9, section 3, the symmetric tensors $(\partial\varepsilon_{\alpha\beta}/\partial T)^*_{\Theta \text{ or } L}$, $(\partial\eta_{\alpha\beta}/\partial T)^*_{\Lambda}$ must be invariant to the transformations of the crystal group. For such a tensor, which we shall call $\psi_{\alpha\beta}$, we see that if the c-axis (x_3-axis) is a 2-fold axis of symmetry, $\psi_{\alpha\beta}$ must be invariant to transformations induced by

$$x_1 \to -x_1, \qquad x_2 \to -x_2, \qquad x_3 \to +x_3. \tag{14.3.2}$$

Thus, we must have

$$\begin{bmatrix} \psi_{11} & \psi_{12} & \psi_{13} \\ \psi_{12} & \psi_{22} & \psi_{23} \\ \psi_{13} & \psi_{23} & \psi_{33} \end{bmatrix} = \begin{bmatrix} \psi_{11} & \psi_{12} & -\psi_{13} \\ \psi_{12} & \psi_{22} & -\psi_{23} \\ -\psi_{13} & -\psi_{23} & \psi_{33} \end{bmatrix}, \tag{14.3.3}$$

and so $\psi_{13} = \psi_{23} = 0$. If, further, the c-axis is a 4-fold axis of symmetry, then, since a 180°-rotation is still a symmetry operation, (14.3.3) holds. In addition, $\boldsymbol{\psi}$ must be invariant to the transformation induced by a 90° rotation about the c-axis, namely,

$$x_1 \to x_2; \qquad x_2 \to -x_1; \qquad x_3 \to x_3. \tag{14.3.4}$$

Hence the tensor elements must satisfy

$$\begin{bmatrix} \psi_{11} & \psi_{12} & 0 \\ \psi_{12} & \psi_{22} & 0 \\ 0 & 0 & \psi_{33} \end{bmatrix} = \begin{bmatrix} \psi_{22} & -\psi_{12} & 0 \\ -\psi_{12} & \psi_{11} & 0 \\ 0 & 0 & \psi_{33} \end{bmatrix}. \tag{14.3.5}$$

Thus, we have the additional symmetry requirements that $\psi_{12}=0$; $\psi_{22}=\psi_{11}$. We can go back to our abbreviated notation and say that, for a crystal with a 4-fold axis of symmetry, i.e., a crystal of the point group class 4 (C_4), the thermal expansion coefficients $\hat{\boldsymbol{m}}$, $\mathring{\boldsymbol{m}}$, $\boldsymbol{m}$ all have the symmetry scheme

$$\begin{bmatrix} m_1 & 0 & 0 \\ 0 & m_1 & 0 \\ 0 & 0 & m_3 \end{bmatrix}.$$

This method of applying the induced transformations of the crystal point group is very well described in the book of Bhagavantam (1966). Since a recently developed different method of dealing with the effects of symmetry in thermodynamics will be used, the above method will not be pursued further. However, in the summary of equations at the end of this chapter, the symmetry scheme for $\hat{\boldsymbol{m}}$, $\mathring{\boldsymbol{m}}$, $\boldsymbol{m}$ will be listed for the different crystal classes.

The symmetry scheme for these quantities depends on the symmetry of the state at which they are evaluated. For example

$$\mathring{\boldsymbol{M}} = (\partial\boldsymbol{\eta}/\partial T)_{\boldsymbol{\Lambda}} \tag{14.3.6}$$

may be evaluated at a state which is not the reference state, but which has a certain symmetry. $\mathring{\boldsymbol{M}}$ will then have the symmetry scheme appropriate to that symmetry.

Summary of important equations

A. *Relations between thermal expansion coefficients*

1. General relations

$$\boldsymbol{m} = \hat{\boldsymbol{m}} + \hat{\boldsymbol{s}}^{(T)}\{\boldsymbol{\Theta}^*(\beta\boldsymbol{E} - \boldsymbol{m}^{\mathrm{t}})\}, \tag{14.1.6}$$

$$\hat{\boldsymbol{m}} = \boldsymbol{m} - \boldsymbol{s}^{(T)}\{\boldsymbol{\Theta}^*(\hat{\beta}\boldsymbol{E} - \hat{\boldsymbol{m}}^{\mathrm{t}})\}. \tag{14.1.7}$$

$$\beta = (\partial J/\partial T)^*_{\boldsymbol{\Theta}} = m_1 + m_2 + m_3. \tag{14.1.5}$$

$$\hat{\boldsymbol{s}}^{(T)}\{\boldsymbol{\Theta}^*(\beta\boldsymbol{E} - \boldsymbol{m}^{\mathrm{t}})\} = \boldsymbol{s}^{(T)}\{\boldsymbol{\Theta}^*(\hat{\beta}\boldsymbol{E} - \hat{\boldsymbol{m}}^{\mathrm{t}})\}. \tag{14.1.8}$$

$$\boldsymbol{m} = \mathring{\boldsymbol{m}} + \mathring{\boldsymbol{s}}^{(T)}(\beta\boldsymbol{\Theta}^* - \boldsymbol{m}\boldsymbol{\Theta}^* - \boldsymbol{\Theta}^*\boldsymbol{m}^{\mathrm{t}}). \tag{14.2.2}$$

$$\mathring{\boldsymbol{m}} = \boldsymbol{m} - \boldsymbol{s}^{(T)}(\mathring{\beta}\boldsymbol{\Theta}^* - \mathring{\boldsymbol{m}}\boldsymbol{\Theta}^* - \boldsymbol{\Theta}^*\mathring{\boldsymbol{m}}^{\mathrm{t}}). \tag{14.2.3}$$

Thermal expansion coefficients

$$\mathring{\mathbf{s}}^{(T)}(\beta\boldsymbol{\Theta}^* - \boldsymbol{m}\boldsymbol{\Theta}^* - \boldsymbol{\Theta}^*\boldsymbol{m}^{\mathrm{t}}) = \mathbf{s}^{(T)}(\mathring{\beta}\boldsymbol{\Theta}^* - \mathring{\boldsymbol{m}}\boldsymbol{\Theta}^* - \boldsymbol{\Theta}^*\mathring{\boldsymbol{m}}^{\mathrm{t}}). \tag{14.2.4}$$

2. Cubic or isotropic materials

$$\hat{\beta} = \beta/(1 - \tfrac{2}{3}\kappa^{(T)}p^*). \tag{14.1.13}$$

$$\mathring{\beta} = \beta/(1 - \tfrac{1}{3}\kappa^{(T)}p^*). \tag{14.2.9}$$

$$\mathring{\beta} = \hat{\beta}\{(1 - \tfrac{2}{3}\kappa^{(T)}p^*)/(1 - \tfrac{1}{3}\kappa^{(T)}p^*)\}. \tag{14.2.10}$$

B. Symmetry scheme for $\hat{\boldsymbol{m}}$, $\mathring{\boldsymbol{m}}$, $\boldsymbol{m}$

Crystal symmetry	Number of independent elements M	Independent elements
Triclinic	6	$\begin{matrix} M_1 & M_6 & M_5 \\ & M_2 & M_4 \\ & & M_3 \end{matrix}$
Monoclinic	4	$\begin{matrix} M_1 & M_6 & 0 \\ & M_2 & 0 \\ & & M_3 \end{matrix}$
Orthorhombic	3	$\begin{matrix} M_1 & 0 & 0 \\ & M_2 & 0 \\ & & M_3 \end{matrix}$
Tetragonal	2	$\begin{matrix} M_1 & 0 & 0 \\ & M_1 & 0 \\ & & M_3 \end{matrix}$
Trigonal and hexagonal	2	$\begin{matrix} M_1 & 0 & 0 \\ & M_1 & 0 \\ & & M_3 \end{matrix}$
Cubic and isotropic	1	$\begin{matrix} M & 0 & 0 \\ & M & 0 \\ & & M \end{matrix}$

Examples

1. Discuss the possible change of symmetry of a monoclinic crystal under thermal expansion at constant stress.
   Prove that $\kappa^{(T)}$, the isothermal volume compressibility, is given by $E_\alpha s^{(T)}_{\alpha\mu} E_\mu$. (Hint: for an infinitesimal deformation $\mathrm{d}J = e_1 + e_2 + e_3 = E_\alpha e_\alpha$.)

3. Look up some typical compressibilities $\kappa^{(T)}$ for solids and discuss the order of magnitude of the various types of thermal expansion coefficients (and of the specific heats discussed in the next chapter).
4. For a tetragonal reference state, the symmetry requirements lead to the following permitted $\boldsymbol{\Theta}$ and $\hat{\boldsymbol{m}}, \acute{\boldsymbol{m}}, \boldsymbol{m}$:

$$\boldsymbol{\Theta}^* = \begin{bmatrix} -p_1 & 0 & 0 \\ 0 & -p_1 & 0 \\ 0 & 0 & -p_3 \end{bmatrix}; \qquad \hat{\boldsymbol{m}}, \acute{\boldsymbol{m}}, \boldsymbol{m}: \begin{bmatrix} M_1 & 0 & 0 \\ 0 & M_1 & 0 \\ 0 & 0 & M_3 \end{bmatrix},$$

and $s_{\alpha\beta}^{(T)}$ is such that $s_{11}^{(T)} = s_{22}^{(T)}$; $s_{13}^{(T)} = s_{23}^{(T)}$.

(*a*) Prove that $\kappa^{(T)} = (2s_{11}^{(T)} + s_{33}^{(T)} + 2s_{12}^{(T)} + 4s_{13}^{(T)})$, and $\beta = 2m_1 + m_3$.

(*b*) Use (14.1.7) to prove that

$$m_1 = \hat{m}_1[1 - p_1(s_{11}^{(T)} + s_{12}^{(T)}) - 2s_{13}^{(T)}p_3] - (s_{11}^{(T)} + s_{12}^{(T)})p_1\hat{m}_3,$$

$$m_3 = -2\hat{m}_1[s_{13}^{(T)}p_1 + s_{33}^{(T)}p_3] + (1 - 2s_{13}^{(T)}p_1)\hat{m}_3.$$

(*c*) Verify that this reduces to the cubic case (14.1.12), by setting $m_1 = m_3$, $\hat{m}_1 = \hat{m}_3$, $p_1 = p_3$. (Hint: Multiply the first equation by 2 and add to the second. *Note:* for cubic crystals $s_{11}^{(T)} = s_{33}^{(T)}$; $s_{12}^{(T)} = s_{23}^{(T)} = s_{31}^{(T)}$, but it is not necessary to assume this for this example.)

15

Specific heats

The relations between the specific heats for the different strain co-ordinate systems are derived. These relations involve the elastic properties of the material.

15.1. Relation between $\hat{c}_{\boldsymbol{L}}$ and $c_{\boldsymbol{\Theta}}$

Specific heats can be measured under a variety of conditions. We could, for example, measure a specific heat, $C_{\gamma_1,\gamma_2,\gamma_3,\tau_4,\tau_5,\tau_6}$, under conditions where the first three strain co-ordinates and the last three stress elements are held fixed. Such 'mixed' specific heats are not a great deal of use. But $C_{\boldsymbol{\gamma}}$, $C_{\boldsymbol{\tau}}$ are of importance, and are the analogue of c_p and c_v in hydrostatics. Thus, we must consider the following pairs of specific heats evaluated at the reference state: $\hat{c}_{\boldsymbol{e}}$, $\hat{c}_{\boldsymbol{L}}$; $\mathring{c}_{\boldsymbol{\eta}}$, $\mathring{c}_{\boldsymbol{\Lambda}}$; $c_{\boldsymbol{\Theta}}$, $c_{\boldsymbol{\epsilon}}$. A general relation between each pair of specific heats (at constant strain and constant stress respectively) has been given in example 6, chapter 11, and relates their ratio to that of the determinants of the adiabatic and isothermal stiffness matrices. Equation (11.3.35) expresses the difference $C_{\boldsymbol{\tau}} - C_{\boldsymbol{\gamma}}$ in terms of thermal expansion coefficients.

We need now to consider the relations between the specific heats for the different strain co-ordinate systems. Again we will see that such a relation may involve the elastic properties, because, for example, $\hat{c}_{\boldsymbol{L}}$ and $c_{\boldsymbol{\Theta}}$ are evaluated at different constant stress conditions. First, we note that, as was pointed out in chapter 13, section 1,

$$\hat{c}_{\boldsymbol{e}} = \mathring{c}_{\boldsymbol{\eta}} = c_{\boldsymbol{\epsilon}}. \tag{15.1.1}$$

Now, let us consider first the relation between $\hat{c}_{\boldsymbol{L}}$ and $c_{\boldsymbol{\Theta}}$. We have

$$\mathrm{d}s = (\partial s/\partial T)_{\boldsymbol{L}}\, \mathrm{d}T + (\partial s/\partial L_\mu)\, \mathrm{d}L_\mu, \tag{15.1.2}$$

where we wish to take account of the change $\mathrm{d}\boldsymbol{L}$ induced by the change $\mathrm{d}T$ occurring at constant $\boldsymbol{\Theta}$. We know this from (14.1.2), and we can substitute it in (15.1.2). At the same time, we can use the result of (11.3.29) to change $\partial s/\partial L_\mu$ to $(\partial e_\mu/\partial T)_{\boldsymbol{L}}$, i.e., to $\hat{m}_\mu$. Thus, (15.1.2) becomes

$$(\partial s/\partial T)_{\boldsymbol{\Theta}} = (\partial s/\partial T)_{\boldsymbol{L}} + \hat{m}_\mu \{\boldsymbol{\Theta}^*(\beta \boldsymbol{E} - \boldsymbol{m}^{\mathrm{t}})\}_\mu \tag{15.1.3}$$

and so, since $\mathring{\boldsymbol{m}}$ is a triangular matrix, we can write this equation as

$$c_\Theta = \hat{c}_L + (T/\rho_0)\,\mathrm{trace}\{\hat{\boldsymbol{m}}(\beta \boldsymbol{E} - \boldsymbol{m})\boldsymbol{\Theta}^*\}, \tag{15.1.4}$$

or

$$c_\Theta = \hat{c}_L + (T/\rho_0)\hat{m}_{\alpha\mu}\{(\beta \boldsymbol{E} - \boldsymbol{m})\boldsymbol{\Theta}^*\}_{\mu\alpha}. \tag{15.1.5}$$

We note that, if $\boldsymbol{\Theta}^* = 0$, the two specific heats are equal.

For a hydrostatic reference state this result may be written

$$c_p - \hat{c}_L = (T/\rho_0)p^*[\hat{m}_1 m_1 + \hat{m}_2 m_2 + \hat{m}_3 m_3 - \hat{\beta}\beta]. \tag{15.1.6}$$

For a cubic or isotropic crystal, for which $m_1 = m_2 = m_3$ and $\beta = 3m_1$,

$$\hat{c}_L - c_p = \tfrac{2}{3}(T/\rho_0)p^*\hat{\beta}\beta, \tag{15.1.7}$$

or, using (14.1.13),

$$\hat{c}_L - c_p = \tfrac{2}{3}(T/\rho_0)p^*\beta^2/(1 - \tfrac{2}{3}\kappa^{(T)}p^*). \tag{15.1.8}$$

15.2. Relation between $\mathring{c}_\Lambda$ and c_Θ

The general method to follow is that of section 1. Using (14.2.1) and the thermodynamic relation (11.3.29), we obtain

$$(\partial s/\partial T)_\Theta = (\partial s/\partial T)_\Lambda + (\partial \eta_\mu/\partial T)_\Lambda[\beta\boldsymbol{\Theta}^* - \boldsymbol{m}\boldsymbol{\Theta}^* - \boldsymbol{\Theta}^*\boldsymbol{m}^{\mathrm{t}}]_\mu, \tag{15.2.1}$$

and so

$$\begin{aligned} c_\Theta - \mathring{c}_\Lambda &= (T/\rho_0)\mathring{m}_\mu[\beta\boldsymbol{\Theta}^* - \boldsymbol{m}\boldsymbol{\Theta}^* - \boldsymbol{\Theta}^*\boldsymbol{m}^{\mathrm{t}}]_\mu \\ &= (T/\rho_0)\,\mathrm{trace}\{\mathring{\boldsymbol{m}}[\cdots]\}, \end{aligned} \tag{15.2.2}$$

where we have used abbreviated notation for η_μ and $\mathring{m}_\mu$ (remember that $\eta_4 = 2\eta_{23}$ etc., and $\mathring{m}_4 = 2\mathring{m}_{23}$, etc.), and for the symmetric stress tensor increment $[\cdots]$ (remember that for stress tensors the abbreviated notation is such that $\Theta_4 = \Theta_{23}$ not $2\Theta_{23}$!).

If the reference state is hydrostatic, (15.2.2) simplifies to

$$\begin{aligned} c_p - \mathring{c}_\Lambda = (T/\rho_0)p^*[&\mathring{\beta}\beta - 2(\mathring{m}_1 m_1 + \mathring{m}_2 m_2 + \mathring{m}_3 m_3) \\ &- (\mathring{m}_4 m_4 + \mathring{m}_5 m_5 + \mathring{m}_6 m_6)]. \end{aligned} \tag{15.2.3}$$

For cubic or isotropic materials, this reduces to

$$\mathring{c}_\Lambda - c_p = \tfrac{1}{3}(T/\rho_0)p^*\mathring{\beta}\beta \tag{15.2.4}$$

$$= \tfrac{1}{3}(T/\rho_0)p^*\beta^2/(1 - \tfrac{1}{3}\kappa^{(T)}p^*), \tag{15.2.5}$$

on using (14.2.9). On subtracting (15.2.5) from (15.1.8) we obtain

$$\hat{c}_L - \mathring{c}_\Lambda = \tfrac{1}{3}(T/\rho_0)p^*\beta^2/(1-\tfrac{1}{3}\kappa^{(T)}p^*)(1-\tfrac{2}{3}\kappa^{(T)}p^*). \quad (15.2.6)$$

15.3. Ratios of specific heats

From example 6 of chapter 11, we may obtain the following relations for the ratios of specific heats, which, since $\hat{c}_{\boldsymbol{e}} = \mathring{c}_{\boldsymbol{\eta}} = c_{\boldsymbol{e}}$, may be written

$$\hat{c}_L/c_{\boldsymbol{e}} = \det(\hat{\boldsymbol{c}}^{(s)})/\det(\hat{\boldsymbol{c}}^{(T)}) \quad (15.3.1)$$

$$c_{\boldsymbol{\Theta}}/c_{\boldsymbol{e}} = \det(\boldsymbol{c}^{(s)})/\det(\boldsymbol{c}^{(T)}) \quad (15.3.2)$$

$$\mathring{c}_\Lambda/c_{\boldsymbol{e}} = \det(\mathring{\boldsymbol{c}}^{(s)})/\det(\mathring{\boldsymbol{c}}^{(T)}), \quad (15.3.3)$$

where $\hat{\boldsymbol{c}}$, $\mathring{\boldsymbol{c}}$, $\boldsymbol{c}$ are the stiffness matrices in the notation described in chapter 13. From these equations we may easily find the ratios of $c_{\boldsymbol{\Theta}}$, $\hat{c}_L$ and $\mathring{c}_\Lambda$ as required.

15.4. The relation between c_v, in hydrostatics, and $c_{\boldsymbol{e}}$, in infinitesimal strain theory

When the temperature of a solid is raised at constant $\boldsymbol{e}$ at, for example, the reference state, not only its volume but its shape must remain constant. In general, the stress tensor $\boldsymbol{\Theta}$ must change in order to allow a temperature change to occur at constant $\boldsymbol{e}$. If the stress tensor is originally hydrostatic, after the change the stress tensor may not be hydrostatic.

On the other hand, if one is restricted to the hydrostatic situation, heating at constant volume will result normally in an increase of pressure – but the shape will not necessarily be constant, except for cubic and isotropic materials. Thus, c_v and $c_{\boldsymbol{e}}$, both measured from the same hydrostatic reference state, are generally not equal – although for cubic and isotropic materials they are equal.

To determine the difference we must find the change $\mathrm{d}\boldsymbol{e}$ which occurs when the temperature of the solid is increased from T to $T+\mathrm{d}T$, at constant volume and *at hydrostatic pressures*. To simplify the discussion we shall use infinitesimal strain theory. We thus have, for $\boldsymbol{\Theta} = -p\boldsymbol{E}$,

$$\mathrm{d}e_\alpha = m_\alpha\,\mathrm{d}T - s_{\alpha\beta}^{(T)}E_\beta\,\mathrm{d}p. \quad (15.4.1)$$

Therefore, if the volume is held constant,

$$\mathrm{d}J = 0 = E_\alpha\,\mathrm{d}e_\alpha = (E_\alpha m_\alpha)\,\mathrm{d}T - (E_\alpha s_{\alpha\beta}^{(T)}E_\beta)\,\mathrm{d}p. \quad (15.4.2)$$

Hence, for the change $\mathrm{d}T$ to occur at constant volume, a pressure increment must be applied, such that

$$\mathrm{d}p = \{(E_\alpha m_\alpha)/\kappa^{(T)}\}\,\mathrm{d}T, \quad (15.4.3)$$

since $E_\alpha s^{(T)}_{\alpha\beta} E_\beta$ is the isothermal compressibility $\kappa^{(T)}$. On using (15.4.3) in (15.4.1), we obtain

$$de_\alpha = [m_\alpha - \{(s^{(T)}_{\alpha\beta} E_\beta)(E_\mu m_\mu)/\kappa^{(T)}\}]\, dT, \tag{15.4.4}$$

which describes the change in shape when the temperature changes at constant volume. It is easy to verify that this satisfies (15.4.2).

We are now in a position to evaluate the resulting entropy change, which is given by

$$ds = (\partial s/\partial T)_{\bullet}\, dT + (\partial s/\partial e_\alpha)\, de_\alpha, \tag{15.4.5}$$

where the de_α are given in (15.4.4).

Thus, on using the thermodynamic relation of example 7, chapter 11, we obtain

$$c_v = c_{\bullet} - (T/\rho_0)(\partial \Theta_\alpha/\partial T)_{\bullet}[m_\alpha - \{(s^{(T)}_{\alpha\beta} E_\beta)(E_\mu m_\mu)/\kappa^{(T)}\}], \tag{15.4.6}$$

and, on using the result of example 4, chapter 11, we have

$$\begin{aligned} c_v &= c_{\bullet} + (T/\rho_0)(c^{(T)}_{\alpha\nu} m_\nu)[\cdots] \\ &= c_{\bullet} + (T/\rho_0)[c^{(T)}_{\alpha\nu} m_\alpha m_\nu - (E_\mu m_\mu)^2/\kappa^{(T)}], \end{aligned} \tag{15.4.7}$$

where the final form results from the fact that $c^{(T)}_{\alpha\nu} s^{(T)}_{\alpha\beta} = \delta_{\nu\beta}$ (since $\boldsymbol{c}^{(T)}$ and $\boldsymbol{s}^{(T)}$ are reciprocal symmetric matrices in the hydrostatic case).

Equation (15.4.7) gives the difference between c_v and $c_{\bullet}$ for a hydrostatic reference state. It is not difficult to find the following expression for the ratio

$$c_v/c_{\bullet} = \{\kappa^{(s)}/\kappa^{(T)}\}\{\det(\boldsymbol{c}^{(s)})/\det(\boldsymbol{c}^{(T)})\}. \tag{15.4.8}$$

This result follows directly from (11.1.24) and from the result of example 6, chapter 11, on noting that, since the state is hydrostatic, $c_p = c_{\Theta}$.

As mentioned earlier, c_v and $c_{\bullet}$ are equal for materials of cubic or isotropic symmetry. To demonstrate this we need the following results for such materials, namely,

$$m_\alpha = m E_\alpha, \tag{15.4.9}$$

(see chapter 14),

$$\kappa^{(T)} = 3(s^{(T)}_{11} + 2 s^{(T)}_{12}) \tag{15.4.10}$$

$$= 3/(c^{(T)}_{11} + 2 c^{(T)}_{12}), \tag{15.4.11}$$

and

$$c^{(T)}_{11} = c^{(T)}_{22} = c^{(T)}_{33}, \qquad c^{(T)}_{12} = c^{(T)}_{23} = c^{(T)}_{31}, \tag{15.4.12}$$

where these last three equations will be derived in chapter 17. From (15.4.9) and (15.4.12), we can show that

$$c_{\alpha\nu}^{(T)} m_\alpha m_\nu = 3m^2(c_{11}^{(T)} + 2c_{12}^{(T)}), \tag{15.4.13}$$

and

$$(E_\mu m_\mu)^2 = 9m^2. \tag{15.4.14}$$

On substituting (15.4.11), (15.4.13) and (15.4.14) into (15.4.7), we see that the difference term vanishes, and so $c_v = c_\bullet$.

Having proved this equality for cubic or isotropic materials we may now derive an interesting consequence of this from (15.4.8). Since, for such materials, the ratio in (15.4.8) is one, we deduce that the ratio of the determinants appearing in the right-hand side of (15.4.8) is equal to the ratio, $\kappa^{(T)}/\kappa^{(s)}$, of the compressibilities. This result may be derived also from results obtained in chapter 16.

Equations (15.4.7) and (15.4.8) will be of importance in considering the behaviour of the specific heats at coherent phase transitions. This topic is discussed in chapter 24.

Finally it should perhaps be emphasised that infinitesimal strain theory has been used throughout this section.

Summary of important equations

A. *Specific heat relations for an arbitrary reference state*

$$c_{\boldsymbol{\Theta}} = \hat{c}_L + (T/\rho_0)\,\text{trace}\{\hat{\boldsymbol{m}}(\beta \boldsymbol{E} - \boldsymbol{m})\boldsymbol{\Theta}^*\}, \tag{15.1.4}$$

where $\boldsymbol{m}$ and $\hat{\boldsymbol{m}}$ are triangular matrices, that is,

$$\hat{m}_{\mu\nu} = m_{\mu\nu} = 0, \quad \mu > \nu,$$

and

$$\beta = (\partial J/\partial T)_{\boldsymbol{\Theta}}.$$

$$c_{\boldsymbol{\Theta}} = \mathring{c}_\Lambda + (T/\rho_0)\,\text{trace}\{\hat{\boldsymbol{m}}[\beta\boldsymbol{\Theta}^* - \boldsymbol{m}\boldsymbol{\Theta}^* - \boldsymbol{\Theta}^*\boldsymbol{m}^t]\}. \tag{15.2.2}$$

$$C_\tau/C_\gamma = \det\{\boldsymbol{C}^{(s)}\}/\det\{\boldsymbol{C}^{(T)}\}, \qquad \text{(chapter 11, example 6)}$$

where $\boldsymbol{C}^{(s)}$ and $\boldsymbol{C}^{(T)}$ are the adiabatic and isothermal stiffness matrices.

B. *Specific heat relations for a hydrostatic reference state*

1. For a general anisotropic material

$$c_p - \hat{c}_L = (T/\rho_0)p^*[\hat{m}_1 m_1 + \hat{m}_2 m_2 + \hat{m}_3 m_3 - \hat{\beta}\beta], \tag{15.1.6}$$

Examples

$$c_p - \mathring{c}_\Lambda = (T/\rho_0)p^*[\mathring{\beta}\beta - 2(\mathring{m}_1 m_1 + \mathring{m}_2 m_2 + \mathring{m}_3 m_3) - (\mathring{m}_4 m_4 + \mathring{m}_5 m_5 + \mathring{m}_6 m_6)]. \tag{15.2.3}$$

2. For cubic or isotropic materials

$$\hat{c}_L - c_p = \tfrac{2}{3}(T/\rho_0)p^*\beta^2/(1 - \tfrac{2}{3}\kappa^{(T)}p^*), \tag{15.1.8}$$

$$\mathring{c}_\Lambda - c_p = \tfrac{1}{3}(T/\rho_0)p^*\beta^2/(1 - \tfrac{1}{3}\kappa^{(T)}p^*), \tag{15.2.5}$$

$$c_L - \mathring{c}_\Lambda = \tfrac{1}{3}(T/\rho_0)p^*\beta^2/(1 - \tfrac{1}{3}\kappa^{(T)}p^*)(1 - \tfrac{2}{3}\kappa^{(T)}p^*). \tag{15.2.6}$$

C. *Using infinitesimal strain theory*

$$c_v = c_\bullet + (T/\rho_0)[c_{\alpha\nu}^{(T)} m_\alpha m_\nu - (E_\mu m_\mu)^2/\kappa^{(T)}], \tag{15.4.7}$$

$$= c_\bullet \text{ for cubic or isotropic materials,}$$

where the isothermal compressibility is given by

$$\kappa^{(T)} = E_\alpha s_{\alpha\beta}^{(T)} E_\beta = -(\partial J/\partial p)_T,$$

and for cubic and isotropic materials

$$\kappa = 3(s_{11} + 2s_{12}),$$

for the adiabatic or isothermal compressibility.

Examples

1. Show, using finite strain theory, that one has

$$(\hat{c}_L/c_p)(c_v/c_\bullet) = \{\kappa^{(s)}/\kappa^{(T)}\}\{\det(\hat{\boldsymbol{c}}^{(s)})/\det(\hat{\boldsymbol{c}}^{(T)})\}$$

and that using (15.1.8), one may obtain a correction to (15.4.8).

2. Prove that the compressibility κ is equal to the sum of those elements $s_{\alpha\beta}$ of the compliance matrix for which α and β are $\leqslant 3$.
3. The symmetry scheme for the stiffness matrix for cubic materials is

$$\boldsymbol{c} = \begin{bmatrix} c_{11} & c_{12} & c_{12} & 0 & 0 & 0 \\ c_{12} & c_{11} & c_{12} & 0 & 0 & 0 \\ c_{12} & c_{12} & c_{11} & 0 & 0 & 0 \\ 0 & 0 & 0 & c_{44} & 0 & 0 \\ 0 & 0 & 0 & 0 & c_{44} & 0 \\ 0 & 0 & 0 & 0 & 0 & c_{44} \end{bmatrix}$$

and is the same for the compliance matrix. For isotropic materials, $c_{44} = \frac{1}{2}(c_{11} - c_{12})$, and $s_{44} = 2(s_{11} - s_{12})$. Use the facts that $\boldsymbol{c}$ and $\boldsymbol{s}$ are reciprocal matrices to prove that, for both cubic and isotropic materials,

$$c_{11} + 2c_{12} = 1/(s_{11} + 2s_{12}),$$

$$c_{11} - c_{12} = 1/(s_{11} - s_{12}), \qquad c_{44} = 1/s_{44}.$$

16

Elastic stiffnesses and compliances

The relations between the elastic moduli are discussed. It is shown that the infinitesimal stiffness matrix $\partial\Theta_\alpha/\partial e_\beta$ is not symmetric, unless the reference state is hydrostatic. For the infinitesimal theory to be a good approximation, not only must the strains be infinitesimal but the stresses must be sufficiently small deviations from a hydrostatic state.

The differences between the adiabatic and isothermal quantities are derived, and discussed.

16.1. Relations between the elastic matrices $\boldsymbol{c}$, $\hat{\boldsymbol{c}}$

The stiffness matrices $\hat{\boldsymbol{c}}$, $\acute{\boldsymbol{c}}$ and $\boldsymbol{C}$ are symmetric matrices. This follows from the fact that their elements are second derivatives of a thermodynamic function. For isothermal stiffness elements this function is f, while for adiabatic elements it is u. To prove this, for example for $\boldsymbol{C}$, one notes first the definition of its elements given in (13.1.7), and second the expression of τ_α as an adiabatic or isothermal derivative of u or f, given in equations (10.2.6) or (10.2.9). By identifying the strain co-ordinates $\boldsymbol{\gamma}$, in turn, with $\boldsymbol{e}$ and $\boldsymbol{\eta}$, the symmetry of $\hat{\boldsymbol{c}}$ and $\acute{\boldsymbol{c}}$ follows.

The elements of the stiffness matrix $\boldsymbol{c}$ of the infinitesimal regime cannot be expressed exactly as second derivatives of a thermodynamic function. Hence it may not be a symmetric matrix. We shall show in this section that it is symmetric *if and only if the reference state is hydrostatic.* For a non-hydrostatic reference state, the degree of asymmetry will depend on the magnitudes of $\Theta_1^*-\Theta_2^*$, $\Theta_2^*-\Theta_3^*$, Θ_4^*, Θ_5^*, Θ_6^*, quantities which all vanish if, and only if, the reference state is hydrostatic. For typical solids, this effect will be of the order of magnitude of 1 per cent for values of the above stress tensor quantities of the order of 1–10 kilobar.

We shall then see that if the infinitesimal treatment is to be satisfactory in application, not only must the deformations considered be infinitesimal, but the stress at the reference state must be limited in magnitude or restricted to small deviations from a hydrostatic stress.

We shall demonstrate the possible asymmetry of $\partial\Theta_\alpha/\partial e_\beta$ by deriving the difference $\hat{\boldsymbol{c}}-\boldsymbol{c}$ and by showing that, in general, this is not symmetric. Then, since $\hat{\boldsymbol{c}}$ is symmetric, we may conclude that $\boldsymbol{c}$ is not necessarily symmetric. Beginning with (13.1.2) we write

$$L_{\alpha\beta} = J\{\boldsymbol{\Theta}(\boldsymbol{H}^{t})^{-1}\}_{\alpha\beta}; \quad \alpha \leqslant \beta, \tag{16.1.1}$$

and so, on taking differentials and setting $J = 1$, $\boldsymbol{H} = \boldsymbol{E}$ for the reference state, we find that

$$\mathrm{d}L_{\alpha\beta} = \{(\mathrm{d}e_{11} + \mathrm{d}e_{22} + \mathrm{d}e_{33})\boldsymbol{\Theta}^* - \boldsymbol{\Theta}^* \,\mathrm{d}\boldsymbol{e}^{t}\}_{\alpha\beta} + \mathrm{d}\Theta_{\alpha\beta}; \quad \alpha \leqslant \beta, \tag{16.1.2}$$

since at the reference state $\mathrm{d}J = \mathrm{d}e_{\alpha\alpha}$.

Thus

$$\mathrm{d}\Theta_{\mu} = \hat{c}_{\mu\nu}\,\mathrm{d}e_{\nu} - \{(E_{\alpha}\,\mathrm{d}e_{\alpha})\boldsymbol{\Theta}^* - \boldsymbol{\Theta}^*\,\mathrm{d}\boldsymbol{e}^{t}\}_{\mu}, \tag{16.1.3}$$

where we have reverted to abbreviated notation. On multiplying out the matrix $\boldsymbol{\Theta}^*\,\mathrm{d}\boldsymbol{e}^{t}$ in the correction term, and expressing it in abbreviated notation we obtain the following result

$$\boldsymbol{c} - \hat{\boldsymbol{c}} = \begin{bmatrix} 0 & -\Theta_1^* & -\Theta_1^* & 0 & \Theta_5^* & \Theta_6^* \\ -\Theta_2^* & 0 & -\Theta_2^* & \Theta_4^* & 0 & 0 \\ -\Theta_3^* & -\Theta_3^* & 0 & 0 & 0 & 0 \\ -\Theta_4^* & -\Theta_4^* & 0 & 0 & 0 & 0 \\ -\Theta_5^* & -\Theta_5^* & 0 & 0 & 0 & 0 \\ -\Theta_6^* & 0 & -\Theta_6^* & \Theta_5^* & 0 & 0 \end{bmatrix}. \tag{16.1.4}$$

If the reference state is hydrostatic

$$\begin{aligned} c_{\alpha\beta} - \hat{c}_{\alpha\beta} &= p^*, \ \alpha\beta = 12, 21; 23, 32; 31, 13; \\ &= 0, \text{ otherwise.} \end{aligned} \tag{16.1.5}$$

We note also that $c_{\alpha\beta}$ is symmetric if and only if the reference state is hydrostatic. Equation (16.1.5) is true for both the adiabatic and isothermal stiffness matrices.

Since the typical elastic moduli $c_{\alpha\beta}$ for solids are of the order 10^5–10^6 bar, we see that the correction terms in (16.1.4) and (16.1.5) are again of the order of 1 per cent for pressures in the range 1–10 kilobar. However, near a structural phase transition for a solid, some stiffness element, or some combination of such elements, may vanish. For example, the bulk modulus $c_{11}^{(T)} + 2c_{12}^{(T)}$ for a cubic crystal may vanish. The difference in the values for this quantity in the two regimes, finite and infinitesimal, is $2p^*$, which could thus be of considerable significance in the study of the transition at high pressures. In any case, experiments are possible at pressures of the order of 100 kilobars, and so care is necessary in knowing and stating which elastic moduli are used.

16.2 The relation between $\hat{c}$ and $\mathring{c}$

This relation is most easily obtained by considering $\hat{c}$ and $\mathring{c}$ as second order derivatives of a thermodynamic function. That is,

$$\hat{c}_{\alpha\beta}^{(T)} = (\partial^2 f/\partial e_\alpha \partial e_\beta)_T^*; \qquad \mathring{c}_{\alpha\beta}^{(T)} = (\partial^2 f/\partial \eta_\alpha \partial \eta_\beta)_T^*, \tag{16.2.1}$$

$$\hat{c}_{\alpha\beta}^{(s)} = (\partial^2 u/\partial e_\alpha \partial e_\beta)_s^*; \qquad \mathring{c}_{\alpha\beta}^{(s)} = (\partial^2 u/\partial \eta_\alpha \partial \eta_\beta)_s^*, \tag{16.2.2}$$

where f is regarded as a function of T and the strain co-ordinates $\boldsymbol{e}$ or $\boldsymbol{\eta}$, and u as a function of s and $\boldsymbol{e}$ or $\boldsymbol{\eta}$.

Now we know that if the restriction $\boldsymbol{V} = \boldsymbol{E}$ is applied,

$$2\boldsymbol{\eta} = \boldsymbol{e} + \boldsymbol{e}^{\mathrm{t}} + \boldsymbol{e}^{\mathrm{t}}\boldsymbol{e}, \tag{16.2.3}$$

and, for the second derivatives of f,

$$\frac{\partial^2 f}{\partial e_\alpha \, \partial e_\beta} = \frac{\partial^2 f}{\partial \eta_\mu \, \partial \eta_\nu} \frac{\partial \eta_\mu}{\partial e_\alpha} \frac{\partial \eta_\nu}{\partial e_\beta} + \frac{\partial f}{\partial \eta_\mu} \frac{\partial^2 \eta_\mu}{\partial e_\alpha \, \partial e_\beta}, \tag{16.2.4}$$

where $\partial \eta_\mu/\partial e_\alpha$ and $\partial^2 \eta_\mu/\partial e_\alpha \, \partial e_\beta$ can be obtained from (16.2.3).

If $\boldsymbol{\Lambda} = 0$, the last term in (16.2.4) vanishes ($\Lambda_\mu = \partial f/\partial \eta_\mu$), and, if we evaluate at the reference state, $\partial \eta_\mu/\partial e_\alpha = \delta_{\mu\alpha}$, and so

$$\hat{\boldsymbol{c}} = \mathring{\boldsymbol{c}} \quad \text{if } \boldsymbol{\Lambda}^* = \boldsymbol{\Theta}^* = 0. \tag{16.2.5a}$$

If this is not the case, we see from (16.2.4) that

$$\hat{c}_{\alpha\beta} = \mathring{c}_{\alpha\beta} + \Theta_\mu^* (\partial^2 \eta_\mu/\partial e_\alpha \, \partial e_\beta)^*, \tag{16.2.5b}$$

where $(\partial^2 \eta_\mu/\partial e_\alpha \, \partial e_\beta)^*$ is a matrix with constant elements which may be evaluated by differentiating (16.2.3) twice. It is then straightforward to present the result as

$$\hat{\boldsymbol{c}} - \mathring{\boldsymbol{c}} = \begin{bmatrix} \Theta_1^* & 0 & 0 & 0 & \Theta_5^* & \Theta_6^* \\ 0 & \Theta_2^* & 0 & \Theta_4^* & 0 & 0 \\ 0 & 0 & \Theta_3^* & 0 & 0 & 0 \\ 0 & \Theta_4^* & 0 & \Theta_3^* & 0 & 0 \\ \Theta_5^* & 0 & 0 & 0 & \Theta_3^* & \Theta_4^* \\ \Theta_6^* & 0 & 0 & 0 & \Theta_4^* & \Theta_2^* \end{bmatrix}, \tag{16.2.6}$$

which we verify is symmetric as it should be.

If the reference state is hydrostatic (as it must be, for instance, for a cubic or isotropic material), then we can write

$$\mathring{\boldsymbol{c}} - \hat{\boldsymbol{c}} = p^* \boldsymbol{E}_6, \tag{16.2.7}$$

where $\boldsymbol{E}_6$ is the unit 6×6 matrix.

Since we know $\boldsymbol{c}-\hat{\boldsymbol{c}}$ from (16.1.4), we may obtain $\boldsymbol{c}-\mathring{\boldsymbol{c}}$ which, for a hydrostatic reference state, becomes

$$c_{\alpha\beta}-\mathring{c}_{\alpha\beta}=p^*, \quad \alpha\beta=12,21;23,32;31,13;$$
$$=-p^*, \quad \alpha\beta=11,22,33,44,55,66,$$
$$=0, \quad \text{otherwise.} \tag{16.2.8}$$

It will be shown later that the crystal symmetry of the reference state can restrict the possible form of $\boldsymbol{\Theta}^*$, this has been mentioned for the cases of cubic and isotropic symmetry where $\boldsymbol{\Theta}^*$ must be hydrostatic. Thus, the correction matrix (16.2.6) may have a simpler form for a particular crystal class as one or more of the Θ^*_α may be zero or equal to another.

Since, as may be shown (see chapter 18, examples 4 and 5), $\hat{\boldsymbol{c}}$ and $\mathring{\boldsymbol{c}}$ have the same symmetry scheme, we see that the correction matrix must conform with it. For example, if the symmetry scheme requires two elements of $\hat{\boldsymbol{c}}$ (or $\mathring{\boldsymbol{c}}$) to be equal, then the two corresponding elements of $\boldsymbol{\Theta}^*$ must be equal.

16.3. The relation between the adiabatic and isothermal elastic matrices *C*, *S*

For the strain co-ordinates and corresponding stress elements we shall use the generalised notation γ_α and τ_α, $\alpha=1,2,\ldots,6$. Thus, for thermodynamic quantities we shall use upper-case symbols, such as $\boldsymbol{C}$, $\boldsymbol{S}$, C_τ, C_γ for stiffness and compliance matrices, and specific heats, as described in chapter 13.

The relation between the isothermal and adiabatic stiffness and compliance matrices is of some practical importance since, as mentioned earlier, the adiabatic quantities are measured by experimental methods involving acoustic wave propagation, while the isothermal elements are measured by static experimental methods.

We shall discuss first the relation between the compliances since it is somewhat simpler to express in terms of familiar quantities than the relation between the stiffnesses.

Using Jacobians, we may write

$$S^{(s)}_{\alpha\beta}=\left(\frac{\partial\gamma_\alpha}{\partial\tau_\beta}\right)_{s,\bar{\tau}_\beta}=\frac{\partial(s,\gamma_\alpha,\bar{\tau}_\beta)}{\partial(s,\tau_\beta,\bar{\tau}_\beta)}, \tag{16.3.1}$$

where here and in the following equation, the Jacobians are evaluated at the reference state. On using the cancelling rule (11.2.3) we write this equation as

$$S^{(s)}_{\alpha\beta}=\frac{\partial(T,\boldsymbol{\tau})}{\partial(s,\boldsymbol{\tau})}\frac{\partial(s,\gamma_\alpha,\bar{\tau}_\beta)}{\partial(T,\tau_\beta,\bar{\tau}_\beta)}. \tag{16.3.2}$$

We note that the first Jacobian is equal to $(T/\rho_0 C_\tau)$. In the second Jacobian the numerator and denominator have five variables $\bar{\tau}_\beta$ in common, and so it will reduce to the following 2×2 determinant,

$$\begin{vmatrix} (\partial s/\partial T)^*_\tau & (\partial s/\partial \tau_\beta)^*_{T,\bar{\tau}_\beta} \\ (\partial \gamma_\alpha/\partial T)^*_\tau & (\partial \gamma_\alpha/\partial \tau_\beta)^*_{T,\bar{\tau}_\beta} \end{vmatrix}. \tag{16.3.3}$$

On using the thermodynamic relation (11.3.29), or by recognising that

$$(\partial s/\partial \tau_\beta)_{T,\bar{\tau}_\beta}=(\partial\gamma_\beta/\partial T)_\tau=-(\partial^2 g^\tau/\partial T\,\partial\tau_\beta), \tag{16.3.4}$$

where g^τ is defined in (13.1.13), we see that the off-diagonal terms are equal respectively to M_α and M_β, where

$$M_\alpha=(\partial\gamma_\alpha/\partial T)^*_\tau. \tag{16.3.5}$$

The leading diagonal term in (16.3.3) is recognised to be $(\rho_0 C_\tau/T)$, and the second diagonal term is $S^{(T)}_{\alpha\beta}$. Hence (16.3.2) becomes

$$S^{(s)}_{\alpha\beta}=S^{(T)}_{\alpha\beta}-(T/\rho_0 C_\tau)M_\alpha M_\beta, \tag{16.3.6}$$

or, in matrix notation,

$$\boldsymbol{S}^{(s)}=\boldsymbol{S}^{(T)}-(T/\rho_0 C_\tau)\boldsymbol{MM}. \tag{16.3.7}$$

In an exactly similar way, by interchanging the roles of $\boldsymbol{\tau}$ and $\boldsymbol{\gamma}$, we would obtain, after using the thermodynamic relation of chapter 11, example 7:

$$C^{(s)}_{\alpha\beta}=C^{(T)}_{\alpha\beta}+(T/\rho_0 C_\gamma)(\partial\tau_\alpha/\partial T)^*_\gamma(\partial\tau_\beta/\partial T)^*_\gamma, \tag{16.3.8}$$

and, on using (11.3.34),

$$C^{(s)}_{\alpha\beta}=C^{(T)}_{\alpha\beta}+(T/\rho_0 C_\gamma)C^{(T)}_{\alpha\mu}C^{(T)}_{\beta\nu}M_\mu M_\nu. \tag{16.3.9}$$

An interesting property is discovered if we attempt to obtain (16.3.9) directly from (16.3.7) by multiplying this equation on the left by $\boldsymbol{C}^{(T)}$ and, then, on the right by $\boldsymbol{C}^{(s)}$. Since these matrices are the reciprocals of $\boldsymbol{S}^{(T)}$ and $\boldsymbol{S}^{(s)}$ respectively, one obtains

$$C^{(s)}_{\alpha\beta}=C^{(T)}_{\alpha\beta}+(T/\rho_0 C_\tau)C^{(T)}_{\alpha\mu}C^{(s)}_{\beta\nu}M_\mu M_\nu, \tag{16.3.10}$$

which would agree with (16.3.9) if

$$(C^{(s)}_{\beta\nu}M_\nu)/C_\tau=(C^{(T)}_{\beta\nu}M_\nu)/C_\gamma, \tag{16.3.11}$$

or, written in matrix language,

$$\boldsymbol{S}^{(T)}\boldsymbol{C}^{(s)}\boldsymbol{M}=(C_\tau/C_\gamma)\boldsymbol{M}, \tag{16.3.12}$$

that is, the six components of the thermal expansion matrix $\boldsymbol{M}$ form an eigenvector of the 6×6 matrix $\boldsymbol{S}^{(T)}\boldsymbol{C}^{(s)}$ with an eigenvalue equal to C_τ/C_γ. We shall give a more direct proof of (16.3.12) below.

It follows from example 6 of chapter 11 that C_τ/C_γ is equal to the determinant of $\boldsymbol{S}^{(T)}\boldsymbol{C}^{(s)}$. Thus, since the determinant of a matrix equals the product of its eigenvalues, we have the interesting result that the product of the other five eigenvalues of the matrix is equal to unity!

As an example, consider in the infinitesimal regime a cubic or isotropic crystal. On using the symmetry scheme for $\boldsymbol{c}$ given in chapter 15, example 3, and that for $\boldsymbol{m}$ given in chapter 14, we may show that five eigenvalues of $\boldsymbol{c}^{(s)}$ are equal respectively to five of $\boldsymbol{c}^{(T)}$, and hence that five eigenvalues of $\boldsymbol{s}^{(T)}\boldsymbol{c}^{(s)}$ are equal to unity. From (16.3.12) we may deduce for the sixth eigenvalues that

$$s_{11}^{(T)}+2s_{12}^{(T)}=(c_\Theta/c_\bullet)\{s_{11}^{(s)}+2s_{12}^{(s)}\}, \tag{16.3.13}$$

or

$$\kappa^{(T)}=(c_\Theta/c_\bullet)\kappa^{(s)}=(c_p/c_v)\kappa^{(s)}, \tag{16.3.14}$$

as we know from chapter 15, section 4.

For tetragonal, trigonal, and hexagonal crystal classes, four eigenvalues of $\boldsymbol{c}^{(T)}$ and $\boldsymbol{c}^{(s)}$ are equal, and it can be shown that the relation given in example 6, chapter 11, reduces to

$$\{c_{11}^{(s)}+c_{12}^{(s)}\}c_{33}^{(s)}-2c_{13}^{(s)2}=(c_\Theta/c_\bullet)[\{c_{11}^{(T)}+c_{12}^{(T)}\}c_{33}^{(T)}-2c_{13}^{(T)2}]. \tag{16.3.15}$$

This equation is a relation between specific heats and stiffness elements alone, which does not involve $\boldsymbol{M}$.

Direct proof of (16.3.12)

Consider

$$\mathrm{d}\tau_\beta=\left(\frac{\partial\tau_\beta}{\partial s}\right)_\gamma \mathrm{d}s+\left(\frac{\partial\tau_\beta}{\partial\gamma_\alpha}\right)_{s,\bar{\gamma}_\alpha}\mathrm{d}\gamma_\alpha. \tag{16.3.16}$$

If we set $\mathrm{d}\tau_\beta=0$, $\beta=1,2,\ldots,6$, we obtain

$$\left(\frac{\partial\tau_\beta}{\partial s}\right)_\gamma=-\left(\frac{\partial\tau_\beta}{\partial\gamma_\alpha}\right)_{s,\bar{\gamma}_\alpha}\left(\frac{\partial\gamma_\alpha}{\partial s}\right)_\tau. \tag{16.3.17}$$

Now, using Jacobians or otherwise, one may prove that this is the same as

$$\frac{1}{C_\gamma}\left(\frac{\partial \tau_\beta}{\partial T}\right)_\gamma = -C^{(s)}_{\beta\alpha} M_\alpha / C_\tau, \tag{16.3.18}$$

and so

$$(\partial \tau_\beta / \partial T)_\gamma = -(C_\gamma / C_\tau) C^{(s)}_{\beta\alpha} M_\alpha. \tag{16.3.19}$$

In chapter 11, example 4, it has been shown how we may prove that

$$(\partial \tau_\beta / \partial T)_\gamma = -C^{(T)}_{\beta\alpha} M_\alpha. \tag{16.3.20}$$

Thus, on comparing these two equations, (16.3.12) is obtained.

Summary of important equations

A. *Relations between the stiffness matrices* $\mathring{\boldsymbol{c}}$, $\hat{\boldsymbol{c}}$ *and* $\boldsymbol{c}$

$$\boldsymbol{c} - \hat{\boldsymbol{c}} = \begin{bmatrix} 0 & -\Theta_1^* & -\Theta_1^* & 0 & \Theta_5^* & \Theta_6^* \\ -\Theta_2^* & 0 & -\Theta_2^* & \Theta_4^* & 0 & 0 \\ -\Theta_3^* & -\Theta_3^* & 0 & 0 & 0 & 0 \\ -\Theta_4^* & -\Theta_4^* & 0 & 0 & 0 & 0 \\ -\Theta_5^* & -\Theta_5^* & 0 & 0 & 0 & 0 \\ -\Theta_6^* & 0 & -\Theta_6^* & \Theta_5^* & 0 & 0 \end{bmatrix}, \tag{16.1.4}$$

which for a hydrostatic reference state reduces to

$$\begin{aligned} c_{\alpha\beta} - \hat{c}_{\alpha\beta} &= p^*, \quad \alpha\beta = 12, 21; 23, 32; 31, 13; \\ &= 0, \quad \text{otherwise.} \end{aligned} \tag{16.1.5}$$

$$\hat{\boldsymbol{c}} - \mathring{\boldsymbol{c}} = \begin{bmatrix} \Theta_1^* & 0 & 0 & 0 & \Theta_5^* & \Theta_6^* \\ 0 & \Theta_2^* & 0 & \Theta_4^* & 0 & 0 \\ 0 & 0 & \Theta_3^* & 0 & 0 & 0 \\ 0 & \Theta_4^* & 0 & \Theta_3^* & 0 & 0 \\ \Theta_5^* & 0 & 0 & 0 & \Theta_3^* & \Theta_4^* \\ \Theta_6^* & 0 & 0 & 0 & \Theta_4^* & \Theta_2^* \end{bmatrix}, \tag{16.2.6}$$

which for a hydrostatic reference state reduces to

$$\mathring{\boldsymbol{c}} - \hat{\boldsymbol{c}} = p^* \boldsymbol{E}_6. \tag{16.2.7}$$

$$\begin{aligned} c_{\alpha\beta} - \mathring{c}_{\alpha\beta} &= p^*, \quad \alpha\beta = 12, 21; 23, 32; 31, 13; \\ &= -p^*, \quad \alpha\beta = 11, 22, 33, 44, 55, 66, \\ &= 0, \quad \text{otherwise.} \end{aligned} \tag{16.2.8}$$

B. *General relations for stiffness and compliance matrices*

$$S^{(s)}_{\alpha\beta} = S^{(T)}_{\alpha\beta} - (T/\rho_0 C_\tau) M_\alpha M_\beta. \quad (16.3.6)$$

$$C^{(s)}_{\alpha\beta} = C^{(T)}_{\alpha\beta} + (T/\rho_0 C_\gamma) C^{(T)}_{\alpha\mu} C^{(T)}_{\beta\nu} M_\mu M_\nu. \quad (16.3.9)$$

$$\boldsymbol{S}^{(T)}\boldsymbol{C}^{(s)}\boldsymbol{M} = (C_\tau/C_\gamma)\boldsymbol{M}. \quad (16.3.12)$$

Examples

1. Show that the correction term in (16.2.5b) may be written directly, for a hydrostatic reference state, as

$$-\tfrac{1}{2}p^*\{\partial^2(e_\mu e_\mu)/\partial e_\alpha\,\partial e_\beta\} = -p^*\delta_{\alpha\beta},$$

thus arriving directly and simply at (16.2.7).

2. For a cubic crystal the Helmholtz free energy, up to terms quadratic in $\boldsymbol{\eta}$, is of the form

$$f = f_0(T) - p^*(\eta_1 + \eta_2 + \eta_3)$$
$$+ \tfrac{1}{2}\{a(\eta_1 + \eta_2 + \eta_3)^2 + b(\eta_1^2 + \eta_2^2 + \eta_3^2) + c(\eta_4^2 + \eta_5^2 + \eta_6^2)\}.$$

(*a*) Express the stiffness elements $\mathring{c}_{\alpha\beta}$ in terms of a, b and c.
(*b*) Express f in terms of $\boldsymbol{e}$ up to second order in the components of $\boldsymbol{e}$, using (16.2.3), and hence obtain the result (16.2.7).

3. Show that, in tensor notation,

$$\hat{c}_{\alpha\beta\mu\nu} = c_{\alpha\beta\mu\nu} + \Theta^*_{\alpha\beta}\delta_{\mu\nu} - \Theta^*_{\alpha\nu}\delta_{\beta\mu},$$
$$= \mathring{c}_{\alpha\beta\mu\nu} + \Theta^*_{\beta\nu}\delta_{\alpha\mu},$$

for $\alpha \leqslant \beta$, $\mu \leqslant \nu$.

Hence show that, for the hydrostatic case,

$$\hat{c}_{\alpha\beta\mu\nu} = c_{\alpha\beta\mu\nu} - p^*(\delta_{\alpha\beta}\delta_{\mu\nu} - \delta_{\alpha\nu}\delta_{\beta\mu}),$$
$$= \mathring{c}_{\alpha\beta\mu\nu} - p^*\delta_{\alpha\mu}\delta_{\beta\nu},$$

for $\alpha \leqslant \beta$, $\mu \leqslant \nu$.

17

Tensorial forms for the elastic stiffness and compliance matrices

The tensor forms of the elastic matrices are discussed. The relationships of tensor elastic moduli and compliances to those already defined and used are derived. The symmetry schemes for elastic stiffness and compliance matrices are given at the end of the chapter, as are a set of examples designed to develop the theory of acoustic propagation in crystals.

17.1. Tensor forms for stress and strain

In chapter 12 we introduced the abbreviated notation for the stress and strain tensors, $\mathbf{\Lambda}$ and $\boldsymbol{\eta}$, as being usually the most convenient method of enumerating the independent variables in non-hydrostatic thermodynamics. However, especially when studying the effects of symmetry on the properties of a crystal, the expression of physical quantities in tensor notation has many advantages. These advantages arise from the linear nature of the transformation of tensors under orthogonal transformations of the co-ordinate axes.

The connection between the abbreviated and full tensor notations may be expressed, for the strain and stress elements, as follows,

$$\eta_1 = \eta_{11}, \quad \eta_2 = \eta_{22}, \quad \eta_3 = \eta_{33}, \quad \eta_4 = \eta_{23} + \eta_{32},$$
$$\eta_5 = \eta_{31} + \eta_{13}, \quad \eta_6 = \eta_{12} + \eta_{21}, \tag{17.1.1}$$

and

$$\Lambda_1 = \Lambda_{11}, \quad \Lambda_2 = \Lambda_{22}, \quad \Lambda_3 = \Lambda_{33}, \quad \Lambda_4 = \tfrac{1}{2}(\Lambda_{23} + \Lambda_{32}),$$
$$\Lambda_5 = \tfrac{1}{2}(\Lambda_{31} + \Lambda_{13}), \quad \Lambda_6 = \tfrac{1}{2}(\Lambda_{12} + \Lambda_{21}). \tag{17.1.2}$$

For $\boldsymbol{\varepsilon}$, the infinitesimal strain tensor, and $\mathbf{\Theta}$, the Cauchy stress tensor, the connections are exactly as for $\boldsymbol{\eta}$ and $\mathbf{\Lambda}$ respectively, except that in the abbreviated notation, the symbol e is used instead of ε, as in (7.3.8). We may also note that the expression for the work done, in an infinitesimal strain variation, is (per unit reference volume)

$$d'w = \Lambda_{\alpha\beta}\, d\eta_{\beta\alpha} = \Lambda_\mu\, d\eta_\mu, \tag{17.1.3}$$
$$= \Theta_{\alpha\beta}\, d\varepsilon_{\beta\alpha} = \Theta_\mu\, de_\mu, \tag{17.1.4}$$

where (17.1.4) applies to the infinitesimal regime.

The stiffness matrix elements, in tensor notation, are defined by

$$\mathring{c}_{\alpha\beta\gamma\delta} = (\partial\Lambda_{\alpha\beta}/\partial\eta_{\gamma\delta})^*, \tag{17.1.5}$$

$$c_{\alpha\beta\gamma\delta} = (\partial\Theta_{\alpha\beta}/\partial\varepsilon_{\gamma\delta})^*, \tag{17.1.6}$$

with corresponding definitions for the tensor compliance elements. We wish to relate such tensorial elements to the corresponding stiffness and compliance elements expressed in abbreviated notation. The use of the relations (17.1.1) and (17.1.2) enables this to be done.

Stiffness elements are second derivatives, with respect to the strain elements, of the thermodynamic functions f or u, where the choice depends on whether the isothermal or adiabatic elements are required. Compliance elements are second derivatives, with respect to the stress elements of the generalized Gibbs free energy density or of the generalised enthalpy density.

For example, in abbreviated notation, the isothermal element

$$\mathring{c}_{\mu\nu} = (\partial^2 f/\partial\eta_\mu\partial\eta_\nu)^*. \tag{17.1.7}$$

If we regard f as a function of the $\eta_{\alpha\beta}$, then on using (17.1.3) we have

$$df = -s\,dT + \Lambda_{\alpha\beta}\,d\eta_{\beta\alpha}, \tag{17.1.8}$$

and so

$$\mathring{c}_{\alpha\beta\gamma\delta} = (\partial^2 f/\partial\eta_{\alpha\beta}\,\partial\eta_{\gamma\delta})^*. \tag{17.1.9}$$

From (17.1.1) we see that

$$(\partial^2 f/\partial\eta_{\alpha\beta}\,\partial\eta_{\gamma\delta}) = (\partial^2 f/\partial\eta_\mu\,\partial\eta_\nu), \tag{17.1.10}$$

where μ and ν correspond, respectively, to the pairs $\alpha\beta$ and $\gamma\delta$, and so

$$\mathring{c}_{\alpha\beta\gamma\delta} = \mathring{c}_{\mu\nu}. \tag{17.1.11}$$

When we come to the compliances, however, we shall see that because of the factor $\frac{1}{2}$ in the last three relations of (17.1.2), a relation is obtained which is not so simple as that of (17.1.11).

From (13.1.12), the definition of $\mathring{g}$, we have

$$d\mathring{g} = -s\,dT - \eta_\mu\,d\Lambda_\mu = -s\,dT - \eta_{\alpha\beta}\,d\Lambda_{\beta\alpha}, \tag{17.1.12}$$

where the last term is justified by reference to the schemes of (17.1.1) and (17.1.2). Thus, we have

$$\eta_\mu = -(\partial\mathring{g}/\partial\Lambda_\mu)^*, \qquad \eta_{\alpha\beta} = -(\partial\mathring{g}/\partial\Lambda_{\alpha\beta})^*. \tag{17.1.13}$$

Because, from (17.1.2),

$$\frac{\partial}{\partial \Lambda_{\alpha\beta}} = (2-\delta_{\alpha\beta})^{-1}\frac{\partial}{\partial \bar{\Lambda}_{\mu}}, \tag{17.1.14}$$

where μ corresponds to $\alpha\beta$, we see that (17.1.13) is consistent with (17.1.1). If we differentiate (17.1.13) with respect to stress elements, we obtain

$$\mathring{s}_{\mu\nu} = -(\partial^2 \mathring{g}/\partial \Lambda_\mu \, \partial \Lambda_\nu)^*, \qquad \mathring{s}_{\alpha\beta\gamma\delta} = -(\partial^2 \mathring{g}/\partial \Lambda_{\alpha\beta} \, \partial \Lambda_{\gamma\delta})^*. \tag{17.1.15}$$

Thus, since

$$\frac{\partial^2}{\partial \Lambda_{\alpha\beta} \, \partial \Lambda_{\gamma\delta}} = \{(2-\delta_{\alpha\beta})(2-\delta_{\gamma\delta})\}^{-1}\frac{\partial^2}{\partial \Lambda_\mu \, \partial \Lambda_\nu}, \tag{17.1.16}$$

where μ and ν correspond, respectively, to $\alpha\beta$ and $\gamma\delta$,

$$\mathring{s}_{\alpha\beta\gamma\delta} = \{(2-\delta_{\alpha\beta})(2-\delta_{\gamma\delta})\}^{-1}\mathring{s}_{\mu\nu}. \tag{17.1.17}$$

In the infinitesimal regime the relations which correspond to (17.1.11) and (17.1.17) can easily be seen to hold.

These relations are true for both adiabatic or isothermal compliances or stiffnesses. The corresponding results for adiabatic quantities may be obtained as above by substituting u for f and $\mathring{h}$ for $\mathring{g}$.

The present treatment may be compared with that of Murnaghan (1951).

17.2. Symmetry schemes for compliances and stiffnesses

For a certain crystal symmetry the tensors $\mathring{\boldsymbol{c}}$, $\mathring{\boldsymbol{s}}$ (and $\boldsymbol{c}$, $\boldsymbol{s}$) are tensors which must be invariant to the symmetry transformations of the axis-system. This point has been discussed in chapter 9. Using this approach, the symmetry schemes for the stiffness and compliance tensors and matrices may be obtained, and this method has been well presented in the book by Bhagavantam (1966). These schemes are listed in the summary at the end of this chapter. It will be seen that the factor $[(2-\delta_{\alpha\beta})(2-\delta_{\gamma\delta})]^{-1}$ in (17.1.17) does not lead to different symmetry schemes for the matrices $\boldsymbol{c}$ and $\boldsymbol{s}$, *except* in the case of the trigonal, hexagonal and isotropic systems, where in all these three systems we have the relations

$$c_{66} = \tfrac{1}{2}(c_{11}-c_{12}); \qquad s_{66} = 2(s_{11}-s_{12}). \tag{17.2.1}$$

In the trigonal system, there are the relations

$$c_{46} = -c_{15}; \qquad c_{56} = c_{14},$$

whereas

$$s_{46} = -2s_{15}; \qquad s_{56} = 2s_{14}. \tag{17.2.2}$$

(Of course, for the isotropic case $c_{44} = c_{55} = c_{66}$.)

In the method of dealing with symmetry properties used in this book, these properties of the symmetry schemes will appear very naturally. The method to be used will be based on the fact that the thermodynamic energy functions u, f, g, h, s, etc., are invariant functions of the crystal group. When this method has been presented, the reader will be able to derive easily the symmetry schemes, *using abbreviated notation only*, for the stiffness and compliance matrices, and indeed for the third order and higher elastic constants if required. However, for convenience, the symmetry schemes are tabulated at the end of this chapter.

17.3. The measurement of *c* and *s*

The *stiffness* elements are sometimes called the *elastic moduli*, and the compliance elements are sometimes known as *elastic constants*.

The measurement of these quantities is of considerable importance in solid state physics. For example, the most important first quantitative step in theoretical solid state physics was made by Max Born, when he related the elastic moduli to the repulsive forces in an ionic crystal.

Static methods would result in the measurement of isothermal quantities. However, the experimental difficulties of static methods are considerable and dynamical methods of measurement are used. These methods employ the properties of the propagation of acoustic waves in solids, and result in the measurement of *adiabatic stiffnesses*. For a description of such methods and a discussion of the practical difficulties, reference may be made to Huntingdon (1958), in which are also given many references. The well-known textbook of Nye (1957) has an excellent discussion on the elastic constants of crystals. The examples at the end of this chapter are designed to develop the theory of acoustic propagation in crystals and the relation between the wave velocity and the elastic moduli.

Summary of important equations

A. *Abbreviated notation for strain and stress tensors*

$$\eta_1 = \eta_{11}, \qquad \eta_2 = \eta_{22}, \qquad \eta_3 = \eta_{33}, \qquad \eta_4 = \eta_{23} + \eta_{32},$$

$$\eta_5 = \eta_{31} + \eta_{13}, \qquad \eta_6 = \eta_{12} + \eta_{21}, \tag{17.1.1}$$

with a corresponding relation for the correspondence in infinitesimal strain theory,

$$e_\mu \to \varepsilon_{\alpha\beta}.$$

$$\Lambda_1 = \Lambda_{11}, \qquad \Lambda_2 = \Lambda_{22}, \qquad \Lambda_3 = \Lambda_{33}, \qquad \Lambda_4 = \tfrac{1}{2}(\Lambda_{23} + \Lambda_{32}),$$

$$\Lambda_5 = \tfrac{1}{2}(\Lambda_{31} + \Lambda_{13}), \qquad \Lambda_6 = \tfrac{1}{2}(\Lambda_{12} + \Lambda_{21}), \tag{17.1.2}$$

with a corresponding relation for the correspondence, in infinitesimal strain theory,

$$\Theta_\mu \to \Theta_{\alpha\beta}.$$

B. *Abbreviated notation for stiffness and compliance tensors*

$$\mathring{c}_{\alpha\beta\gamma\delta} = (\partial\Lambda_{\alpha\beta}/\partial\eta_{\gamma\delta})^* = \mathring{c}_{\mu\nu}, \qquad \begin{cases}(17.1.5)\\(17.1.11)\end{cases}$$

$$\mathring{s}_{\alpha\beta\gamma\delta} = \{(2-\delta_{\alpha\beta})(2-\delta_{\gamma\delta})\}^{-1}\mathring{s}_{\mu\nu}. \tag{17.1.17}$$

Similar relations hold for the stiffness and compliance elements of the infinitesimal strain regime.

C. *Elastic stiffness coefficients for crystal systems*

(*a*) Triclinic system: the most general stiffness matrix for this system is the symmetric 6×6 matrix with no element necessarily zero.

(*b*) Monoclinic system

$$\begin{matrix} c_{11} & c_{12} & c_{13} & 0 & 0 & c_{16} \\ & c_{22} & c_{23} & 0 & 0 & c_{26} \\ & & c_{33} & 0 & 0 & c_{36} \\ & & & c_{44} & c_{45} & 0 \\ & & & & c_{55} & 0 \\ & & & & & c_{66}. \end{matrix}$$

(*c*) Orthorhombic system

$$\begin{matrix} c_{11} & c_{12} & c_{13} & 0 & 0 & 0 \\ & c_{22} & c_{23} & 0 & 0 & 0 \\ & & c_{33} & 0 & 0 & 0 \\ & & & c_{44} & 0 & 0 \\ & & & & c_{55} & 0 \\ & & & & & c_{66}. \end{matrix}$$

(*d*) (i) Tetragonal system: classes 4, $\bar{4}$, $4/m$

$$\begin{matrix} c_{11} & c_{12} & c_{13} & 0 & 0 & c_{16} \\ & c_{11} & c_{13} & 0 & 0 & -c_{16} \\ & & c_{33} & 0 & 0 & 0 \\ & & & c_{44} & 0 & 0 \\ & & & & c_{44} & 0 \\ & & & & & c_{66}. \end{matrix}$$

(*d*) (ii) Tetragonal system: classes $4mm$, $\bar{4}2m$, 422, $4/mmm$

$$\begin{matrix} c_{11} & c_{12} & c_{13} & 0 & 0 & 0 \\ & c_{11} & c_{13} & 0 & 0 & 0 \\ & & c_{33} & 0 & 0 & 0 \\ & & & c_{44} & 0 & 0 \\ & & & & c_{44} & 0 \\ & & & & & c_{66}. \end{matrix}$$

(*e*) (i) Trigonal system: classes 3, $\bar{3}$

$$\begin{matrix} c_{11} & c_{12} & c_{13} & c_{14} & c_{15} & 0 \\ & c_{11} & c_{13} & -c_{14} & -c_{15} & 0 \\ & & c_{33} & 0 & 0 & 0 \\ & & & c_{44} & 0 & -c_{15} \\ & & & & c_{44} & c_{14} \\ & & & & & \frac{1}{2}(c_{11}-c_{12}). \end{matrix}$$

(*e*) (ii) Trigonal system: classes $3m$, 32, $\bar{3}m$

$$\begin{matrix} c_{11} & c_{12} & c_{13} & c_{14} & 0 & 0 \\ & c_{11} & c_{13} & -c_{14} & 0 & 0 \\ & & c_{33} & 0 & 0 & 0 \\ & & & c_{44} & 0 & 0 \\ & & & & c_{44} & c_{14} \\ & & & & & \frac{1}{2}(c_{11}-c_{12}). \end{matrix}$$

(*f*) Hexagonal system

$$\begin{matrix} c_{11} & c_{12} & c_{13} & 0 & 0 & 0 \\ & c_{11} & c_{13} & 0 & 0 & 0 \\ & & c_{33} & 0 & 0 & 0 \\ & & & c_{44} & 0 & 0 \\ & & & & c_{44} & 0 \\ & & & & & \frac{1}{2}(c_{11}-c_{12}). \end{matrix}$$

(*g*) Cubic systems

$$\begin{array}{cccccc} c_{11} & c_{12} & c_{12} & 0 & 0 & 0 \\ & c_{11} & c_{12} & 0 & 0 & 0 \\ & & c_{11} & 0 & 0 & 0 \\ & & & c_{44} & 0 & 0 \\ & & & & c_{44} & 0 \\ & & & & & c_{44}. \end{array}$$

(*h*) Isotropic medium

$$\begin{array}{cccccc} c_{11} & c_{12} & c_{12} & 0 & 0 & 0 \\ & c_{11} & c_{12} & 0 & 0 & 0 \\ & & c_{11} & 0 & 0 & 0 \\ & & & \frac{1}{2}(c_{11}-c_{12}) & 0 & 0 \\ & & & & \frac{1}{2}(c_{11}-c_{12}) & 0 \\ & & & & & \frac{1}{2}(c_{11}-c_{12}). \end{array}$$

The scheme is unchanged for **s** except that in the trigonal, hexagonal and isotropic systems $s_{66}=2(s_{11}-s_{12})$, and in the trigonal system $s_{46}=-2s_{15}$, and $s_{56}=2s_{14}$.

Examples (mainly to indicate the theory of acoustic wave propagation)

1. From the tables of elastic stiffnesses in the summary, solve for the compliance matrix by taking the reciprocal of the stiffness matrix in the case of an isotropic medium, the hexagonal system, the tetragonal systems and any others you wish.
2. Derive the symmetry scheme for the stiffness elements for the trigonal and hexagonal systems, using the method described in chapter 9 (and Bhagavantam, 1966).

Note. The following examples are designed to indicate the theory of acoustic wave propagation in elastic media, which are not necessarily isotropic. From these examples it should be seen how the adiabatic stiffnesses may be obtained from measurements of the phase velocities of acoustic waves. For an account of the modern experimental methods, reference may be made to, e.g., Huntingdon, 1958.

3.(*a*) Prove that the equation of continuity for an elastic medium may be written as

$$\partial(\rho U_\alpha)/\partial X_\alpha + \partial\rho/\partial t = 0,$$

where $\boldsymbol{U}$ is the vector function of position $\boldsymbol{X}$ and time t, which describes the velocity field of the medium (see chapter 5).

(*b*) Using the equation of continuity and neglecting terms quadratic in strain and velocity components, prove that (5.4.2) leads to an equation of motion which may be written, to this approximation, as

$$-(\partial^2\rho/\partial t^2) = \partial^2\Theta_{\alpha\beta}/\partial X_\alpha\,\partial X_\beta,$$

gravity and other body forces being neglected.

4.(*a*) If $\boldsymbol{\Theta} = -p\boldsymbol{E}$, i.e., *if the stress remains hydrostatic*, show that the equation of motion of example 3(b) reduces to

$$(\partial^2\rho/\partial t^2) = \partial^2 p/\partial X_\alpha\,\partial X_\alpha = \nabla^2 p.$$

If the processes in the time-dependent behaviour are assumed to be adiabatic, prove that for a perfect gas

$$\nabla^2 p = (\gamma p/\rho)\nabla^2\rho,$$

and hence that the speed of sound in a perfect gas is given by $\surd(\gamma p/\rho)$, where p and ρ are the static undisturbed values of the pressure and density; $\gamma = c_p/c_v$, the ratio of the specific heats.

(*b*) Prove that the adiabatic bulk modulus $-V(\partial p/\partial V)_s$ is γp for a perfect gas.

5. For the conditions of example 4(*a*) prove that the velocity of sound in any elastic medium, *if the stress remains hydrostatic in the acoustic disturbance*, is given by $\surd(n^{(s)}/\rho)$, where $n^{(s)}$ is the adiabatic bulk modulus, (a stiffness) $-V(\partial p/\partial V)_s$.

Note. In examples 4 and 5, the assumption that the stress remains hydrostatic in a wave disturbance cannot be sustained in crystalline materials, and the methods discussed really only apply to fluids. In the following examples, a method is developed which is applicable to such materials. This method applies to waves of infinitesimal amplitude so that quantities higher than first order in strain and particle velocity components are neglected. In effect, in the following examples 6–16, infinitesimal strain theory is used and results are expressed in terms of the corresponding stiffness matrix $\mathbf{c}$.

6. Assume that a plane acoustic wave is being propagated, such that the displacements of the particles from their reference positions are given by

$$\boldsymbol{u} = \boldsymbol{u}^{(w)} \exp\{\mathrm{i}(\boldsymbol{k}\cdot\boldsymbol{x} - \omega t)\},$$

where $\boldsymbol{u}^{(w)}$ is the amplitude (considered infinitesimal) of the wave, and is independent of position. The time and space variation of other quantities linearly related to $\boldsymbol{u}$ is also given by the same exponential factor.
Prove that the wave variation of $\boldsymbol{D}$ is given by

$$d_{\alpha\beta} = D_{\alpha\beta} - \delta_{\alpha\beta} = -U_\alpha k_\beta/\omega = -U_\alpha n_\beta/v_{\mathrm{ph}},$$

where $U_\alpha = \partial u_\alpha/\partial t$ is the particle velocity, $\boldsymbol{k} = (2\pi/\lambda)\boldsymbol{n}$, $\boldsymbol{n}$ being the unit vector in the direction of propagation, λ the wavelength, $v_{\mathrm{ph}} = (\omega/2\pi)\lambda$ being the phase velocity.

7. Show that the equations of motion (5.4.2), with body forces such as gravity neglected, reduce to

$$\rho_0\omega U_\alpha + k_\beta\theta_{\alpha\beta} = 0,$$

i.e.,

$$v_{\mathrm{ph}}\rho_0 U_\alpha + n_\beta\theta_{\alpha\beta} = 0,$$

where $\boldsymbol{\theta}$, the wave part of $\boldsymbol{\Theta} = \boldsymbol{\Theta}^* + \boldsymbol{\theta}$, being linearly related to $\boldsymbol{d}$, thus has the time and space dependence given by

$$\boldsymbol{\theta} = \boldsymbol{\theta}^{(w)} \exp\{\mathrm{i}(\boldsymbol{k} \cdot \boldsymbol{x} - \omega t)\}.$$

Note: when differentiating with respect to $\boldsymbol{X}$, the difference between $\boldsymbol{X}$ and $\boldsymbol{x}$ in $\boldsymbol{k} \cdot \boldsymbol{x}$ is to be ignored; further, ρ is to be taken as ρ_0.

8. Use the symmetry properties of $c_{\alpha\beta\mu\nu}$ to prove that

$$\theta_{\alpha\beta} = \tfrac{1}{2} c_{\alpha\beta\mu\nu}(d_{\mu\nu} + d_{\nu\mu}) = c_{\alpha\beta\mu\nu} d_{\mu\nu}.$$

9.(*a*) Using the result of example 8 in the equation of motion obtained in example 7, and using the result of example 6 for $d_{\alpha\beta}$, show that there are three equations of motion as follows.

$$(\Omega_{\alpha\nu} - a\delta_{\alpha\nu})U_\nu = 0, \quad \alpha = 1, 2, 3,$$

or

$$(\boldsymbol{\Omega} - a\boldsymbol{E})\boldsymbol{U} = 0,$$

where

$$\Omega_{\alpha\nu} = c_{\alpha\beta\mu\nu} n_\beta n_\mu, \quad a = \rho_0 v_{\mathrm{ph}}^2.$$

(*b*) Show that, if $\boldsymbol{c}$ is symmetric, so is $\boldsymbol{\Omega}$ (see Note after example 16), and that in order that there may be non-trivial solutions to the equations we must have

$$\det(\boldsymbol{\Omega} - a\boldsymbol{E}) = 0.$$

The elastic moduli $c_{\alpha\beta\mu\nu}$ are usually taken as adiabatic, since, in acoustic waves, the time-constants of relaxation processes are assumed to be very much greater than the period of oscillation.

(*c*) Show that there are three modes of propagation for a given wave direction $\boldsymbol{n}$, the three phase velocities being determined by the three values of the eigenvalues a, and the three polarisations being determined by the three mutually orthogonal eigenvectors $\boldsymbol{U}$.

(*d*) Show that the modes are not necessarily exactly longitudinal or transverse. (*Note*: it will be shown in succeeding examples that many modes are transverse or longitudinal in materials of high crystal symmetry.)

10. For a cubic crystal, show that, for $\boldsymbol{n} = (1, 0, 0)$,

$$\boldsymbol{\Omega} = \begin{bmatrix} c_{11} & 0 & 0 \\ 0 & c_{44} & 0 \\ 0 & 0 & c_{44} \end{bmatrix},$$

and that the corresponding wave modes consist of two transverse modes of equal phase velocity and one longitudinal mode. Measurement of the phase velocities will determine c_{11} and c_{44}.

11. For a cubic crystal, show that, for $\boldsymbol{n} = (1, 1, 1)/\sqrt{3}$,

$$\boldsymbol{\Omega} = \frac{1}{3}\begin{bmatrix} \alpha & \beta & \beta \\ \beta & \alpha & \beta \\ \beta & \beta & \alpha \end{bmatrix} = \tfrac{1}{3}(\alpha - \beta)\boldsymbol{E} + \tfrac{1}{3}\beta \begin{bmatrix} 1 & 1 & 1 \\ 1 & 1 & 1 \\ 1 & 1 & 1 \end{bmatrix},$$

where $\alpha = c_{11} + 2c_{44}$, $\beta = c_{12} + c_{44}$.

Prove that the corresponding modes consist of two transverse modes of equal phase velocity $\sqrt{\{\frac{1}{3}(c_{11}-c_{12}+c_{44})/\rho_0\}}$ and one longitudinal mode of phase velocity $\sqrt{\{\frac{1}{3}(c_{11}+2c_{12}+4c_{44})/\rho_0\}}$. What would these phase velocities be for the case of an isotropic medium? (*Note*: Measurements of phase velocities for a cubic crystal for wave direction $(1,0,0)$ and $(1,1,1)/\sqrt{3}$ determine c_{11}, c_{44}, c_{12}.)

12.(*a*) Show that the matrix $\boldsymbol{N}$, defined by

$$N_{\alpha\beta}=n_\alpha n_\beta,$$

where $\boldsymbol{n}$ is a unit vector, has the eigenvalues 1, 0, 0, with corresponding orthogonal eigenvectors $\boldsymbol{n}$, $\boldsymbol{l}$, $\boldsymbol{m}$, the latter two being any pair of orthogonal (unit) vectors which are orthogonal to $\boldsymbol{n}$.

(*b*) Show that, if $\boldsymbol{\Omega}$ is of the form $x\boldsymbol{N}+y\boldsymbol{E}$, there are two transverse modes of equal phase velocity and one longitudinal mode. Note that the matrix $\boldsymbol{\Omega}$ of Example 11 is of this form.

13.(*a*) For the tetragonal classes 4, $\bar{4}$, $4/m$, show that for $\boldsymbol{n}=(1,0,0)$,

$$\boldsymbol{\Omega}=\begin{bmatrix} c_{11} & c_{16} & 0 \\ c_{16} & c_{66} & 0 \\ 0 & 0 & c_{44} \end{bmatrix}.$$

Discuss the phase velocities and polarisation of the corresponding three independent modes.

(*b*) Is there a direction of propagation for these classes, such that there are one longitudinal and two transverse modes?

14.(*a*) Show that for a cubic crystal and for a general direction of propagation,

$$\boldsymbol{\Omega}=(c_{12}+c_{44})\boldsymbol{N}+\boldsymbol{Y},$$

where $\boldsymbol{N}$ is defined in example 12, and $\boldsymbol{Y}$ is the diagonal matrix whose diagonal elements are

$$Y_{ii}=c_{44}+n_i^2(c_{11}-c_{12}-2c_{44}),\quad i=1,2,3.$$

(*b*) Hence show that, if the additional condition of isotropy, i.e., $c_{44}=\frac{1}{2}(c_{11}-c_{12})$, is satisfied, there are, for any wave direction $\boldsymbol{n}$, two transverse modes of equal phase velocity and one longitudinal mode.

15. If the isotropy condition of example 14(*b*) is not satisfied, show that for a cubic crystal, there are two transverse modes of equal phase velocity and one longitudinal mode for each of the following wave directions: $\boldsymbol{n}=(1,0,0)$; $(0,1,0)$; $(0,0,1)$; $(\pm1,\pm1,\pm1)/\sqrt{3}$.

16. In the treatment of acoustic propagation given in the above examples, show that there is no dispersion for a given mode, i.e., for a given mode, v_{ph} is independent of frequency.

Note: *According to chapter 16,* $\mathbf{c}$ *is not symmetric in general, so a rigorous treatment should be given. In fact, the equation of motion of chapter 6 example 7(c) may be used, again neglecting quadratic velocity terms in* $\mathrm{D}U/\mathrm{D}t$. *It may be shown that* $\boldsymbol{\Omega}$ *in example 9 must be given by*

$$\Omega_{\alpha\nu}=\mathring{c}_{\alpha\beta\mu\nu}n_\beta n_\mu+(\Theta^*_{\beta\mu}n_\beta n_\mu)\delta_{\alpha\nu}=(\partial B_{\beta\alpha}/\partial D_{\mu\nu})n_\beta n_\mu.$$

For a hydrostatic reference state

$$\Omega_{\alpha\nu} = \mathring{c}_{\alpha\beta\mu\nu} n_\beta n_\mu - p^* \delta_{\alpha\nu} = c_{\alpha\beta\mu\nu} n_\beta n_\mu .$$

This correction could be important at high non-hydrostatic stresses or near structural phase transitions where elastic softening may occur. We note from the last equation that the result of infinitesimal strain theory is regained for a hydrostatic case.

18

The effects of symmetry on the thermodynamic properties of crystals

It is shown how the development of the symmetry properties of crystals in thermodynamics may be based on the invariance of thermodynamic energy functions. Such functions can be developed in terms of polynomials of the strain co-ordinates, which can be expressed as polynomials of a finite set of invariant polynomials known as an integrity basis. Thus, elastic constants of any order can be described and related. This method simplifies the derivation of the relations between the different systems of co-ordinates and also the transformations of elastic stiffnesses, etc., with an arbitrary orthogonal transformation of the axis system.

18.1. The integrity basis of a finite transformation group

Suppose, at first, that we consider finite symmetry groups defined as transformations on a space of three dimensions, which will be described by an orthonormal co-ordinate system with co-ordinates x, y, z (or x_1, x_2, x_3). A simple example of such a transformation group is the crystal point group m. The elements of this group are $\boldsymbol{E}$, the identity element, and the reflection operator m defining reflection in a plane which, in this case, we can choose as the XOY plane. The reflection operator is then commonly designated m_z. We shall regard this operation, as with all others, as transforming the co-ordinate axis system from a system where the co-ordinates of a point are x, y, z, to that where the co-ordinates of the same point are x', y', z'. Thus m_z has the representation, in this case, given by

$$x' \to x; \qquad y' \to y; \qquad z' \to -z, \tag{18.1.1}$$

i.e., it is represented by the matrix

$$m_z = \begin{bmatrix} 1 & 0 & 0 \\ 0 & 1 & 0 \\ 0 & 0 & -1 \end{bmatrix}. \tag{18.1.2}$$

Similarly

$$m_y = \begin{bmatrix} 1 & 0 & 0 \\ 0 & -1 & 0 \\ 0 & 0 & 1 \end{bmatrix}; \qquad m_x = \begin{bmatrix} -1 & 0 & 0 \\ 0 & 1 & 0 \\ 0 & 0 & 1 \end{bmatrix}. \tag{18.1.3}$$

Both these matrices are of det(−1). We see that the product $m_x m_y$ (which is equal to $m_y m_x$) is given by

$$m_x m_y = \begin{bmatrix} -1 & 0 & 0 \\ 0 & -1 & 0 \\ 0 & 0 & +1 \end{bmatrix} \tag{18.1.4}$$

of determinant +1.

It is easy to see that $m_x m_y$ represents a rotation of π about the z-axis, and this is a particular case of a general result that any rotation ϕ about an axis $\boldsymbol{n}$ (unit vector) is identical with the product of successive reflections in two planes whose line of intersection is given by $\boldsymbol{n}$, and such that the dihedral angle between the two planes is $\phi/2$. The reader may easily verify this by drawing suitable diagrams.

Thus, for example, the crystal point group C_2 (International symbol: 2) which consists conventionally of $\boldsymbol{E}$ and 2_z, the π rotation about the z-axis, has two elements which can be taken as consisting of the elements $\boldsymbol{E}$, $m_x m_y$.

Since $2_z 2_z = \boldsymbol{E}$, C_2 can be said to be generated by the element 2_z in the sense that every element of the group is a power of the generator 2_z and every such power of 2_z is a group element.

The group C_2 is a subgroup of the group C_{2v} (International symbol: $2mm$) which has the elements

$$\boldsymbol{E},\ m_x,\ m_y,\ m_x m_y.$$

C_{2v} is generated by the two elements m_x and m_y, in the sense that every element of the group is a product of powers of m_x and m_y, and, conversely, that every such product is an element of the group. This follows since $m_x^2 = m_y^2 = \boldsymbol{E}$, $m_x m_y = m_y m_x$. We note here that the set of generators is not unique. The group C_{2v} could also be generated by the elements m_x and $m_x m_y$.

It is a general property of a finite group that such a set of generators may be chosen. (For further discussion of this and other topics, the reader may consult a wealth of books applying group theory to physics and to solid state physics, in particular Heine (1963) is especially recommended.)

A finite group such as C_{2v} which may be generated by reflection operators is known as a finite reflection group, and nowadays is often termed a Coxeter group. The crystal point groups may be divided into two sets, one of which consists of Coxeter groups, and the other which consists of sub-groups of Coxeter groups.

Coxeter groups have several important properties relevant to invariant polynomials of the co-ordinates of the space on which the group is defined as a set of transformations. For a Coxeter group of transformations on a space of n dimensions, $x_1, x_2, \ldots, x_n$, we have the important property that

all invariants of a Coxeter group which are polynomials of $x_1, x_2, \ldots, x_n$ may be expressed as polynomials of n integrity basis polynomials.

For example, for the Coxeter group C_{2v} in three dimensions, it may be shown that the integrity basis has three members: x^2, y^2, z. This means that any polynomial invariant of C_{2v} can be expressed as a polynomial of these three polynomials. Another important property of Coxeter groups is that the product of orders m_i, $i = 1, 2, \ldots, n$ of the integrity basis polynomials is equal to g, the order of the group (the number of elements of the group):

$$\prod_{i=1}^{n} m_i = g. \tag{18.1.5}$$

In our example of C_{2v} we have,

$$1 \times 2 \times 2 = 4, \tag{18.1.6}$$

agreeing with the order 4 of the group.

The integrity bases for all the crystal point groups have been derived and listed by Killingbeck (1972).

For a complete discussion of Coxeter groups, reference may be made to Coxeter & Moser (1972).

We may conclude this section by mentioning the full rotation group, the symmetry group for isotropic substances. This is the group of all rotations and the inversion operator $\bar{\boldsymbol{E}}$. It is defined in three dimensions and it may be shown that it has an integrity basis of one member, namely, $x^2 + y^2 + z^2$. It is a continuous group, not a Coxeter group.

18.2. The integrity basis for polynomial invariant functions of the elements of a second rank symmetric tensor

As has been discussed in chapter 9, the co-ordinate axis transformations of 3-dimensional space induce corresponding transformations on the space described by the six independent elements of a symmetric second rank tensor such as $\boldsymbol{\eta}$, $\boldsymbol{\varepsilon}$, $\boldsymbol{\Lambda}$ or $\boldsymbol{\Theta}$. A crystal point group will thus *induce* a transformation group on this space of six dimensions. The group induced may be isomorphic (i.e. one-to-one) with the corresponding point group, but frequently it will be a sub-group. For example for the group C_i consisting of $\boldsymbol{E}$, $\bar{\boldsymbol{E}}$, (International symbol $\bar{1}$), it is easy to see that every element of $\boldsymbol{\eta}$, say, is invariant to this group and so the group induced is the identity group $\boldsymbol{E}$, which of course is a sub-group of C_i. For the group C_2 the induced group contains the identity element and an element of order 2 which may be represented by

$$\begin{aligned} &\eta'_{xx} \rightarrow \eta_{xx}; \quad &&\eta'_{yy} \rightarrow \eta_{yy}; \quad &&\eta'_{zz} \rightarrow \eta_{zz}; \\ &\eta'_{yz} \rightarrow -\eta_{yz}; \quad &&\eta'_{zx} \rightarrow -\eta_{zx}; \quad &&\eta'_{xy} \rightarrow \eta_{xy}; \end{aligned} \tag{18.2.1}$$

in abbreviated notation, this element may be expressed as

$$\begin{matrix} \eta'_1 \\ \eta'_2 \\ \eta'_3 \\ \eta'_4 \\ \eta'_5 \\ \eta'_6 \end{matrix} = \begin{bmatrix} 1 & 0 & 0 & 0 & 0 & 0 \\ 0 & 1 & 0 & 0 & 0 & 0 \\ 0 & 0 & 1 & 0 & 0 & 0 \\ 0 & 0 & 0 & -1 & 0 & 0 \\ 0 & 0 & 0 & 0 & -1 & 0 \\ 0 & 0 & 0 & 0 & 0 & 1 \end{bmatrix} \begin{matrix} \eta_1 \\ \eta_2 \\ \eta_3 \\ \eta_4 \\ \eta_5 \\ \eta_6 \end{matrix} \tag{18.2.2}$$

A reflection operator in six dimensions would be one which leaves a 5-dimensional sub-space invariant and changes the sign of the sixth dimension, when the axis system is suitably chosen. That is, it is represented by a matrix, of whose eigenvalues five are +1 and one −1. We see that the element of the induced group, represented by (18.2.2), is the product of two reflection operators whose representations are diagonal matrices with diagonal elements (1, 1, 1, −1, 1, 1) and (1, 1, 1, 1, −1, 1) respectively. Thus the induced group is not a Coxeter group in six dimensions. The reader may show that the group induced by C_{2v} is also *not* a Coxeter group in six dimensions. In general, none of the groups so induced by the crystal point groups are Coxeter groups, but are sub-groups.

McLellan (1974) has recently derived the integrity bases for the groups induced on the above six-dimensional space by the crystal point groups. The method used is general, and consists of first finding the integrity basis members for the Coxeter group in six dimensions, of which the induced group is a sub-group. Apart from these there are additional members of the integrity basis of the induced group which are not invariant to the enclosing Coxeter group, but transform according to certain irreducible representations of this group.

The mathematical method enables all these integrity basis invariants to be identified.

It should be pointed out that there is no unique integrity basis for a particular group. For example, the octahedral group (in three dimensions), O_h, is a Coxeter group and has three integrity basis members, which may be chosen as

$$x^2+y^2+z^2, \qquad x^4+y^4+z^4, \qquad x^6+y^6+z^6. \tag{18.2.3}$$

But we could equivalently choose polynomials of these of the same order, e.g.,

$$x^2+y^2+z^2, \qquad x^2y^2+y^2z^2+z^2x^2, \qquad x^2y^2z^2, \tag{18.2.4}$$

as may readily be verified.

A transformation induced by the operation 2_z in three dimensions will be exactly the same as that induced by m_z, since $m_z = 2_z\bar{\boldsymbol{E}}$ and since a second rank tensor is invariant to $\bar{\boldsymbol{E}}$. Similarly, that induced by 3_z will be the same as that induced by the improper rotation $\bar{3}_z(=3_z\bar{\boldsymbol{E}})$. So the induced group will be identical for several point groups. For example, the transformations induced by the groups D_2, D_{2h}, C_{2v} of the orthorhombic lattice system are identical.

In the cases of the triclinic, monoclinic and orthorhombic lattice systems, all of the point groups belonging to the same lattice system induce an identical group on the η_α. In the cases of the tetragonal, trigonal, hexagonal and cubic systems, the point groups belonging to the same system separate into two sets, such that all groups in a set induce the same transformation group.

These sets are

Tetragonal (*a*) C_4, S_4, C_{4h} (International symbols: 4, $\bar{4}$, $4/m$)
Tetragonal (*b*) D_4, C_{4v}, D_{2d}, D_{4h} (422, $4mm$, $\bar{4}2m$, $4/mmm$)
Trigonal (*a*) C_3, S_6 (or C_{3i}), (3, $\bar{3}$)
Trigonal (*b*) C_{3v}, D_3, D_{3d} ($3m$, 32, $\bar{3}m$)
Hexagonal (*a*) C_6, C_{6h}, C_{3h} (6, $6/m$, $\bar{6}$)
Hexagonal (*b*) C_{6v}, D_6, D_{6h}, D_{3h} ($6mm$, 622, $6/mmm$, $\bar{6}m2$)
Cubic (*a*) T, T_h (23, $m3$)
Cubic (*b*) O, O_h, T_d (432, $m3m$, $\bar{4}3m$).

In every case the Coxeter group in six dimensions, of which an induced group is a sub-group, is a direct product of two Coxeter groups in three dimensions. For example, for both the induced groups of the tetragonal class, this Coxeter group is $m_d \times D_{4h}$, i.e., the direct product of the Coxeter groups in three dimensions, m_d and D_{4h}. The elements η_1, η_2, η_3 are transformed by m_d which is the reflection operator for the plane containing the η_3-axis which is equally inclined to η_1 and η_2-axes, i.e., it interchanges η_1 and η_2, thus

$$m_d\colon \eta_1 \leftrightarrow \eta_2. \tag{18.2.5}$$

Under the transformations induced by the group D_{4h}, the tensor elements η_4, η_5, η_6 transform in exactly the same way as do the vector components x, y, z.

For each induced group it has been shown (McLellan, 1974) that the polynomial invariants of the group are 'generated' by a generating function of the form

$$I(t)=\frac{1+t(N_1^{(1)}+\cdots+N_{\mu_1}^{(1)})+t^2(N_1^{(2)}+\cdots+N_{\mu_2}^{(2)}) +\cdots+t^k(N_1^{(k)}+\cdots+N_{\mu_k}^{(k)})}{\prod_{\alpha=1}^{6}(1-t^{m_\alpha}I_\alpha)} \tag{18.2.6}$$

That is, the coefficient of t^a in the expansion contains all the linearly independent polynomial invariants of order a. The numerator contains a finite number of terms. The $N_j^{(i)}$, $j=1,\ldots,\mu_i$, are called numerator invariants and are of order i. The I_α (each of order m_α) are homogeneous polynomials (as are the N's) and are called denominator invariants. Not every power from 1 to k need occur in the numerator.

Equation (18.2.6) corresponds to the generating function

$$K_0(t)=\frac{1+\mu_1 t+\mu_2 t^2+\cdots+\mu_k t^k}{\prod_{\alpha=1}^{6}(1-t^{m_\alpha})} \tag{18.2.7}$$

which is such that the coefficient of t^a gives the number of linearly independent homogeneous polynomial invariants of the η_α of order a.

If the group were a Coxeter group, the numerator in both (18.2.6) and (18.2.7) would be unity, and for the Coxeter group which contains the induced group as a sub-group the generating functions (18.2.6) would be

$$I(t)=\prod_{\alpha=1}^{6}(1-t^{m_\alpha}I_\alpha)^{-1}. \tag{18.2.8}$$

In other words, the I_α, $\alpha=1,2,\ldots,6$, are the six integrity basis members for the enclosing group, as, in the expansion of (18.2.8), we obtain all polynomial invariants of the Coxeter group, and we see that they are polynomials of the I_α.

The numerator invariants for the sub-group, i.e., the $N_j^{(i)}$, are not invariants of the Coxeter group, but transform according to an irreducible representation, other than the identity, of this group.

Thus, the invariants of an induced group fall into two sets, such that the denominator invariants can occur to any power in an invariant polynomial, but the numerator invariants occur to *at most* the first power. This can be seen when the expression (18.2.6) is expanded in powers of t.

To illustrate this property, let us go back to three dimensions and consider the invariants of the crystal point group O in relation to those of its enclosing Coxeter group O_h.

For O_h, the generating function of polynomial invariants is

$$1/(1-t^2 I_2)(1-t^4 I_4)(1-t^6 I_6), \tag{18.2.9}$$

where $I_2 = x^2 + y^2 + z^2$, $I_4 = x^4 + y^4 + 2^4$, $I_6 = x^2 y^2 z^2$. We use here a suffix for each I to denote its order. For O, the generating function is

$$I(t) = (1 + t^9 N_9)/D(O_h), \tag{18.2.10}$$

where $D(O_h)$ is the denominator of (18.2.9). The polynomial

$$N_9 = xyz(x^2 - y^2)(y^2 - z^2)(z^2 - x^2) \tag{18.2.11}$$

is an invariant of O, but transforms according to the irreducible representation A_{1u} of O_h, i.e., it is invariant to the elements of det(+1) in O_h but changes sign under transformation by those of det(−1) in O_h. Thus, we see that, for example, the invariant polynomials of O are

$$\begin{aligned} &I_2;\ I_4, I_2^2;\ I_4 I_2, I_2^3, I_6;\ I_4^2, I_4 I_2^2, I_2^4, I_6 I_2; \\ &N_9;\ I_2^5, I_2 I_4^2, I_2^3 I_4, I_2^2 I_6, I_4 I_6, N_9 I_2, \text{etc.} \end{aligned} \tag{18.2.12}$$

in ascending powers of homogeneous polynomials. In fact N_9^2 is equal to a polynomial of I_2, I_4, I_6, as it must be, since it is an invariant of O_h and therefore expressible as a polynomial of the integrity basis of this Coxeter group. It is unnecessary to take higher powers of N_9.

The most general invariant polynomial is of the form

$$J + \sum_i N_i J_i, \tag{18.2.13}$$

where J and each J_i are polynomials constructed from denominator invariants only. Thus all the N's of (18.2.6) have the same property: they occur singly to – *at most* – first order. This important property of some of the integrity basis members of a sub-group of a Coxeter group does not appear to be very well known. In the context of this book, it should be particularly noted, since if this property is not recognised and used in formulating invariant thermodynamic functions, unnecessary parameters could be introduced as coefficients. For example, for tetragonal (b) crystals, the fourth-order term $\{(\eta_1 - \eta_2)\eta_6\}^2$, the square of a numerator invariant, may not be recognised as identical to the product of two denominator invariants. It may be included in the thermodynamic function f, with a separate coefficient, together with the above product and another coefficient, whereas of course only one such term should be included.

18.3. More about the integrity basis for invariant polynomials of the η_α

At the end of this chapter are tabulated the integrity bases of the groups induced on the η_α by the crystal point groups. This table contains other information, and in this section some further details are given in explanation. This section, however, is not really necessary for an understanding of how to construct invariant polynomials from the listed integrity bases.

As mentioned, the induced group is enclosed in a Coxeter group which is always a direct product of two Coxeter groups of transformations both in three dimensions.

For example, as discussed in section 2, the two induced groups of the tetragonal class have an enclosing Coxeter group

$$m_d \times D_{4h}$$

where η_1, η_2, η_3 transform under m_d and η_4, η_5, η_6 under D_{4h}.

The irreducible representations of $m_d \times D_{4h}$ are products of the irreducible representations of m_d and D_{4h} respectively. If m_d has the irreps (common abbreviation for *irreducible* representations) Γ'_i, $i = 1, 2, \ldots,$ (actually $i = 1, 2$) and D_{4h} has the irreps Γ_j, $j = 1, 2, \ldots,$ (actually $j = 1, 2, \ldots, 10$), then $m_d \times D_{4h}$ has irreps $\Gamma'_i\Gamma_j$, where, in this case, there are 20 such product irreps. For example, the identity irrep is $\Gamma'_1\Gamma_1$. In the usual notation this would be denoted by $A'_{1g}A_{1g}$ (see Heine, 1963). In this notation, A, E and T denote respectively, 1-, 2- and 3-dimensional irreps. The suffix g (for the German *gerade* = even) denotes an even irrep, i.e., one invariant to the inversion operator $\bar{E}$, while the suffix u (for *ungerade*) denotes an odd irrep (which changes sign under inversion).

In the table, each numerator invariant is listed together with the irrep according to which it transforms under the enclosing Coxeter group. The product space of this group is shown, as well as its order g – a useful number because the powers m_α of the denominator invariants obey (18.1.5).

For example, for the orthorhombic crystal class the Coxeter group is $E_3 \times D_{2h}$, whose product space is $(\eta_1, \eta_2, \eta_3) \times (\eta_4, \eta_5, \eta_6)$. η_1, η_2, η_3 are all invariant to the induced group and, in fact, they are denominator invariants. η_4, η_5, η_6 transform according to D_{2h} ($= 222 \times \bar{E}$) under the Coxeter group. At most, they change sign, leading to $\eta_4^2, \eta_5^2, \eta_6^2$ as denominator invariants. Under the Coxeter group, the product $\eta_4\eta_5\eta_6$

transforms as the irrep A_{1u} of D_{2h}, but it is invariant under the transformations of the subgroup which is induced by the crystal point groups of the orthorhombic crystal class. Thus, it is a numerator invariant and is listed as

$$|A_1' A_{1u}\rangle : \eta_4\eta_5\eta_6.$$

In the table, too, are listed the numerator $N_0(t)$ and the denominator D of $K_0(t)$ of the induced groups (see (18.2.7)).

18.4. The trigonal and hexagonal crystal classes and the isotropic case

The sets η_1, η_2, η_3 and η_4, η_5, η_6 occur symmetrically in all the induced groups except those of the trigonal and hexagonal classes. Where they occur symmetrically, it is immaterial which of the following conventions we use:

$$\psi_\mu = 2\psi_{(\alpha\beta)_\mu} \quad \text{or} \quad \psi_{(\alpha\beta)_\mu}, \quad \mu = 4, 5, 6. \tag{18.4.1}$$

There is no normalising, so the factor 2 is not important in the integrity basis members. Thus $\boldsymbol{\eta}$ and $\boldsymbol{\Lambda}$, or $\boldsymbol{\varepsilon}$ and $\boldsymbol{\Theta}$, can be treated in a similar way in the sense that $\boldsymbol{\eta}$ with the first convention of (18.4.1) has the same integrity basis as $\boldsymbol{\Lambda}$ with the second convention of (18.4.1).

We shall show how this leads simply in these cases to the same symmetry scheme for $\mathring{\boldsymbol{c}}$ and $\mathring{\boldsymbol{s}}$ (or $\boldsymbol{c}$ and $\boldsymbol{s}$).

To avoid building up complicated notations, we shall do this by taking the orthorhombic class as an example. The second-order part of the Helmholtz free energy, in this case, can be written as

$$f^{(2)} = \tfrac{1}{2}(\tilde{c}_{\alpha\beta}\eta_\alpha\eta_\beta + c_4'\eta_4^2 + c_5'\eta_5^2 + c_6'\eta_6^2), \tag{18.4.2}$$

on consulting the table of integrity bases. Here the coefficients $\tilde{c}_{\alpha\beta}$ may be non-zero only when $1 \leqslant \alpha \leqslant 3$ *and* $1 \leqslant \beta \leqslant 3$. Since we may define $\tilde{c}_{\alpha\beta} = \tilde{c}_{\beta\alpha}$, $f^{(2)}$ is determined by nine constants. For the stiffness elements (at the reference state), we see, on taking the second order derivatives of (18.4.2), that, for example,

$$\mathring{c}_{12} = \tilde{c}_{12}, \qquad \mathring{c}_{14} = 0, \qquad \mathring{c}_{44} = c_4', \qquad \mathring{c}_{55} = c_5', \qquad \mathring{c}_{66} = c_6'. \tag{18.4.3}$$

That is, there are nine elastic constants exactly as given in chapter 17.

The same scheme, in terms of the Λ_α, may be used for $g^{(2)}$, the second-order part of the Gibbs free energy $\mathring{g}$, and so we find that the $\mathring{s}_{\alpha\beta} = -(\partial^2 g^{(2)}/\partial\Lambda_\alpha\partial\Lambda_\beta)^*$ satisfy the same symmetry scheme.

The reader may wish to verify, in the above way, that the symmetry schemes for $\mathring{c}$, for all the crystal classes, are as listed in chapter 17, and that except for the hexagonal, trigonal and isotropic classes, the symmetry schemes for $\mathring{s}$ are the same as those for $\mathring{c}$.

In the case of hexagonal and trigonal classes a complication arises which results in different schemes for $\mathring{c}$ and $\mathring{s}$. This is because there is a lack of symmetry in the integrity basis elements, such that η_6 is 'mixed' in linearly with η_1 and η_2. We note that among the second order denominator invariants we have

$$(\eta_1-\eta_2)^2+\eta_6^2=(\eta_{11}-\eta_{22})^2+4\eta_{12}^2, \tag{18.4.4}$$

and so, for the $\mathbf{\Lambda}$ scheme, we would have the denominator invariant

$$(\Lambda_1-\Lambda_2)^2+4\Lambda_6^2. \tag{18.4.5}$$

In the trigonal cases only, we also have a second-order numerator invariant for trigonal (a) and (b), of the form

$$\psi_1\psi_4+\psi_2\psi_5+\psi_3\psi_6, \tag{18.4.6}$$

and for trigonal (a) an additional one, of the form

$$\psi_1\psi_5+\psi_2\psi_6+\psi_3\psi_4, \tag{18.4.7}$$

where ψ_1 and ψ_2 (for x_3-axis$\equiv$trigonal axis) are given by $\eta_2-\eta_1+\sqrt{3}\eta_6$ and $\eta_2-\eta_1-\sqrt{3}\eta_6$. For the $\mathbf{\Lambda}$ scheme, we would have

$$\psi_1=\Lambda_2-\Lambda_1+2\sqrt{3}\Lambda_6 \tag{18.4.8}$$

and

$$\psi_2=\Lambda_2-\Lambda_1-2\sqrt{3}\Lambda_6. \tag{18.4.9}$$

It can easily be verified that no other differences occur in the first or second order terms. Wherever η_6 (or Λ_6) occurs in constructing $f^{(2)}$ (or $g^{(2)}$) we shall have a term $2\Lambda_6$ for η_6 which *does affect* the symmetry scheme. In the symmetry schemes, we must replace

$$\begin{aligned}&\mathring{c}_{\alpha 6} \text{ by } 2\mathring{s}_{\alpha 6}, \quad \alpha\neq 6\\ &\mathring{c}_{66} \text{ by } 4\mathring{s}_{66}\end{aligned} \tag{18.4.10}$$

and

$$\mathring{c}_{\alpha\beta} \text{ by } \mathring{s}_{\alpha\beta} \quad \text{for all other values of } \alpha \text{ and } \beta.$$

In the isotropic case, there is a similar complication arising from the 'mixing' which occurs in the denominator invariant,

$$2(\eta_1^2+\eta_2^2+\eta_3^2)+(\eta_4^2+\eta_5^2+\eta_6^2), \tag{18.4.11}$$

which is the only second-order term (or lower) which contains η_4, η_5, η_6. It is a simple matter to show that (18.4.10) is again required.

Finally, we should note an important point in the case of the trigonal groups 32, $3m$, $\bar{3}m$. This concerns the convention used for crystal axes. In this section we have quoted invariant polynomials which are based on the convention that the two-fold axis is the x_1 axis. The tables of invariants for these trigonal groups, given at the end of this chapter, use this convention, which is then consistent with the usual symmetry schemes for the stiffness and compliance matrices given in chapter 17. The notation 32_1, $3m_1$, $\bar{3}m_1$, used in the tables indicates this choice of x_1 axis. McLellan (1974) used the convention that the two-fold axis was the x_2 axis.

18.5. The thermodynamic functions

To avoid complicated notations, we shall again use examples. The general method will then be obvious.

For the cubic class we have six denominator invariants, five numerator invariants for cubic (b) and an additional six for cubic (a).

The free energy f can be written in terms of the integrity basis. Up to, and including, second order terms it is the same for cubic (a) and cubic (b):

$$f = f_0(T) - p^*(\eta_1+\eta_2+\eta_3) + \tfrac{1}{2}\{\mathring{c}_{12}^{(T)}(\eta_1+\eta_2+\eta_3)^2 + (\mathring{c}_{11}^{(T)} - \mathring{c}_{12}^{(T)})(\eta_1^2+\eta_2^2+\eta_3^2) + \mathring{c}_{44}^{(T)}(\eta_4^2+\eta_5^2+\eta_6^2)\} + \cdots \tag{18.5.1}$$

Because of the extra numerator invariants, the higher order terms for cubic (a) will include terms additional to those for cubic (b).

We see from (18.5.1) that a reference state of the cubic class must be hydrostatic, and that there are three independent elastic moduli. The Helmholtz free energy of the reference state, $f_0(T)$, depends on the temperature, as do p^* and the other coefficients $\mathring{c}_{\alpha\beta}$ and those of higher-order terms.

Similarly,

$$u = u_0(s) - p^*(\eta_1+\eta_2+\eta_3) + \tfrac{1}{2}\{\mathring{c}_{12}^{(s)}(\eta_1+\eta_2+\eta_3)^2 + \cdots\} + \cdots, \tag{18.5.2}$$

where we have adiabatic stiffness quantities, and

$$\mathring{g} = \mathring{g}_0(T) - \tfrac{1}{2}\{\mathring{s}_{12}^{(T)}(\Lambda_1+\Lambda_2+\Lambda_3)^2 + (\mathring{s}_{11}^{(T)} - \mathring{s}_{12}^{(T)})(\Lambda_1^2+\Lambda_2^2+\Lambda_3^2) + \mathring{s}_{44}^{(T)}(\Lambda_4^2+\Lambda_5^2+\Lambda_6^2)\} + \cdots. \tag{18.5.3}$$

There is no first-order term in $\mathring{g}$ because

$$\eta_\alpha = -(\partial\mathring{g}/\partial\Lambda_\alpha)^* = 0, \tag{18.5.4}$$

i.e., at the reference state, $\boldsymbol{\eta} = 0$.

The enthalpy, $\mathring{h} = u - \Lambda_\alpha\eta_\alpha$, may be written similarly to $\mathring{g}$, with adiabatic compliances replacing the isothermal.

The third-order terms for cubic (b) are

$$\begin{aligned} f^{(3)} = {} & \alpha\eta_1\eta_2\eta_3 + \beta\eta_4\eta_5\eta_6 + (\eta_1+\eta_2+\eta_3)\{\gamma(\eta_1^2+\eta_2^2+\eta_3^2) \\ & + \delta(\eta_4^2+\eta_5^2+\eta_6^2)\} + \varepsilon(\eta_1\eta_4^2+\eta_2\eta_5^2+\eta_3\eta_6^2). \end{aligned} \tag{18.5.5}$$

Thus, for cubic (b) we have five independent third-order elastic moduli $\alpha, \beta, \gamma, \delta, \varepsilon$ (these symbols are not meant to be standard notation). For cubic (a) we see that we have two additional cubic terms

$$\zeta(\eta_1-\eta_2)(\eta_2-\eta_3)(\eta_3-\eta_1) + \nu(\eta_1\eta_5^2+\eta_2\eta_6^2+\eta_3\eta_4^2), \tag{18.5.6}$$

and so we have seven independent third-order elastic moduli (or constants) for cubic (a).

This method of the integrity basis is clearly the simplest method of treating the higher order terms of the thermodynamic functions.

If we wish to use the strain co-ordinates $\boldsymbol{e}$ appropriate to the geometrical restriction $\boldsymbol{V} = \boldsymbol{E}$, we use

$$2\boldsymbol{\eta} = \boldsymbol{e} + \boldsymbol{e}^{\mathrm{t}} + \boldsymbol{e}^{\mathrm{t}}\boldsymbol{e}, \tag{18.5.7}$$

in order to express f or u in terms of the e_α.

Equation (18.5.1) is expressed, of course, in terms of the cubic axes as co-ordinate axes, and it is sometimes necessary, when we have more than one crystal in the system under consideration, to consider a particular crystal in terms of co-ordinate axes not chosen to be its crystallographic axes. Suppose that we transform the axis system to an arbitrary one. We may write (18.5.1), up to second-order terms, as

$$\begin{aligned} f = {} & f_0(T) - p^*(\eta_1+\eta_2+\eta_3) + \tfrac{1}{2}\{\mathring{c}_{12}(\eta_1+\eta_2+\eta_3)^2 \\ & + (\mathring{c}_{11}-\mathring{c}_{12})[\eta_1^2+\eta_2^2+\eta_3^2+\tfrac{1}{2}(\eta_4^2+\eta_5^2+\eta_6^2)] \\ & + [\mathring{c}_{44}-\tfrac{1}{2}(\mathring{c}_{11}-\mathring{c}_{12})][\eta_4^2+\eta_5^2+\eta_6^2]\}, \end{aligned} \tag{18.5.8}$$

where, on referring to the isotropic case in the table of integrity bases, we see that all the terms, except $\eta_4^2+\eta_5^2+\eta_6^2$, are invariant to the arbitrary transformation of axes.

Thus, the stiffness matrix for this arbitrary axis system will be the sum of that for an isotropic system and the matrix $\mathring{\boldsymbol{c}}'$, derived from the quadratic form

$$\tfrac{1}{2}\{\mathring{c}_{44}-\tfrac{1}{2}(\mathring{c}_{11}-\mathring{c}_{12})\}\{S_{2\mu'}S_{2\mu}S_{3\nu'}S_{3\nu}+S_{3\mu'}S_{3\mu}S_{1\nu'}S_{1\nu}$$
$$+S_{1\mu'}S_{1\mu}S_{2\nu'}S_{2\nu}\}\eta_{\mu'\nu'}\eta_{\mu\nu}, \qquad (18.5.9)$$

where $\boldsymbol{S}$ represents the transformation of (9.1.5) describing the change of axes. Equation (18.5.9) cannot be simplified further until we put in the actual form of $\boldsymbol{S}$ in any particular case. The element $\mathring{c}'_{\mu'\nu'\mu\nu}$ of the additional stiffness matrix is twice the coefficient of $\eta_{\mu'\nu'}\eta_{\mu\nu}$ in (18.5.9).

Integrity bases for invariant functions of the elements of $\boldsymbol{\eta}$

$\eta_1=\eta_{11};\ \eta_2=\eta_{22};\ \eta_3=\eta_{33};\ \eta_4=2\eta_{23};\ \eta_5=2\eta_{31};\ \eta_6=2\eta_{12}.$

1. Triclinic: C_i, C_1. (International Symbols: $1, \bar{1}$)

Coxeter group	*space*	*order*
E_6	$\eta_i,\ i=1,\ldots,6$	1

Denominator invariants $\eta_i,\ i=1,\ldots,6$

2. Monoclinic: C_{2h}, C_2, C_s. $(2/m, 2, m)$

Coxeter group	*space*	*order*
$E_3\times C_{2v}$	$(\eta_1,\eta_2,\eta_3)\times(\eta_4,\eta_5,\eta_6)$	4

Numerator invariant $|A_1'A_2\rangle:\eta_4\eta_5$

Denominator invariants $\eta_1,\eta_2,\eta_3,\eta_6,\eta_4^2,\eta_5^2$

$K_0(t):(1+t^2)/(1-t)^4(1-t^2)^2$

3. Orthorhombic: C_{2v}, D_2, D_{2h}. $(2mm, 222, mmm)$

Coxeter group	*space*	*order*
$E_3\times D_{2h}$	$(\eta_1,\eta_2,\eta_3)\times(\eta_4,\eta_5,\eta_6)$	8

Numerator invariants $|A_1'A_{1u}\rangle:\eta_4\eta_5\eta_6$

Denominator invariants $\eta_1,\eta_2,\eta_3,\eta_4^2,\eta_5^2,\eta_6^2$

$K_0(t):(1+t^3)/(1-t)^3(1-t^2)^3$

4. Tetragonal: (a) C_4, S_4, C_{4h}. $(4, \bar{4}, 4/m)$
   Tetragonal: (b) D_4, C_{4v}, D_{2d}, D_{4h}. $(422, 4mm, \bar{4}2m, 4/mmm)$

Coxeter group	*space*	*order*
$m_d\times D_{4h}$	$(\eta_1,\eta_2,\eta_3)\times(\eta_4,\eta_5,\eta_6)$	32

Numerator invariants	*Additional numerator invariants*
Tetragonal (b)	Tetragonal (a)
$\|A_2'B_{1g}\rangle:(\eta_1-\eta_2)(\eta_4^2-\eta_5^2)$	$\|A_2'A_{2u}\rangle:(\eta_1-\eta_2)\eta_6$
$\|A_1'B_{1u}\rangle:\eta_4\eta_5\eta_6$	$\|A_1'B_{2u}\rangle:\eta_6(\eta_4^2-\eta_5^2)$
$\|A_2'A_{1u}\rangle:(\eta_1-\eta_2)\eta_4\eta_5\eta_6(\eta_4^2-\eta_5^2)$	$\|A_2'B_{2g}\rangle:(\eta_1-\eta_2)\eta_4\eta_5$
	$\|A_1'A_{2g}\rangle:\eta_4\eta_5(\eta_4^2-\eta_5^2)$

Denominator invariants $\eta_1+\eta_2,\ \eta_3,\ (\eta_1-\eta_2)^2,\ \eta_4^2+\eta_5^2,\ \eta_6^2,\ \eta_4^2\eta_5^2$

$N_0(t): 1+2t^3+t^6;\ 1+t^2+4t^3+t^4+t^6$
$D(1-t)^2(1-t^2)^3(1-t^4)$

5. Cubic: (*a*) T, T_h. $(23, m3)$
   Cubic: (*b*) O, O_h, T_d. $(432, m3m, \bar{4}3m)$

Coxeter group	*space*	*order*
$C_{3v}\times O_h$	$(\eta_1,\eta_2,\eta_3)\times(\eta_4,\eta_5,\eta_6)$	288

Numerator invariants

Cubic (*b*)

$|E'E_g\rangle: \eta_1\eta_4^2+\eta_2\eta_5^2+\eta_3\eta_6^2;$
$\eta_1^2\eta_4^2+\eta_2^2\eta_5^2+\eta_3^2\eta_6^2;$
$\eta_1\eta_4^4+\eta_2\eta_5^4+\eta_3\eta_6^4;$
$\eta_1^2\eta_4^4+\eta_2^2\eta_5^4+\eta_3^2\eta_6^4;$
$|A_2'A_{2g}\rangle: (\eta_1-\eta_2)(\eta_2-\eta_3(\eta_3-\eta_1)$
$\times(\eta_4^2-\eta_5^2)(\eta_5^2-\eta_6^2)(\eta_6^2-\eta_4^2)$

Additional numerator invariants

Cubic (*a*)

$|A_2'A_{1g}\rangle: (\eta_1-\eta_2)(\eta_2-\eta_3)(\eta_3-\eta_1);$
$|E'E_g\rangle: \eta_1\eta_5^2+\eta_2\eta_6^2+\eta_3\eta_4^2;$
$\eta_1^2\eta_5^2+\eta_2^2\eta_6^2+\eta_3^2\eta_4^2;$
$\eta_1\eta_5^4+\eta_2\eta_6^4+\eta_3\eta_4^4;$
$\eta_1^2\eta_5^4+\eta_2^2\eta_6^4+\eta_3^2\eta_4^4;$
$|A_1'A_{2g}\rangle: (\eta_4^2-\eta_5^2)(\eta_5^2-\eta_6^2)(\eta_6^2-\eta_4^2)$

Denominator invariants $\eta_1+\eta_2+\eta_3;\ \eta_1^2+\eta_2^2+\eta_3^2;\ \eta_1\eta_2\eta_3$
$\eta_4^2+\eta_5^2+\eta_6^2;\ \eta_4^4+\eta_5^4+\eta_6^4;\ \eta_4\eta_5\eta_6$

$N_0(t)\ 1+t^3+t^4+t^5+t^6+t^9;\quad 1+3t^3+2t^4+2t^5+3t^6+t^9$
$D(1-t)(1-t^2)^2(1-t^3)^2(1-t^4)$

6. Trigonal: (*a*) C_3, S_6. $(3, \bar{3})$
   Trigonal: (*b*) C_{3v}, D_3, D_{3d}. $(3m_1, 32_1, \bar{3}m_1)$

Coxeter group	*space*	*order*
$C_{3v}\times C_{3v}$	$(\psi_1,\psi_2,\psi_3)\times(\psi_4,\psi_5,\psi_6)$, or $(\eta_1,\eta_2,\eta_6)\times(\eta_3,\eta_4,\eta_5)$.	36

Trigonal axis $= x_3$-axis: $\psi_1=\eta_2-\eta_1+\sqrt{3}\eta_6;$
$\psi_2=\eta_2-\eta_1-\sqrt{3}\eta_6;\quad \psi_3=2(\eta_1-\eta_2)$
$\psi_4=-\eta_4+\sqrt{3}\eta_5;\quad \psi_5=-\eta_4-\sqrt{3}\eta_5;$
$\psi_6=2\eta_4.$

Trigonal axis $=(1,1,1)/\sqrt{3}$: $\psi_i=\eta_i$, $i=1,2,\ldots,6$.

Numerator invariants

Trigonal (*b*)

$|E'E\rangle: \psi_1\psi_4+\psi_2\psi_5+\psi_3\psi_6;$
$\psi_1\psi_4^2+\psi_2\psi_5^2+\psi_3\psi_6^2;$
$\psi_1^2\psi_4+\psi_2^2\psi_5+\psi_3^2\psi_6;$
$\psi_1^2\psi_4^2+\psi_2^2\psi_5^2+\psi_3^2\psi_6^2;$
$|A_2'A_2\rangle: (\psi_1-\psi_2)(\psi_2-\psi_3)(\psi_3-\psi_1)$
$\times(\psi_4-\psi_5)(\psi_5-\psi_6)(\psi_6-\psi_4).$

Additional numerator invariants

Trigonal (*a*)

$|E'E\rangle: \psi_1\psi_5+\psi_2\psi_6+\psi_3\psi_4;$
$\psi_1\psi_5^2+\psi_2\psi_6^2+\psi_3\psi_4^2;$
$\psi_1^2\psi_5+\psi_2^2\psi_6+\psi_3^2\psi_4;$
$\psi_1^2\psi_5^2+\psi_2^2\psi_6^2+\psi_3^2\psi_4^2;$
$|A_1'A_2\rangle: (\psi_4-\psi_5)(\psi_5-\psi_6)(\psi_6-\psi_4);$
$|A_2'A_1\rangle: (\psi_1-\psi_2)(\psi_2-\psi_3)(\psi_3-\psi_1).$

Examples

Denominator invariants

Trigonal axis $= x_3$-axis: $\eta_3, \eta_1+\eta_2, (\eta_1-\eta_2)^2+\eta_6^2; \eta_4^2+\eta_5^2,$

$(\eta_1-\eta_2)\{(\eta_1-\eta_2)^2-3\eta_6^2\}; \eta_4(\eta_4^2-3\eta_5^2).$

Trigonal axis $= (1, 1, 1)/\sqrt{3}$: $\eta_1+\eta_2+\eta_3; \eta_4+\eta_5+\eta_6; \eta_1^2+\eta_2^2+\eta_3^2$

$\eta_4^2+\eta_5^2+\eta_6^2; \eta_1\eta_2\eta_3; \eta_4\eta_5\eta_6.$

$N_0(t):(1+t^2+2t^3+t^4+t^6); \qquad 1+2t^2+6t^3+2t^4+t^6$

$D(1-t)^2(1-t^2)^2(1-t^3)^2$

7. Hexagonal: (*a*) C_6, C_{6h}, C_{3h}. $(6, 6/m, \bar{6})$

Hexagonal: (*b*) $C_{6v}, D_6, D_{6h}, D_{3h}$. $(6mm, 622, 6/mmm, \bar{6}m2)$

Hexagonal axis $= x_3$-axis

Coxeter group	*space*	*order*
$C_{3v}\times C_{6v}$	$(\psi_1, \psi_2, \psi_3)\times(\psi_4, \psi_5, \psi_6)$	72
	or	
	$(\eta_1, \eta_2, \eta_6)\times(\eta_3, \eta_4, \eta_5)$	

The ψ_i *are as defined for trigonal axis* $= x_3$*-axis.*

Numerator invariants

Hexagonal (*b*)

$|E'E_2\rangle: \psi_1\psi_4^2+\psi_2\psi_5^2+\psi_3\psi_6^2;$

$\psi_1^2\psi_4^2+\psi_2^2\psi_5^2+\psi_3^2\psi_6^2;$

$\psi_1\psi_4^4+\psi_2\psi_5^4+\psi_3\psi_6^4;$

$\psi_1^2\psi_4^4+\psi_2^2\psi_5^4+\psi_3^2\psi_6^4;$

$|A_2'A_2\rangle: (\psi_1-\psi_2)(\psi_2-\psi_3)(\psi_3-\psi_1)$

$\times(\psi_4^2-\psi_5^2)(\psi_5^2-\psi_6^2)(\psi_6^2-\psi_4^2).$

Additional numerator invariants

Hexagonal (*a*)

$|E'E_2\rangle: \psi_1\psi_5^2+\psi_2\psi_6^2+\psi_3\psi_4^2;$

$\psi_1^2\psi_5^2+\psi_2^2\psi_6^2+\psi_3^2\psi_4^2;$

$\psi_1\psi_5^4+\psi_2\psi_6^4+\psi_3\psi_4^4;$

$\psi_1^2\psi_5^4+\psi_2^2\psi_6^4+\psi_3^2\psi_4^4;$

$|A_1'A_2\rangle: (\psi_4^2-\psi_5^2)(\psi_5^2-\psi_6^2)(\psi_6^2-\psi$

$|A_2'A_1\rangle: (\psi_1-\psi_2)(\psi_2-\psi_3)(\psi_3-\psi_1$

Denominator invariants $\eta_3; \eta_1+\eta_2; (\eta_1-\eta_2)^2+\eta_6^2; \eta_4^2+\eta_5^2;$

$(\eta_1-\eta_2)\{(\eta_1-\eta_2)^2-3\eta_6^2\}; \eta_5^2(\eta_5^2-3\eta_4^2)^2$

$N_0(t):(1+t^3+t^4+t^5+t^6+t^9); \qquad (1+3t^3+2t^4+2t^5+3t^6+t^9)$

$D:(1-t)^2(1-t^2)^2(1-t^3)(1-t^6).$

8. Isotropic material (full rotation group)

Denominator invariants $\eta_1+\eta_2+\eta_3; 2(\eta_1^2+\eta_2^2+\eta_3^2)+(\eta_4^2+\eta_5^2+\eta_6^2);$

$4\eta_1\eta_2\eta_3+\eta_4\eta_5\eta_6-(\eta_1\eta_4^2+\eta_2\eta_5^2+\eta_3\eta_6^2).$

Examples

1.(*a*) Show that the second order contribution to f in the isotropic case consists of two terms I_1^2 and I_2, and the third order consists of three terms I_1^3, I_1I_2, I_3. Show that there are 4 fourth-order terms, 5 fifth-order, but 7 sixth-order.

(*b*) Discuss the elastic moduli of the second and third orders for an elastic isotropic medium.

2. Write down all the terms in the fourth- and fifth-order contributions to f for the cubic (*b*) case.

3. Consider all the crystal classes and, taking the x_3-axis, where appropriate, as the principal symmetry axis, show that

(*a*) $\boldsymbol{\Theta}^*$ may have off-diagonal elements only for the triclinic class, when all elements may be non-zero and unequal.
(*b*) For monoclinic and orthorhombic classes $\boldsymbol{\Theta}^* = \text{diag}(\Theta_1^*, \Theta_2^*, \Theta_3^*)$, i.e., a diagonal tensor where no two elements need be equal.
(*c*) For tetragonal, trigonal and hexagonal classes $\boldsymbol{\Theta}^* = \text{diag}(\Theta_1^*, \Theta_1^*, \Theta_3^*)$ where Θ_3^* need not equal Θ_1^*.
(*d*) For cubic classes, $\boldsymbol{\Theta}^* = -p\boldsymbol{E}$, i.e., it must be hydrostatic.

4. Use example 3 and (16.2.6) to show that $\hat{\boldsymbol{c}}$ has the same symmetry scheme as $\acute{\boldsymbol{c}}$ (or $\boldsymbol{c}$) for all crystal classes.

19

Equilibrium and stability conditions for thermodynamic systems

The conditions of equilibrium and stability of systems, of possibly more than one phase, are introduced, and the conditions of equilibrium are discussed, in terms of the properties of thermodynamic functions, for such processes as phase transitions, diffusion, and chemical reactions. In order to illustrate the principles and methods, discussion is largely limited to hydrostatic systems, before application is made, in succeeding chapters, to non-hydrostatic systems.

19.1. The entropy principle

In many books on thermodynamics (e.g., Zemansky, 1968), the principle of the increase of entropy is stated thus: The change of the entropy of the universe as a result of any kind of process is given by

$$\Delta S \geqslant 0, \tag{19.1.1}$$

where the equality sign refers to reversible processes, and the inequality sign to irreversible processes.

A system, adiabatically isolated from the rest of the universe, will proceed spontaneously to equilibrium in such a way that in the processes occurring, (19.1.1) will hold where S is the entropy of this system only. When equilibrium is finally reached, S will be at a maximum.

The condition of equilibrium of an isolated thermodynamic system is that its entropy cannot increase. This may be expressed in the following well-known way: for any virtual variation of the system at the equilibrium state under study, we must have

$$\delta^1 S = 0, \tag{19.1.2}$$

for

$$\delta W = \delta U = 0, \tag{19.1.3}$$

where this latter equation expresses the isolation of the system, such that no work is done on the system and no heat transfer occurs. The above argument also implies that for stability of the system,

$$\delta^2 S \leqslant 0, \tag{19.1.4}$$

under the conditions (19.1.3). Here

$$\delta S = \delta^1 S + \delta^2 S + \cdots, \tag{19.1.5}$$

where δS is the variation of S, and $\delta^1 S$, $\delta^2 S$ are terms respectively first- and second-order in the parameters determining the variation. The linear term $\delta^1 S$ is mathematically equivalent to $\mathrm{d}S$, but we shall see that it is often clearer to use the above notation for a variation from an equilibrium state. We may note, in passing, that we shall use the symbol Δ to denote increments occurring in a natural process, and δ to denote increments occurring in a virtual variation at an equilibrium state. The differential $\mathrm{d}S$, etc., will be reserved for use in integration and to display mathematical relationships between a function and its derivatives. Work done in a variation is, of course, not a change in a thermodynamic function, but the use of δW makes clear the meaning of the term.

To return to the conditions of equilibrium and stability, (19.1.2) together with (19.1.3) is usually known as a condition of equilibrium, while (19.1.4) is known as a condition of stability. If the inequality in (19.1.4) alone holds, the system is said to be strongly stable. If the equality holds, the system is said to be weakly stable, but in fact in this case the system may be stable, unstable or at neutral equilibrium, a matter which can be settled only by considering higher terms. However, whichever holds, (19.1.4) is a necessary (but not sufficient) condition of stability.

In hydrostatics, the conditions (19.1.3) may be written $\delta V = \delta U = 0$.

The above conditions of equilibrium and stability apply to an isolated system. It is, however, often convenient to consider the conditions appropriate to a system which is not isolated. For example, we shall be able to prove that the condition of equilibrium, under the accessory conditions

$$\delta T = \delta W = 0, \tag{19.1.6}$$

is that the first-order virtual change of the Helmholtz free energy

$$\delta^1 F = 0. \tag{19.1.7}$$

This is a very useful form of the equilibrium condition.

To derive this, and other conditions of equilibrium, we consider the system in diathermal contact with a large heat reservoir, i.e., it can exchange heat with the reservoir. The reservoir and system are now regarded as a composite system.

We imagine the system as originally isolated adiabatically from the heat reservoir, and at a slightly different temperature from it. If a diathermal partition is substituted for the adiabatic partition between system and reservoir, heat will be exchanged and the system will come to equilibrium with the reservoir. In this process, the heat exchanged will be so small compared with the heat capacity of the reservoir (which may be considered as large as necessary) that the temperature of the reservoir remains effectively unchanged.

If ΔQ is the heat *absorbed* by the system from the reservoir, then the change of entropy of the reservoir will be $-\Delta Q/T$. Equation (19.1.1) will become

$$\Delta S - \Delta Q/T \geqslant 0, \tag{19.1.8}$$

where S is now the entropy of the system. The left-hand side represents the change of entropy of the universe in the process. On using the first law equation

$$\Delta U = \Delta Q + \Delta W, \tag{19.1.9}$$

(19.1.8) can be written

$$T\,\Delta S - \Delta U + \Delta W \geqslant 0. \tag{19.1.10}$$

From this equation we may now obtain equilibrium conditions for a variety of accessory conditions.

First of all, we note that if the system is adiabatically isolated, $\Delta U = \Delta W = 0$, and (19.1.10) reverts to (19.1.1). The equilibrium condition (19.1.2) and the stability condition (19.1.4) are again obtained.

If the accessory conditions are that the temperature is held constant and $\Delta W = 0$, then (19.1.10) becomes

$$\Delta(TS - U) \geqslant 0,$$

or

$$\Delta F \leqslant 0. \tag{19.1.11}$$

Thus, the Helmholtz free energy F tends to a minimum in this case, and the equilibrium condition is

$$\delta^1 F = 0, \quad \text{when } \delta T = \delta W = 0, \tag{19.1.12}$$

and the stability condition is

$$\delta^2 F \geqslant 0, \tag{19.1.13}$$

for the same accessory conditions.

If the entropy is held constant and $\Delta W = 0$, then (19.1.10) becomes

$$\Delta U \leqslant 0, \tag{19.1.14}$$

and so the equilibrium condition is

$$\delta^1 U = 0, \quad \text{when } \delta S = \delta W = 0, \tag{19.1.15}$$

and the stability condition is

$$\delta^2 U \geqslant 0. \tag{19.1.16}$$

Finally, we note that we have used (19.1.10) to show how, under certain accessory conditions, thermodynamic functions such as U and F must have minimum values at a state of stable equilibrium. We shall now derive, from (19.1.10), a general condition for stable equilibrium, which will be useful when dealing with the stability of phases under non-hydrostatic stresses.

Suppose we have an equilibrium state. Consider a variation from that state. If it is stable, there will be a natural path back to it, along which (19.1.10) holds. Thus, the changes in the thermodynamic quantities which occur in the variation will be given by $\delta U = -\Delta U$, $\delta S = -\Delta S$, $\delta W = -\Delta W$, that is, they are the negative of those changes which occur in the natural process back to the equilibrium state. Hence, the condition of stable equilibrium is

$$\delta U - T\delta S - \delta W \geqslant 0, \tag{19.1.17}$$

or, from the first law of thermodynamics,

$$\delta Q \geqslant T\delta S. \tag{19.1.18}$$

The reader may see how the conditions obtained in terms of U and F may be derived from (19.1.17) when the appropriate accessory conditions are applied.

We have derived, from (19.1.10), the conditions of equilibrium under various sets of accessory conditions. It should be noted that the temperature T in this equation is that of the large heat reservoir. Thus, in applying (19.1.17), or any of the equilibrium conditions derived from this equation, we are restricted to variations such that T must be held at a fixed uniform value, at least over that part of the external surface of the system which is in diathermal contact with the reservoir. This restriction raises no difficulties in later discussions of equilibrium and stability.

19.2. Examples of the conditions of equilibrium

In this section we shall illustrate the methods used in applying these conditions, and, for simplicity, hydrostatic examples will be considered.

First we note a remark of Gibbs (1906) which may be found on page 45:

> It will be observed that the supposition of a rigid and non-conducting envelope enclosing the mass under discussion involves no real loss of generality, for if any

mass of matter is in equilibrium, it would also be so, if the whole or any part of it were enclosed in an envelope as supposed; therefore the conditions of equilibrium for a mass thus enclosed are the general conditions which must be satisfied in case of equilibrium.

This quotation refers to the case of adiabatic isolation which was our first example of the conditions of equilibrium in section 1. In this case the system must be supposed to be enclosed in a rigid non-conducting envelope. Gibbs in the above quotation is pointing out, however, that the conditions of equilibrium so obtained are general, and are not confined to systems so enclosed.

The same consideration would apply in the case where we wish to consider a system or part of a system contained in a rigid envelope (hence $\delta W = 0$) which can conduct heat so that the processes considered may be taken as isothermal. The conditions of equilibrium are general. Let us consider such an example consisting of two phases in contact over a surface. If the system is hydrostatic, we shall prove that the pressure is the same in both phases. Let part of the system be enclosed in a rigid diathermal envelope, which encloses volumes V_a and V_b of phases a and b. Then, we imagine an isothermal infinitesimal variation to occur, such that the volumes change by δV_a and δV_b, respectively. Now, in an isothermal variation (see (11.3.2)),

$$\delta^1 F_a = -p_a\,\delta V_a, \tag{19.2.1}$$

which is, simply, the work done on the phase a. Thus

$$\delta^1 F = \delta^1 F_a + \delta^1 F_b = (p_b - p_a)\,\delta V_a = 0 \tag{19.2.2}$$

is the condition of equilibrium (19.1.7), in which we have used the accessory condition

$$\delta V = \delta V_a + \delta V_b = 0. \tag{19.2.3}$$

Thus, we have proved that

$$p_b = p_a. \tag{19.2.4}$$

In a similar manner, we may discuss the interesting example of a spherical bubble in a fluid, and find the well-known relation between the surface tension of the bubble surface and the pressures inside and outside the bubble. We imagine the fluid enclosed in a rigid diathermal envelope – the bubble not being in contact with the envelope (gravity must be neglected, or the system must be in a state of 'free fall'). If we consider an isothermal variation involving changes of volume of bubble and fluid, the bubble remaining spherical, then

$$\delta^1 F = -p_b\,\delta V_b - p_f\,\delta V_f + \gamma\,\delta A, \tag{19.2.5}$$

with the accessory condition

$$\delta V = \delta V_b + \delta V_f = 0, \tag{19.2.6}$$

where V_f and V_b are the volumes of fluid and bubble respectively, p_f and p_b the respective pressures; γ and A are the surface tension and area of the spherical bubble surface. It is a simple matter to show that

$$\delta A = (2/3)(A/V_b)\,\delta V_b = (2/r)\,\delta V_b, \tag{19.2.7}$$

where r is the radius of the bubble. On using this and (19.2.6) in (19.2.5), we have

$$\delta^1 F = \{p_f - p_b + (2\gamma/r)\}\,\delta V_b = 0, \tag{19.2.8}$$

and so we obtain the well-known condition of equilibrium

$$p_b = p_f + (2\gamma/r). \tag{19.2.9}$$

If, in the first example of two phases, the stresses are not necessarily hydrostatic, then we can find the mechanical conditions to be satisfied by the stress tensors in each phase at the surface dividing the two phases. We assume a small region containing both phases to be enclosed in a rigid diathermal envelope small enough so that in each part of the phases enclosed the stress tensor may be considered to be uniform.

Thus in an infinitesimal isothermal virtual change at equilibrium

$$\delta^1 F = \int (\Theta^{(a)}_{\alpha\beta}\,\delta X^{(a)}_\alpha - \Theta^{(b)}_{\alpha\beta}\,\delta X^{(b)}_\alpha)\,\mathrm{d}\Sigma^{(ab)}_\beta = 0, \tag{19.2.10}$$

where the integral is over the dividing surface $\Sigma^{(ab)}$ between the two phases. We note that the variations satisfy $\delta\boldsymbol{X}^{(a)} = \delta\boldsymbol{X}^{(b)} = 0$ everywhere at the rigid envelope. The minus sign in (19.2.10) denotes that $\mathrm{d}\Sigma^{(ab)}$ is referred to the normal drawn outward from phase a. We now demand that the variation be continuous at the dividing surface and so

$$\delta\boldsymbol{X}^{(a)} = \delta\boldsymbol{X}^{(b)} \tag{19.2.11}$$

at this surface. Thus, (19.2.10) may be written

$$\delta^1 F = \int (\Theta^{(a)}_{\alpha\beta} - \Theta^{(b)}_{\alpha\beta})\,\delta X^{(a)}_\alpha N_\beta\,\mathrm{d}\Sigma^{(ab)} = 0, \tag{19.2.12}$$

where $\boldsymbol{N}$ is the unit normal at the surface element $\mathrm{d}\Sigma^{(ab)}$ outward drawn from phase a.

Since $\delta \boldsymbol{X}^{(a)}$ is arbitrary at the dividing surface, we must have

$$\Theta^{(a)}_{\alpha\beta} N_\beta = \Theta^{(b)}_{\alpha\beta} N_\beta. \tag{19.2.13}$$

We see, on referring to (6.1.2), that at equilibrium the surface force density exerted by phase a on phase b at any point of the dividing surface must be equal to that exerted by phase b on phase a at the same point.

As another example, in hydrostatics this time, we can prove that for any system in equilibrium, the temperature and pressure must be uniform throughout the system. We use the conditions (19.1.15).

Let us imagine a variation of entropy and density in the system, where δs is the change in entropy density (per unit volume), and δJ is the change in $J = \det(\boldsymbol{D})$ at the volume element $\mathrm{d}v$. We assume that the chemical composition remains constant.

The first-order change in U is then

$$\delta^1 U = \int \delta u \, \mathrm{d}v = \int \{T\,\delta s - p\,\delta J\}\,\mathrm{d}v, \tag{19.2.14}$$

where integration is over the whole reference region v.

The accessory conditions are

$$\delta S = \int \delta s \, \mathrm{d}v = 0, \tag{19.2.15}$$

$$\delta V = \int \delta J \, \mathrm{d}v = 0, \tag{19.2.16}$$

where the latter equation signifies that no external work is done, as the system is, in this case, supposed enclosed in a rigid adiabatic container. T, p, δs, δJ are all possibly functions of position.

If we multiply the equations (19.2.15) and (19.2.16) by, respectively, the undetermined parameters T_0 and p_0, and if we then subtract both these equations from (19.2.14) we obtain

$$\delta^1 U = \int \{(T - T_0)\delta s - (p - p_0)\,\delta J\}\,\mathrm{d}v = 0, \tag{19.2.17}$$

for equilibrium. δs and δJ are arbitrary functions of $\boldsymbol{x}$ except that (19.2.15) and (19.2.16) must be satisfied. It is well known that, in order for (19.2.17) to be satisfied, we must have

$$T = T_0, \qquad p = p_0, \tag{19.2.18}$$

that is that T and P must be uniform over the whole (possibly multiphase) hydrostatic system. However, a simple proof of this important result is given now, as this type of argument will be freely used hereafter.

First, we consider variations in which δJ is zero throughout v – this of course satisfies (19.2.16)! To prove, with such variations, that (19.2.17) has the consequence that T must be uniform, we proceed as follows. For variations in the entropy we choose

$$\delta s = \delta s' + a\,\delta s_0, \tag{19.2.19}$$

where δs_0 is some fixed function of position. $\delta s'$ may be chosen completely arbitrarily, because we may choose the parameter a, for any choice of $\delta s'$, so that the accessory condition (19.2.15) is satisfied. Once we have chosen the function δs_0, a will depend on our choice of $\delta s'$.

Equation (19.2.17) may thus be written as

$$\int_v (T - T_0)\,\delta s'\,\mathrm{d}v + a\int (T - T_0)\,\delta s_0\,\mathrm{d}v = 0. \tag{19.2.20}$$

We may choose T_0 so that the second integral in this equation vanishes. This choice of T_0 depends only on the functions T and δs_0. To satisfy, now, the condition of equilibrium, the first integral of (19.2.20) must vanish for all arbitrary choices of the function $\delta s'$. This requires $T = T_0$ throughout v. For, if $T \neq T_0$, we may choose $\delta s'$ to be positive in regions of v where $T - T_0$ is positive, and negative in regions where $T - T_0$ is negative. Thus the integrand would be positive everywhere in v, and so the integral would not vanish. Thus T must be uniform ($= T_0$).

By choosing δs zero throughout v, we may consider variations δJ, and follow through an identical argument to show that the pressure must be uniform, and equal to the undetermined parameter p_0.

In the examples of this section, we have dealt with the equilibrium conditions for the thermal and mechanical state of a system. We have assumed that, in the variations, the chemical composition has been constant. The results obtained, nevertheless, apply to systems in which it is possible for the composition to change either by diffusion or by chemical reactions. The equilibrium state considered may, in fact, have been reached from some initial non-equilibrium state, by means of such processes. The variations which we have employed in the examples form a sub-set of all possible variations. Thus, the conditions derived are necessary for the equilibrium of the system. By considering variations which involve a change of chemical composition further equilibrium conditions will be obtained. These conditions will govern the diffusion processes or chemical reactions which provide the mechanism for composition changes.

If the chemical composition can vary by the diffusion of a mobile component, we may introduce the concept of chemical potential, and show that the chemical potential of the mobile component must be a constant through the whole system, which may possibly be multi-phase. We may also introduce the chemical potential of a phase, and show that, if this phase can coexist with other phases of the same chemical composition, the chemical potentials of the coexisting phases must be the same for equilibrium.

We can go further and show that, if phases of different chemical compositions can coexist, certain chemical potentials exist which must satisfy certain relations – one relation for each possible chemical process.

A discussion of these chemical processes, for hydrostatics, will be given in sections 4 and 5.

19.3. Comments on methods using the Gibbs free energy

In hydrostatic thermodynamics, chemical processes are most simply dealt with by using the Gibbs free energy, $G = U - TS + pV$. It may be quickly shown that, from (19.1.17), the conditions of equilibrium and stability are

$$\delta^1 G = 0, \qquad \delta^2 G \geq 0, \tag{19.3.1}$$

under the accessory conditions,

$$\delta T = \delta p = 0. \tag{19.3.2}$$

The considerable utility of this function G lies wholly in the fact that it is a condition of equilibrium in a *hydrostatic* system that the pressure be uniform throughout. Thus, if we are considering a system of several phases then p, the pressure, is a common quantity and it makes sense to use (19.3.1).

We shall consider various examples of chemical processes from this point of view, but a complete generalisation of this function G cannot be achieved, in the sense that we cannot define a function for a multi-phase system of the form

$$G = U - TS - \tau_\alpha \Gamma_\alpha, \tag{19.3.3}$$

where τ_α, $\alpha = 1, 2, \ldots, 6$, are intensive stress elements, and the Γ_α are extensive mechanical co-ordinates. The reason is that, in general, the stress elements cannot be defined as common to all parts of the system. While we may have, for example, Cauchy stress tensors which are uniform in each phase they need not have the same value in different phases; all that is required is that at the surface between two phases the mechanical equilibrium condition (19.2.13) be satisfied. An obvious example is that of a fluid phase in contact with a solid which is non-hydrostatically stressed. If the Cauchy stress tensor in the solid is non-diagonal it cannot be equal to that in the fluid, which is $-p\boldsymbol{E}$, where p is the pressure.

In the next section, the use of (19.3.1) and (19.3.2) will be illustrated by some examples, while in section 5, a method will be used for hydrostatics which can be carried over to non-hydrostatics.

19.4. The Gibbs free energy for equilibrium conditions in hydrostatics

For a closed hydrostatic thermodynamic system, we have

$$dG = -S\,dT + V\,dp, \tag{19.4.1}$$

from the definition

$$G = U - TS + pV, \tag{19.4.2}$$

and using (11.1.1).

Thus, for an infinitesimal variation at an equilibrium state, we see that $\delta^1 G = 0$, if $\delta T = \delta p = 0$.

For a homogeneous phase G is regarded as a function of T, p and of the composition of the phase expressed in terms of the masses of the chemical components. That is,

$$G = G(T, p, M_1, M_2, \ldots, M_\kappa), \tag{19.4.3}$$

where $M_1, M_2, \ldots, M_\kappa$ are the masses of the κ chemical components. Now U, S, and V are extensive in the sense that, if we keep T and p constant, and increase all the masses by the factor λ, then

$$U(T, p, \lambda M_i) = \lambda U(T, p, M_i), \tag{19.4.4}$$

$$S(T, p, \lambda M_i) = \lambda S(T, p, M_i), \tag{19.4.5}$$

$$V(T, p, \lambda M_i) = \lambda V(T, p, M_i), \tag{19.4.6}$$

where M_i is short-hand for $M_1, M_2, \ldots, M_\kappa$, i.e.,

$$\lambda M_i \to \lambda M_1, \lambda M_2, \lambda M_3, \ldots, \lambda M_\kappa. \tag{19.4.7}$$

Therefore G satisfies exactly the same property. The above equations mean that U, S, V, (and G) are all homogeneous functions of the first order in the masses M_i. We may thus apply Euler's theorem which states that for a function f of variables $x_1, x_2, \ldots, x_\kappa$, which is homogeneous of nth order in these variables, i.e.,

$$f(\lambda x_1, \lambda x_2, \ldots, \lambda x_\kappa) = \lambda^n f(x_1, x_2, \ldots, x_\kappa), \tag{19.4.8}$$

then

$$nf = x_\alpha \frac{\partial f}{\partial x_\alpha}. \tag{19.4.9}$$

This is easily proved by differentiating the equation (19.4.8) with respect to λ:

$$\mathrm{d}f(\lambda x_i)/\mathrm{d}\lambda = x_\alpha\{\partial f(\lambda x_i)/\partial(\lambda x_\alpha)\} = n\lambda^{n-1}f(x_i). \tag{19.4.10}$$

On setting $\lambda = 1$, the result (19.4.9) follows. Applying this theorem to U, S, V and G, we obtain

$$U(T, p, M_i) = M_1\tilde{u}_1 + M_2\tilde{u}_2 + \cdots, M_\kappa\tilde{u}_\kappa = M\tilde{u}, \tag{19.4.11}$$

$$S(T, p, M_i) = M_1\tilde{s}_1 + M_2\tilde{s}_2 + \cdots, M_\kappa\tilde{s}_\kappa = M\tilde{s}, \tag{19.4.12}$$

$$V(T, p, M_i) = M_1\tilde{V}_1 + M_2\tilde{V}_2 + \cdots, M_\kappa\tilde{V}_\kappa = M\tilde{V}, \tag{19.4.13}$$

$$G(T, p, M_i) = M_1\tilde{g}_1 + M_2\tilde{g}_2 + \cdots, M_\kappa\tilde{g}_\kappa = M\tilde{g}, \tag{19.4.14}$$

where M is the total mass and, for example, $\tilde{u}_1 = (\partial U/\partial M_1)_{T,p,\bar{M}_1}$.

The quantities $\tilde{u}_\alpha$ are known as the partial specific energies, *or, if the masses are expressed in moles, as the partial molar energies.* The quantities $\tilde{s}_\alpha$, $\tilde{V}_\alpha$, $\tilde{g}_\alpha$ are also respectively the partial specific or molar entropies, volumes, Gibbs free energies, and $\tilde{u}$, $\tilde{s}$, $\tilde{V}$, $\tilde{g}$ are the specific or molar energy, entropy, volume and the Gibbs free energy.

The $\tilde{g}_\alpha$ are often written as μ_α and are known as the chemical potentials, and in conformity with this custom we shall use the symbol μ for a chemical potential. We note from above, then, that, for a phase,

$$\mu_\alpha = (\partial G/\partial M_\alpha)_{T,p,\bar{M}_\alpha} = \tilde{u}_\alpha - T\tilde{s}_\alpha + p\tilde{V}_\alpha. \tag{19.4.15}$$

Thus, for a phase which is not closed as to composition and mass, we have

$$\mathrm{d}G = -S\,\mathrm{d}T + V\,\mathrm{d}p + \mu_\alpha\,\mathrm{d}M_\alpha. \tag{19.4.16}$$

For a multi-phase system, we can define G as in (19.4.2), and we shall show that, at equilibrium, μ_α is a constant throughout the system, if the corresponding chemical component can diffuse throughout the system. Then we shall be able to write the following equation for a multi-phase system at equilibrium,

$$\mathrm{d}G = \sum_j \mathrm{d}G^{(j)} = -S\,\mathrm{d}T + V\,\mathrm{d}p + \mu_\alpha\,\mathrm{d}M_\alpha, \tag{19.4.17}$$

where $\mathrm{d}M_\alpha$ is now the total change of the mass of component α in the system and where $G^{(j)}$ refers to the jth phase. This result is true whether or not chemical reactions can occur. We shall show below that, for each possible reaction, an additional equilibrium condition is required.

Let us consider an infinitesimal variation occurring at constant T and p in a closed multi-phase system, assumed to be at equilibrium. In this variation the only change taking place is that in a certain volume element a, there is an increase δM_i of the chemical component i, while, in another volume element b, there is an increment $-\delta M_i$. Thus, using (19.4.17), the equilibrium condition (19.3.1) reduces to

$$\delta^1 G = \delta M_i(\mu_i^{(a)} - \mu_i^{(b)}) = 0, \tag{19.4.18}$$

that is, to

$$\mu_i^{(a)} = \mu_i^{(b)}. \tag{19.4.19}$$

Hence the equality of the chemical potential at two points depends, simply, on whether the component i can migrate from one point to the other.

Suppose that we generalise the argument by considering a variation of composition, described at every point by $\delta\rho_\alpha$, where ρ_α is the density (mass/unit reference volume) of component α. $\delta\rho_\alpha$ is a function of position, and the variation is to occur at constant T and p. Then, since $\delta^1 g = \delta\rho_\alpha\mu_\alpha$, the condition of equilibrium is

$$\delta^1 G = \int \delta\rho_\alpha\mu_\alpha \, \mathrm{d}v = 0, \tag{19.4.20}$$

where integration is throughout the reference state (which may be the state under study) of the system. The chemical potential μ_α may change from phase to phase – although we will now prove that it does not.

Now, since the condition of equilibrium applies to a closed system, we have the mass conservation restrictions

$$\int \delta\rho_\alpha \, \mathrm{d}v = 0, \quad \alpha = 1, 2, \ldots, \kappa. \tag{19.4.21}$$

Thus $\delta\rho_\alpha$ is not a completely arbitrary variation, but, following the method of the 'undetermined parameters' developed in section 2, we multiply each equation of (19.4.21) by a constant $\mu_\alpha^{(0)}$, $\alpha = 1, 2, \ldots, \kappa$, and subtract from (19.4.20) to obtain

$$\delta^1 G = \int \delta\rho_\alpha(\mu_\alpha - \mu_\alpha^{(0)}) \, \mathrm{d}v = 0. \tag{19.4.22}$$

Therefore we must have

$$\mu_\alpha = \mu_\alpha^{(0)}, \quad \alpha = 1, 2, \ldots, \kappa. \tag{19.4.23}$$

That is, each μ_α is a constant throughout the system. We see from the additive properties of S and V that (19.4.17) applies to the whole system for the case when we add new matter to it.

Next, suppose that we have a chemical reaction possible between three components A_1, A_2, A_3. Then instead of three restrictions, as in (19.4.21), we shall have only two, namely

$$\int \{(\delta\rho_1/a_1)+(\delta\rho_3/a_3)\}\,\mathrm{d}v=0 \qquad (19.4.24)$$

and

$$\int \{(\delta\rho_2/a_2)+(\delta\rho_3/a_3)\,\mathrm{d}v=0, \qquad (19.4.25)$$

where the ρ_i are measured in moles per unit reference volume. The stoichiometric coefficients a_1, a_2, a_3, are defined by the stoichiometric equation

$$a_1A_1+a_2A_2\rightleftarrows a_3A_3. \qquad (19.4.26)$$

Thus, the condition of equilibrium becomes

$$\int [\delta\rho_1(\mu_1-\mu_1^{(0)}/a_1)+\delta\rho_2(\mu_2-\mu_2^{(0)}/a_2)$$
$$+\delta\rho_3\{\mu_3-(\mu_0^{(1)}+\mu_0^{(2)})/a_3\}$$
$$+\delta\rho_\beta(\mu_\beta-\mu_\beta^{(0)})]\,\mathrm{d}v=0, \qquad (19.4.27)$$

where β runs over the other components. Again, each chemical potential has the same value through the system, but we have the additional relation

$$a_3\mu_3=a_1\mu_1+a_2\mu_2, \qquad (19.4.28)$$

as may easily be seen by noting that the coefficients of $\delta\rho_1$, $\delta\rho_2$, $\delta\rho_3$ in (19.4.27) are zero, and by eliminating $\mu_1^{(0)}$, $\mu_2^{(0)}$, the undetermined multipliers for the restrictions (19.4.24) and (19.4.25).

It should be emphasised that for (19.4.28) to hold, the μ_i, $i=1,2,3$, must be molar quantities, otherwise the molecular weights must be introduced.

We have shown, from (19.4.27), that each chemical potential has, at equilibrium, the same value throughout the whole system. Thus, (19.4.17) applies to the whole system. If chemical reactions may occur, we have, for each such reaction, a relation like (19.4.28).

Next, suppose that we have a substance, like quartz, of fixed composition, which can exist in several phases under hydrostatic conditions. The condition for coexistence of such phases is easily obtained by considering a variation in such a system where mass δM of phase β transforms to phase α at constant T and p. The condition of equilibrium is then

$$\delta^1 G = \delta M(\tilde{g}^{(\alpha)} - \tilde{g}^{(\beta)}) = 0, \tag{19.4.29}$$

that is,

$$\tilde{g}^{(\alpha)} = \tilde{g}^{(\beta)}. \tag{19.4.30}$$

Thus, the specific or molar Gibbs functions of the two phases must be equal.

In conclusion, we note that equality of the chemical potential throughout a system has meaning only if the corresponding component can migrate throughout the system. If, for instance, component A_α cannot be mobile in a particular region of the system then we must take, in all variations, $\delta\rho_\alpha = 0$ in that region. The above treatment cannot then prove that the chemical potential of component α in this region is equal to that elsewhere in the system. In general, equality of chemical potential or molar Gibbs function at two points, or for two phases, implies a possible process between these two points or phases.

19.5. Hydrostatic equilibrium conditions not using the Gibbs free energy

As pointed out previously, the use of G is particularly important in hydrostatics since, in equilibrium, p is uniform throughout a multi-phase system. In non-hydrostatics the similar assumption cannot be made, and the most convenient function is the Helmholtz free energy $F = U - TS$. The accessory conditions for the equilibrium and stability conditions (19.1.12) can easily be satisfied by imagining the system enclosed in a rigid diathermal container, so that isothermal infinitesimal variations may occur and so that no external work is done.

Let us consider the case of mobile components as in the previous section and imagine the isothermal variation $\delta\rho_\alpha$, $\alpha = 1, 2, \ldots, \kappa$. Under this variation the volume of each element $\mathrm{d}v$ will change by $\delta J\,\mathrm{d}v$. Now we evaluate $\delta^1 F$ by evaluating the change of free energy in each volume element, and by integrating. This change is evaluated in two steps. First, we imagine the volume element to change its composition at constant T and p. The corresponding first order change of free energy will be

$$\{(\partial F/\partial M_\alpha)_{T,p,\bar{M}_\alpha}\}\delta\rho_\alpha\,\mathrm{d}v = \tilde{f}_\alpha\,\delta\rho_\alpha\,\mathrm{d}v, \tag{19.5.1}$$

where $\tilde{f}_\alpha$ is the partial molar Helmholtz free energy evaluated at the volume element, and the change of its volume will be

$$\{(\partial V/\partial M_\alpha)_{T,p,\bar{M}_\alpha}\}\delta\rho_\alpha\, dv = \tilde{V}_\alpha\, \delta\rho_\alpha\, dv. \tag{19.5.2}$$

For the second step, we imagine that the volume element is deformed isothermally so that it takes up its actual shape after the variation. This deformation is required since the system is in a rigid container. In this second isothermal step the first order change in the free energy of the volume element will be the work done,

$$-p\{\delta J - \tilde{V}_\alpha\, \delta\rho_\alpha\}\, dv. \tag{19.5.3}$$

Thus, after these two steps,

$$\delta^1 F = \int \{\tilde{f}_\alpha\, \delta\rho_\alpha - p(\delta J - \tilde{V}_\alpha\, \delta\rho_\alpha)\}\, dv = 0. \tag{19.5.4}$$

But

$$\delta V = \int \delta J\, dv = 0, \tag{19.5.5}$$

because of the rigid container. Further, since the system is closed,

$$\int \delta\rho_\alpha\, dv = 0, \quad \alpha = 1, 2, \ldots, \kappa. \tag{19.5.6}$$

On using (19.5.5), and on introducing undetermined multipliers μ_α, $\alpha = 1, 2, \ldots, \kappa$, for the restrictions (19.5.6), we obtain

$$\tilde{f}_\alpha + p\tilde{V}_\alpha = \mu_\alpha. \tag{19.5.7}$$

Since

$$\tilde{f}_\alpha = \tilde{u}_\alpha - T\tilde{s}_\alpha, \tag{19.5.8}$$

we have shown that $\mu_\alpha (= \tilde{g}_\alpha)$ is constant throughout the system, at least wherever migration is possible.

A similar derivation of the equilibrium condition may be carried out using the internal energy U. In this case the system would be supposed to be enclosed in a *rigid adiabatic* container. The change of internal energy in a volume element would be calculated in two steps to obtain, to first order

$$\delta^1 U = \delta Q + \int \{\tilde{u}_\alpha + p\tilde{V}_\alpha\}\, \delta\rho_\alpha\, dv = 0, \tag{19.5.9}$$

where δQ is any heat liberated in the variation. In addition to the restrictions (19.5.6) we also have, because of the adiabatic nature of the process, to first order,

$$(\delta Q/T)+\int \tilde{s}_\alpha\,\delta\rho_\alpha\,\mathrm{d}v=0. \tag{19.5.10}$$

On eliminating δQ in (19.5.9) by using this restriction, one obtains, again, the condition

$$\mu_\alpha=\tilde{u}_\alpha-T\tilde{s}_\alpha+p\tilde{V}_\alpha, \tag{19.5.11}$$

where μ_α is independent of position and has been introduced as an undetermined multiplier to deal with the restrictions of (19.5.6).

Summary of important equations

A. *Equilibrium and stability conditions*

$$T\Delta S-\Delta U+\Delta W\geqslant 0. \tag{19.1.10}$$

$$\delta U-T\,\delta S-\delta W\geqslant 0. \tag{19.1.17}$$

$$\delta Q\geqslant T\,\delta S. \tag{19.1.18}$$

$$\delta^1 S=0;\qquad \delta^2 S\leqslant 0\quad \text{for } \delta U=\delta W=0. \tag{19.1.3}$$

$$\delta^1 U=0;\qquad \delta^2 U\geqslant 0\quad \text{for } \delta S=\delta W=0. \tag{19.1.15}$$

$$\delta^1 F=0;\qquad \delta^2 F\geqslant 0\quad \text{for } \delta T=\delta W=0. \tag{19.1.6}$$

$$\delta^1 G=0;\qquad \delta^2 G\geqslant 0\quad \text{for } \delta T=\delta p=0. \tag{19.3.1}$$

B. *Definitions involving extensive quantities in hydrostatics*

$$U(T,p,M_i)=M_\alpha\tilde{u}_\alpha=M\tilde{u},\quad \alpha=1,2,\ldots,\kappa, \tag{19.4.11}$$

$$S(T,p,M_i)=M_\alpha\tilde{s}_\alpha=M\tilde{s}, \tag{19.4.12}$$

$$V(T,p,M_i)=M_\alpha\tilde{V}_\alpha=M\tilde{V}, \tag{19.4.13}$$

$$G(T,p,M_i)=M_\alpha\tilde{g}_\alpha=M_\alpha\mu_\alpha=M\tilde{g}. \tag{19.4.14}$$

$\tilde{u}_\alpha$ = partial specific or molar internal energy.

$\tilde{s}_\alpha$ = partial specific or molar entropy.

$\tilde{u}$ = specific or molar internal energy, etc.

$M=M_1+M_2+\cdots+M_\kappa$ = total mass.

$\tilde{u}_\alpha=(\partial U/\partial M_\alpha)_{T,p,\bar{M}_\alpha}$,

$\tilde{s}_\alpha=(\partial S/\partial M_\alpha)_{T,p,\bar{M}_\alpha}$.

C. *An important thermodynamic equation for a system at equilibrium*

$$dG = -S\,dT + V\,dp + \mu_\alpha\,dM_\alpha, \tag{19.4.17}$$

for a phase or a multi-phase system in hydrostatics.

Examples

1. In the evaluation of (19.5.1) and (19.5.3), that is the two terms in the change of the free energy of the material in the volume element dv, it is important to realise that these are first order in the variations. In these variations, because the container is rigid, p will change by an infinitesimal amount. Show, however, that this effect will produce in δF a term, second order in the variations, which does not contribute to the equilibrium condition (it would be important, of course, in the stability condition).
2. In the restriction (19.5.11) the temperature will differ from T by a first order term, due to the liberation of heat δQ, and because the container is adiabatic. Show that this effect will contribute, in δU, a term, second order in the variations, which will not alter the equilibrium conditions.
3. The partial derivatives on the left hand sides of (19.5.1) and (19.5.2) are to be evaluated at a volume element of reference volume dv. The Helmholtz free energy of the volume element is given by $F = f\,dv$, the mass of component i contained in the volume element is given by $M_i = \rho_i\,dv$, and the volume of the element is given by $V = J\,dv$, where f, ρ_i, J are evaluated at the volume element. Show that the two partial derivatives mentioned above may be written, respectively, as $(\partial f/\partial \rho_\alpha)_{T,p,\bar{\rho}_\alpha}$ and $(\partial J/\partial \rho_\alpha)_{T,p,\bar{\rho}_\alpha}$.

20

Equilibrium conditions for diffusion in phases under non-hydrostatic stresses

The thermodynamics of a phase of variable chemical composition is discussed, and such quantities as molar and partial molar energies, entropies, etc., and chemical potentials are introduced and defined. The equilibrium conditions for diffusion of mobile chemical components in non-hydrostatically stressed solids are derived, and it is shown how the case where there may be a chemical reaction between such components may be treated. These equilibrium conditions may be expressed in terms of the chemical potentials of mobile components throughout a multiphase system.

20.1. The thermodynamics of a phase of variable composition

Phases which sustain non-hydrostatic stresses are necessarily solid, and so the situation as far as diffusidlimitingcerned is as follows. There will be a 'matrix' solid of constant composition, which will be the basic material sustaining the non-hydrostatic stresses, and into this matrix material the mobile components will be able to diffuse. The composition of the phase will vary through the variation of the masses of these mobile components. The mass of the matrix material can only be changed by such incoherent processes as crystallisation or solution which occur at the surfaces of the phase, phenomena which will be studied in the next chapter. In this chapter we shall consider the phenomenon of diffusion into a solid. We shall discuss the two cases, first where the mobile components are chemically inert to each other, and second where chemical reactions may occur between mobile components.

In the process of diffusion of a gas or fluid into a solid, the solid may tend to change its shape as well as its volume, and this effect will play its part in the balancing processes that lead to equilibrium. Also, the entropy may change and this will have an effect on the equilibrium.

We must now discuss the thermodynamic functions of such a solid under non-hydrostatic stresses. They will be functions of such variables as the temperature T, the generalised strain or stress co-ordinates $\boldsymbol{\gamma}$ or $\boldsymbol{\tau}$, and the masses or densities of the mobile components. The mass of the matrix solid may also be a variable – for example, we may imagine a homogeneous solid subdivided into parts of different mass. As in hydrostatics, we shall show that we may introduce specific or partial energy and entropy functions.

Let us, then, imagine the homogeneous solid to be subdivided into different parts. We may write the Helmholtz free energy, for example, of a part, whose reference volume is v_0, as

$$F = F(T, \boldsymbol{\gamma}, M_0, M_1, \ldots, M_\kappa), \tag{20.1.1}$$

where M_0 denotes the mass of the matrix solid in that part, and M_i, $i = 1, 2, \ldots, \kappa$, the masses of the mobile components. As discussed in chapter 4, the strain co-ordinates will be determined from the deformation which is coherent with that of the lattice atoms, molecules or ions of the matrix solid. From the homogeneous nature of the solid, we may assume that

$$F(T, \boldsymbol{\gamma}, \lambda M_0, \lambda M_1, \ldots, \lambda M_\kappa) = \lambda F(T, \boldsymbol{\gamma}, M_0, M_1, \ldots, M_\kappa), \tag{20.1.2}$$

where the left-hand side is the Helmholtz free energy of a part of the solid of reference volume λv_0, and where the masses will therefore be λM_i, $i = 0, 1, 2, \ldots, \kappa$. F is thus a first order homogeneous function of the M_i, and, hence, by Euler's theorem it may be written as

$$F(T, \boldsymbol{\gamma}, M_i) = (\partial F/\partial M_\alpha)M_\alpha, \tag{20.1.3}$$

where the partial derivatives are evaluated with respect to the variables M_i, $i = 0, 1, \ldots, \kappa$.

In hydrostatics, we take the specific or partial functions as functions of T, p and the masses (or densities) of the chemical components. For example,

$$\tilde{f}_\alpha = (\partial F/\partial M_\alpha)_{T,p,\bar{M}_\alpha}. \tag{20.1.4}$$

The corresponding generalisation in non-hydrostatics is

$$\tilde{f}_\alpha = (\partial F/\partial M_\alpha)_{T,\boldsymbol{\tau},M_\alpha}, \tag{20.1.5}$$

on introducing the six stress components as variables, instead of the strain components. If we then regard F, as we may, as a function of the variables $T, \boldsymbol{\tau}, M_i$, $i = 0, 1, \ldots, \kappa$, we would have, similarly to (20.1.2),

$$F(T, \boldsymbol{\tau}, \lambda M_i) = \lambda F(T, \boldsymbol{\tau}, M_i), \tag{20.1.6}$$

where M_i has been written to denote all the masses. Hence, by Euler's theorem, we may write

$$F = M_\alpha(\partial F/\partial M_\alpha)_{T,\boldsymbol{\tau},\bar{M}_\alpha} = \tilde{f}_\alpha M_\alpha. \tag{20.1.7}$$

If we divide by v_0, this equation becomes

$$f = \tilde{f}_\alpha \rho_\alpha, \tag{20.1.8}$$

where

$$\tilde{f}_\alpha = (\partial F/\partial M_\alpha)_{T,\boldsymbol{\tau},\bar{M}_\alpha} = (\partial f/\partial \rho_\alpha)_{T,\boldsymbol{\tau},\bar{\rho}_\alpha}, \tag{20.1.9}$$

and

$$\rho_\alpha = M_\alpha / v_0. \tag{20.1.10}$$

In (20.1.9) f is treated as a function of T, $\boldsymbol{\tau}$, ρ_i, $i = 0, 1, \ldots, \kappa$. By its definition as F/v_0, it is a homogeneous function of zero order in $M_0, M_1, \ldots, M_\kappa$, i.e.,

$$f(T, \boldsymbol{\tau}, \lambda M_i) = f(T, \boldsymbol{\tau}, M_i). \tag{20.1.11}$$

If we put $\lambda = 1/v_0$ in (20.1.11) we obtain

$$f(T, \boldsymbol{\tau}, \rho_i) = f(T, \boldsymbol{\tau}, M_i). \tag{20.1.12}$$

Thus, wherever the M_i appear in f we may replace them by the ρ_i. The $\tilde{f}_\alpha$ may then, as can be seen from (20.1.8), also be regarded as functions of T, $\boldsymbol{\tau}$, and the ρ_i. The $\tilde{f}_\alpha$ are known as the partial Helmholtz free energies. In a similar way we could discuss the other important thermodynamic functions such as U and S, and introduce the $\tilde{u}_\alpha$ and $\tilde{s}_\alpha$ as, respectively, the partial internal energies and entropies. For example,

$$u = \tilde{u}_\alpha \rho_\alpha; \qquad \tilde{u}_\alpha = (\partial u/\partial \rho_\alpha)_{T,\boldsymbol{\tau},\bar{\rho}_\alpha}. \tag{20.1.13}$$

Before we can introduce a generalised Gibbs function we must, briefly, discuss the mechanical co-ordinates used. The strain co-ordinates $\boldsymbol{\gamma}$ are independent of the size of the homogeneous phase. We can define generalised extensive co-ordinates Γ_μ by taking

$$\Gamma_\mu = v_0 \gamma_\mu, \quad \mu = 1, 2, \ldots, 6, \tag{20.1.14}$$

so that on multiplying (10.2.4) by v_0,

$$\mathrm{d}U = T\,\mathrm{d}S + \tau_\mu\,\mathrm{d}\Gamma_\mu. \tag{20.1.15}$$

We may now define a generalised Gibbs function (or free energy) as

$$G^\tau = U - TS - \tau_\mu \Gamma_\mu, \tag{20.1.16}$$

and, so

$$\mathrm{d}G^{\tau} = -S\,\mathrm{d}T - \Gamma_{\mu}\,\mathrm{d}\tau_{\mu}. \tag{20.1.17}$$

Thus, analogously to the Gibbs function of hydrostatics for which $V = (\partial G/\partial p)_T$, here we have

$$\Gamma_{\mu} = -(\partial G^{\tau}/\partial \tau_{\mu})_{T,\bar{\tau}_{\mu}}. \tag{20.1.18}$$

In the sense of (20.1.6) G^{τ} is extensive, and we may define a Gibbs free energy density function g^{τ} (per unit reference volume) and partial Gibbs free energies $\tilde{g}^{\tau}_{\alpha}$, such that

$$g^{\tau} = G^{\tau}/v_0 = \tilde{g}^{\tau}_{\alpha}\rho_{\alpha}; \qquad \tilde{g}^{\tau}_{\alpha} = (\partial g^{\tau}/\partial \rho_{\alpha})_{T,\tau,\bar{\rho}_{\alpha}}. \tag{20.1.19}$$

Since

$$g^{\tau} = u - Ts - \tau_{\mu}\gamma_{\mu} = f - \tau_{\mu}\gamma_{\mu} \tag{20.1.20}$$

then

$$\tilde{g}^{\tau}_{\alpha} = \tilde{f}_{\alpha} - \tau_{\mu}(\partial \gamma_{\mu}/\partial \rho_{\alpha})_{T,\tau,\bar{\rho}_{\alpha}}, \tag{20.1.21}$$

a result which will be useful later.

20.2. Equilibrium conditions for diffusion within a homogeneous phase

In order to emphasise the essential steps in the derivation of the equilibrium conditions for diffusion, they will be derived first for a single homogeneous solid phase. First, we consider such a phase containing mobile components chemically inert to each other, which can diffuse homogeneously throughout the phase.

We assume the phase to be enclosed by rigid diathermal walls so that in an infinitesimal virtual variation,

$$\delta W = 0, \qquad \delta T = 0, \tag{20.2.1}$$

i.e., no work is done externally and the variational process is isothermal. Under these circumstances, the condition of equilibrium is that $\delta^1 F$ be zero to first order in the variation. We consider a variation such that the ρ_i, the densities per unit reference volume of the mobile components, change infinitesimally by the amounts $\delta\rho_i$, $i = 1, \ldots, \kappa$. Since the phase is closed, we have the κ restrictions,

$$\int \delta\rho_i \,\mathrm{d}v = 0, \quad i = 1, 2, \ldots, \kappa, \tag{20.2.2}$$

but, otherwise, the $\delta\rho_i$ are arbitrary functions of position. Integration in

(20.2.2) is throughout the phase. We note that there is no restriction similar to (20.2.2) for the matrix solid material whose density is ρ_0.

Now we must evaluate $\delta^1 F$, the resulting variation in the Helmholtz free energy, by a method analogous to that developed in the previous chapter for hydrostatics.

Consider the material whose reference volume is the volume element $\mathrm{d}v$. We calculate its change of free energy in two steps as follows.

Step 1. We imagine that its composition is changed at constant T and $\boldsymbol{\tau}$, so that the resulting change of free energy, to first order, is by (20.1.7)

$$\tilde{f}_\alpha \delta\rho_\alpha \,\mathrm{d}v. \tag{20.2.3}$$

Step 2. In step 1, the isothermal change in the composition occurs at constant stress. Hence the material in the volume element will change its shape. However the stress cannot remain constant everywhere, as the variation occurs in a rigid container. Thus the material in the volume element must be deformed isothermally to its final shape after the variation. This deformation is step 2, and the corresponding change in the free energy is equal to the work done on the phase in this step. It will be shown below that in step 2 the first order change in the free energy of the whole phase is

$$-\int \{\tau_\mu (\partial\gamma_\mu/\partial\rho_\alpha)_{T,\boldsymbol{\tau},\bar{\rho}_\alpha}\}\,\delta\rho_\alpha \,\mathrm{d}v. \tag{20.2.4}$$

Hence

$$\delta^1 F = \int \{\tilde{f}_\alpha - \tau_\mu(\partial\gamma_\mu/\partial\rho_\alpha)_{T,\boldsymbol{\tau},\bar{\rho}_\alpha}\}\,\delta\rho_\alpha\,\mathrm{d}v = \int \tilde{g}^{\,T}_\alpha\,\delta\rho_\alpha\,\mathrm{d}v, \tag{20.2.5}$$

on using (20.1.21).

On introducing κ undetermined parameters μ_α, $\alpha = 1, 2, 3, \ldots, \kappa$, to take care of the restrictions of (20.2.2) as discussed in the previous chapter, the equilibrium condition $\delta^1 F = 0$ results in

$$\mu_\alpha = \tilde{g}^{\,T}_\alpha,\ \alpha = 1, 2, \ldots, \kappa, \tag{20.2.6}$$

in other words the right-hand side must have the same value throughout the phase.

To prove that (20.2.4) is the work done in step 2, we proceed as follows. Let $\delta\boldsymbol{X}$ be the deformation arising from the variation. Then we may write

$$\delta\boldsymbol{X} = \delta\boldsymbol{X}' + \delta\boldsymbol{X}'', \tag{20.2.7}$$

where $\delta \boldsymbol{X}'$ is the deformation which would arise, at constant $T, \boldsymbol{\tau}$, from the change in composition. That is, $\delta \boldsymbol{X}'$ is the deformation of step 1 and $\delta \boldsymbol{X}''$ is the deformation in step 2. Now the total external work done is zero and therefore, from (6.4.14), to first order,

$$\delta W = \int B_{\alpha\beta}\{\partial(\delta X_\beta)/\partial x_\alpha\}\, dv = 0, \tag{20.2.8}$$

where $B_{\alpha\beta}$ are the elements of the Boussinesq stress tensor. Thus

$$\int B_{\alpha\beta}\{\partial(\delta X''_\beta)/\partial x_\alpha\}\, dv = -\int B_{\alpha\beta}\{\partial(\delta X'_\beta)/\partial x_\alpha\}\, dv \tag{20.2.9}$$

to first order.

The second integral is equal to the work which would be done in step 1 on the whole phase if the change of composition were to occur at constant T and $\boldsymbol{\tau}$. In other words this integral is equal to

$$\int B_{\alpha\beta}\delta D_{\beta\alpha}\, dv = \int \tau_\mu \delta^1 \gamma_\mu\, dv, \tag{20.2.10}$$

where

$$\delta^1 \gamma_\mu = (\partial \gamma_\mu / \partial \rho_\alpha)_{T,\boldsymbol{\tau},\bar{\rho}_\alpha}\, \delta\rho_\alpha . \tag{20.2.11}$$

Hence, the work done on the whole phase in step 2 is given by (20.2.4), as stated, and the derivation of the equilibrium condition (20.2.6) is therefore complete.

Thus, we may introduce the concept of the chemical potential μ_α for a *mobile* chemical component in non-hydrostatic thermodynamics.

If the mobile components can interact chemically, (20.2.5) is again obtained. If we take, as an example of a chemical reaction, the same one as in the previous chapter, (19.4.26), the only change is in the restrictions (20.2.2); for $i = 1, 2, 3$ we have then only two restrictions (19.4.24) and (19.4.25), and, in an exactly similar way we arrive at chemical potentials $\mu_\alpha = \tilde{g}^{\tau}_{\alpha}$, uniform through the phase, for the *mobile components only*, with the additional condition that

$$a_3\mu_3 = a_1\mu_1 + a_2\mu_2, \tag{20.2.12}$$

as in (19.4.28).

Finally, a check through the derivation of equilibrium conditions given in this section shows that the equilibrium conditions (20.2.6) and (20.2.12) would still be obtained even if conditions were not homogeneous in the phase. Quantities such as $\tilde{g}^{\tau}_{\alpha}$, $B_{\alpha\beta}$, τ_α would be functions of position, but all of the essential steps would be exactly the same. Thus, even in an inhomogeneous phase, the concept of the chemical potential of a *mobile* component as a quantity uniform through the phase would still be valid.

20.3. Equilibrium conditions for diffusion within a multi-phase system

We suppose that the multi-phase system is enclosed in a closed rigid diathermal container so that, for a virtual variation, $\delta W = 0$, $\delta T = 0$. The equilibrium condition is

$$\delta^1 F = \sum_{(a)} \delta^1 F^{(a)} = 0, \tag{20.3.1}$$

where the summation is over the phases, and $\delta^1 F^{(a)}$ is the first order variation in the Helmholtz free energy $F^{(a)}$ of phase a.

Again, we consider variations $\delta\rho_\alpha^{(a)}$ of the densities of the mobile components, which must satisfy the restrictions

$$\sum_{(a)} \int_a \delta\rho_\alpha^{(a)}\, dv = 0, \quad \alpha = 1, 2, \ldots, \kappa, \tag{20.3.2}$$

where each integration is over the reference volume of each phase and summation is over all phases in the system.

If there is a chemical reaction of the type (19.4.26), the first three restrictions reduce as before to two, namely,

$$\sum_{(a)} \int_a \{(\delta\rho_1^{(a)}/a_1) + (\delta\rho_3^{(a)}/a_3)\}\, dv = 0, \tag{20.3.3}$$

and

$$\sum_{(a)} \int_a \{(\delta\rho_2^{(a)}/a_2) + (\delta\rho_3^{(a)}/a_3)\}\, dv = 0. \tag{20.3.4}$$

To evaluate the variation in F, we consider two steps, as in the previous section.

In the first step, the first order change in F due to the variation occurring in a volume element at constant T and $\boldsymbol{\tau}$ is seen to be, by comparison with (20.2.3),

$$\sum_{(a)} \int_a \bar{f}_\alpha^{(a)} \delta\rho_\alpha^{(a)}\, dv. \tag{20.3.5}$$

The work done in then deforming each volume element to its final shape is

$$-\sum_{(a)}\int_a \tau_\mu\{(\partial\gamma_\mu/\partial\rho_\alpha)^{(a)}_{T,\tau^{(a)},\bar{\rho}^{(a)}_\alpha}\}\,\delta\rho^{(a)}_\alpha\,\mathrm{d}v, \tag{20.3.6}$$

which we shall now prove.

In proving this, in this case of a multi-phase system, it will be simpler to consider the Cauchy stress tensor and integration over the actual deformed states of the phases rather than, as in the previous section, the Boussinesq stress tensor and integration over the reference states.

As in (20.2.7), we split $\delta\boldsymbol{X}^{(a)}$ into two parts where $\delta X'^{(a)}$ is the deformation in phase a which would result if each volume element in it changed its shape at constant T and $\boldsymbol{\tau}^{(a)}$. $\delta X''^{(a)}$ is the deformation which brings each volume element to its final shape. (*If the phase is inhomogeneous,* $\delta\boldsymbol{X}'^{(a)}$ *would be the change of shape of each volume element with* T *constant, and with* $\boldsymbol{\tau}$ *held constant at its value at the volume element in question.*)

If $\delta\boldsymbol{X}$ is continuous everywhere in the multi-phase system, and, in particular, at interfaces between solid phases, we may write, to first order,

$$\delta W=\int_\Sigma \Theta_{\alpha\beta}\,\delta X_\alpha\,\mathrm{d}\Sigma_\beta \tag{20.3.7}$$

$$=\sum_{(a)}\int_{\Sigma^{(a)}}\Theta^{(a)}_{\alpha\beta}\,\delta X^{(a)}_\alpha\,\mathrm{d}\Sigma^{(a)}_\beta=0, \tag{20.3.8}$$

where Σ denotes the outside surface of the whole system, and $\Sigma^{(a)}$ denotes the surface enclosing phase a. To show that the expressions in (20.3.7) and (20.3.8) are equal, we note that they differ by a sum of integrals, one over each interface between two phases. That over the interface between phases a and b is

$$\int\{\Theta^{(a)}_{\alpha\beta}\,\delta X^{(a)}_\alpha-\Theta^{(b)}_{\alpha\beta}\,\delta X^{(b)}_\alpha\}\,\mathrm{d}\Sigma^{(a)}_\beta. \tag{20.3.9}$$

On using the condition (19.2.13), we see that each such integral is zero, since we have assumed that $\delta\boldsymbol{X}$ is continuous.

If we now substitute (20.2.7) for each phase into (20.3.8), we can show that the first order change in the free energy is given by

$$\sum_{(a)}\int_{\Sigma^{(a)}}\Theta^{(a)}_{\alpha\beta}\,\delta X''^{(a)}_\alpha\,\mathrm{d}\Sigma^{(a)}_\beta=-\sum_{(a)}\int_{\Sigma^{(a)}}\Theta^{(a)}_{\alpha\beta}\,\delta X'^{(a)}_\alpha\,\mathrm{d}\Sigma^{(a)}_\beta. \tag{20.3.10}$$

That is, it is minus the work done in the first step,

$$-\sum_{(a)}\int_a \tau_\mu^{(a)}\delta^1\gamma_\mu^{(a)}\,\mathrm{d}v = -\sum_{(a)}\int_a \tau_\mu^{(a)}(\partial\gamma_\mu/\partial\rho_\alpha)^{(a)}\,\delta\rho_\alpha^{(a)}\,\mathrm{d}v. \tag{20.3.11}$$

Hence, combining (20.3.5) and (20.3.11), we have

$$\delta^1 F = \sum_{(a)}\int_a \tilde{g}_\alpha^{\tau(a)}\,\delta\rho_\alpha^{(a)}\,\mathrm{d}v = 0, \tag{20.3.12}$$

subject to the restrictions (20.3.2), if there are no chemical reactions.

Again, we note that there is nothing in all these steps which would be invalidated if each phase were inhomogeneous. On introducing the μ_α as undetermined multipliers, one has, for the mobile components,

$$\mu_\alpha = \tilde{g}_\alpha^\tau, \tag{20.3.13}$$

i.e., $\tilde{g}_\alpha^\tau$ has the same value throughout the multi-phase system.

We should note that the type of co-ordinate $\boldsymbol{\gamma}$ and its corresponding stress $\boldsymbol{\tau}$ need not be the same for each phase. This is rather surprising, but, in fact, $\tilde{g}_\alpha^\tau$ is independent of the choice of co-ordinates, because $\tau_\mu(\partial\gamma_\mu/\partial\rho_\alpha)$ transforms in the same way as $\tau_\mu\,\mathrm{d}\gamma_\mu$ in the transformations from one set of co-ordinates and stress elements to another, and we know that

$$\mathrm{d}'w = \tau_\mu\,\mathrm{d}\gamma_\mu = \tau'_\mu\,\mathrm{d}\gamma'_\mu, \tag{20.3.14}$$

for two such sets of co-ordinates and stress elements.

As an example, consider a fluid in contact with a solid phase, and a mobile component α. Then $\tilde{g}_\alpha$ for the component in the fluid will be

$$\mu_\alpha^{(f)} = \tilde{f}_\alpha + p\tilde{V}_\alpha, \tag{20.3.15}$$

i.e., the usual hydrostatic expression for the chemical potential in the fluid. For equilibrium against *diffusion* we shall have

$$\mu_\alpha^{(f)} = \tilde{g}_\alpha^\tau, \tag{20.3.16}$$

where the right-hand side refers to the solid phase.

If the mobile component is the fluid itself, one can see, from (20.1.7) and (20.1.9), that $\tilde{f}_\alpha$ becomes $\tilde{f}$, the specific or molar free energy of the fluid.

Finally, if there is a chemical reaction possible, of the type, for example, of (19.4.26), then, exactly as in chapter 19, we will again obtain the relation (19.4.28) between the chemical potentials of the reacting components.

Summary of important equations

A. *The Helmholtz free energy*

$$F = \tilde{f}_\alpha M_\alpha, \tag{20.1.7}$$

$$f = \tilde{f}_\alpha \rho_\alpha, \tag{20.1.8}$$

$$\tilde{f}_\alpha = (\partial F/\partial M_\alpha)_{T,\tau,\bar{M}_\alpha} = (\partial f/\partial \rho_\alpha)_{T,\tau,\bar{\rho}_\alpha}, \tag{20.1.9}$$

where

$$\rho_\alpha = M_\alpha/v_0, \qquad f = F/v_0 \tag{20.1.10}$$

and v_0 is the reference volume of the phase.

B. *The internal energy density*

$$u = \tilde{u}_\alpha \rho_\alpha, \qquad \tilde{u}_\alpha = (\partial u/\partial \rho_\alpha)_{T,\tau,\bar{\rho}_\alpha}. \tag{20.1.13}$$

C. *The generalised Gibbs free energy*

$$G^\tau = U - TS - \tau_\mu \Gamma_\mu, \tag{20.1.16}$$

$$\mathrm{d}G^\tau = -S\,\mathrm{d}T - \Gamma_\mu\,\mathrm{d}\tau_\mu, \tag{20.1.17}$$

$$\Gamma_\mu = v_0 \gamma_\mu = -(\partial G^\tau/\partial \tau_\mu)_{T,\bar{\tau}_\mu}. \qquad \begin{cases}(20.1.14)\\(20.1.18)\end{cases}$$

$$g^\tau = G^\tau/v_0 = \tilde{g}^\tau_\alpha \rho_\alpha, \qquad \tilde{g}^\tau_\alpha = (\partial g^\tau/\partial \rho_\alpha)_{T,\tau,\bar{\rho}_\alpha}. \tag{20.1.19}$$

$$\tilde{g}^\tau_\alpha = \tilde{f}_\alpha - \tau_\mu(\partial \gamma_\mu/\partial \rho_\alpha)_{T,\tau,\bar{\rho}_\alpha}. \tag{20.1.21}$$

D. *Equilibrium conditions for diffusion*

1. $\mu_\alpha = \tilde{g}^\tau_\alpha =$ constant through the system. (20.3.13)

2. $a_1\mu_1 + a_1\mu_2 = a_3\mu_3$, (20.2.12)

for the chemical reaction

$$a_1A_1 + a_2A_2 \leftrightarrows a_3A_3. \tag{19.4.26}$$

21

The equilibrium of a stressed solid in contact with a solution of the solid

It is shown how to derive the equilibrium condition for the case where a stressed solid is in contact with a solution of a solid. The proof does not depend on the stress or strain being uniform. It is proved that, for a stable solid under non-hydrostatic stresses, such equilibrium is metastable in the sense that a seed crystal placed in the solution, and thus in a hydrostatic state, would grow at the expense of the stressed solid. This result depends on conditions of phase stability developed in chapter 23.

21.1. Introductory comments

In hydrostatics, if a solid is in contact and in equilibrium with a solution of the solid, the situation is simplified by the fact that the pressure is uniform throughout the solid and fluid system. Thus, as pointed out in chapter 19, we can use the equilibrium condition $\delta^1 G = 0$ for a virtual change in which $\delta T = \delta p = 0$. We can assume that a mass δM of the solid material either dissolves from or crystallises on to the solid phase.

Then

$$\delta^1 G = \delta M(\mu^{(s)} - \mu^{(L)}) = 0, \quad \text{i.e., } \mu^{(s)} = \mu^{(L)}. \tag{21.1.1}$$

The chemical potential $\mu^{(L)}$ of the solid dissolved in the solution is given by

$$\mu^{(L)} = (\partial G^{(L)}/\partial M_s)_{T,p}, \tag{21.1.2}$$

where M_s is the mass of the solid dissolved in the solution and $G^{(L)}$ is the Gibbs function of the solution. The chemical potential $\mu^{(s)}$ of the solid is $G^{(s)}/M$, where $G^{(s)}$ is the Gibbs function, $U - TS + pV$, of the solid phase of mass M. Because $G^{(s)}$ is a first order homogeneous function of M, we may write

$$\mu^{(s)} = (\partial G^{(s)}/\partial M)_{T,p}. \tag{21.1.3}$$

When a system consisting of a non-hydrostatically stressed solid in contact with a fluid phase is at equilibrium, the mechanical conditions (19.2.13) must be satisfied at the surface between fluid and solid. Since the stress in the fluid is hydrostatic,

$$\Theta_{\alpha\beta}N_{\beta} = -pN_{\alpha}, \quad \text{i.e., } \boldsymbol{\Theta}\boldsymbol{N} = -p\boldsymbol{N}. \tag{21.1.4}$$

p is the pressure in the fluid, and $\boldsymbol{\Theta}$ the Cauchy stress tensor in the solid at a point on the surface. From (21.1.4) we see that $\boldsymbol{N}$, the unit normal to the surface at this point, is an eigenvector of $\boldsymbol{\Theta}$ – in other words $\boldsymbol{N}$ is a 'principal direction' of $\boldsymbol{\Theta}$. For example, if $\boldsymbol{\Theta}$ is uniform in the solid, this places restrictions, for equilibrium, on the shape of the surface between the solid and fluid. Consider, for instance, the case of a uniform stress tensor $\boldsymbol{\Theta}$, whose three eigenvalues are all unequal. Then, it may be seen that the only equilibrium surface possible between the solid and fluid is a plane surface whose normal is an eigenvector of $\boldsymbol{\Theta}$.

Suppose $\boldsymbol{\Theta}$ has two equal eigenvalues Θ, and that $\boldsymbol{N}_1$ and $\boldsymbol{N}_2$ are a pair of orthogonal unit vectors which are both eigenvectors corresponding to this eigenvalue Θ. Then, any linear combination $\alpha\boldsymbol{N}_1+\beta\boldsymbol{N}_2$ is also an eigenvector of $\boldsymbol{\Theta}$ corresponding to Θ. Thus, we see that the surface may be a general cylinder whose axis is perpendicular to $\boldsymbol{N}_1$ and $\boldsymbol{N}_2$, and, of course, we must have $\Theta=-p$. As an example of this general cylindrical shape we could take two intersecting planes whose normals are respectively $\boldsymbol{N}_1$ and $\boldsymbol{N}_2$.

In the case of hydrostatic conditions, any external surface shape can be an equilibrium shape.

In the next section we shall derive the equilibrium conditions for the case of a non-hydrostatically stressed solid in contact with a solution of the solid. These conditions were first derived by Gibbs (1906). (In reading Gibbs' work it is well worth noting and verifying that the stress tensor used by him is identical with the Boussinesq stress tensor as defined in this book.)

21.2. Derivation of the equilibrium conditions

We consider a system consisting of the solid phase and the fluid solution in contact at a surface, not necessarily a plane. For equilibrium at every point of the surface (21.1.4) must hold, even though $\boldsymbol{\Theta}$ may not be uniform.

The system is considered to be enclosed in a rigid diathermal container, so that the equilibrium conditions are

$$\delta^1 F = 0, \tag{21.2.1}$$

under the accessory conditions

$$\delta T = 0, \qquad \delta W = 0, \tag{21.2.2}$$

where, as usual δW is the work done externally on the system.

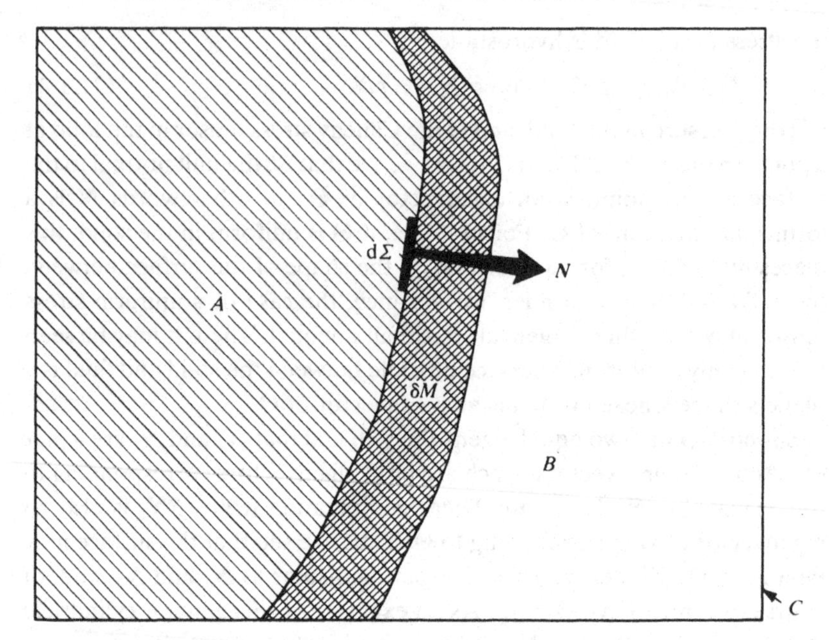

Fig. 15. The container C is a rigid diathermal container; A is the solid phase, δM is the layer of solid crystallised on to A; B is the solution.

Under the conditions (21.2.2) a virtual infinitesimal variation is considered, in which an amount δM of the solid material is crystallised on to the solid phase – see fig. 15. (If δM is negative, $|\delta M|$ has dissolved from the surface of the solid.)

If this process were to occur in such a way that the stress tensor $\mathbf{\Theta}$, at every point in the solid, were held constant and the pressure in the liquid were held constant, there would be an increase of volume given by

$$\left\{\int_{(sL)} \tilde{V}_s\, \delta m\, \mathrm{d}\Sigma\right\} - \tilde{v}_s^{(L)}\, \delta M = \int_{(sL)} (\tilde{V}_s - \tilde{v}_s^{(L)})\, \delta m\, \mathrm{d}\Sigma. \qquad (21.2.3)$$

$\tilde{V}_s$ is the molar volume of the solid phase in its stressed state at the surface element $\mathrm{d}\Sigma$ – it may be a function of position as we have not assumed $\mathbf{\Theta}$ to be uniform. $\delta m\, \mathrm{d}\Sigma$ is that part of δM which has grown on or dissolved from the surface element $\mathrm{d}\Sigma$ on the interface surface, which has been denoted by (sL) in the integral. δm is the mass increment per unit area, and so

$$\int_{(sL)} \delta m\, \mathrm{d}\Sigma = \delta M. \qquad (21.2.4)$$

The partial molar volume of the solid material in the solution is defined by

$$\tilde{v}_s^{(L)} = (\partial V^{(L)}/\partial M_s)_{T,p}. \tag{21.2.5}$$

$V^{(L)}$ and M_s are, respectively, the volume of the liquid, and the mass, in moles, of the solid in solution.

However, because the system is contained in a rigid enclosure, the total change of volume must be zero. Thus, after the variation, the solid will be deformed and the fluid volume changed, relative to the initial state. That is, the strain state of the solid and the fluid volume will be different from those for the equilibrium state being studied. (It is this deformation and fluid volume change, in the variation, which may be avoided in the hydrostatic case by considering the variation to occur at constant T and p, when we can use the Gibbs function). We must include the work done in this deformation and fluid volume change, when we evaluate the change $\delta^1 F$ in the Helmholtz free energy.

There are four contributions to $\delta^1 F$, as follows.

1. The new mass δM added to the solid at constant T and $\boldsymbol{\Theta}$: the corresponding contribution to $\delta^1 F$ is, to first order in the variation,

$$\int_{(sL)} \tilde{f}_s \, \delta m \, d\Sigma, \tag{21.2.6}$$

where $\tilde{f}_s$ is the molar Helmholtz free energy of the solid which may be a function of position. Any change in $\tilde{f}_s$ due to the first order change in $\boldsymbol{\Theta}$ mentioned above will only contribute a second order term to (21.2.6).

2. The isothermal deformation of the solid phase (*A* in fig. 15) to bring it to its final shape: the resulting first order change in the free energy is given by

$$\int_{(sL)} \Theta_{\alpha\beta} \delta X_\alpha N_\beta \, d\Sigma, \tag{21.2.7}$$

since integration over the surface of the rigid container contributes nothing to the work. $\boldsymbol{N}$ is the unit normal, outward drawn from the solid phase, $\delta \boldsymbol{X}$ describes the deformation.

3. The loss δM, *at constant T and p*, of the solid material from the liquid phase (*B* in fig. 15): the resulting first order change in the free energy is given by

$$-\tilde{f}_s^{(L)} \delta M, \tag{21.2.8}$$

where

$$\tilde{f}_s^{(L)} = (\partial F^{(L)}/\partial M_s)_{T,p}, \tag{21.2.9}$$

is the partial molar free energy of the solid material in the liquid solution.

4. The isothermal compression or expansion of the liquid phase to bring it to its final volume: the resulting change in the free energy is equal to the work done on the phase. This, to first order, is given by

$$-p\,\delta V', \tag{21.2.10}$$

where $\delta V'$ is the change in volume required to bring the liquid to its final state from the state resulting from the loss of solid δM at *constant* T and p. That is the total change of volume of the fluid phase is

$$\delta V^{(L)} = -\tilde{v}_s^{(L)}\delta M + \delta V'. \tag{21.2.11}$$

We thus have four contributions to $\delta^1 F$, namely, (21.2.6), (21.2.7), (21.2.8), and (21.2.10). On using (21.1.4) then (21.2.7) becomes equal to

$$-p\delta V'', \tag{21.2.12}$$

where

$$\delta V'' = \int_{(sL)} \delta X_\alpha N_\alpha \mathrm{d}\Sigma. \tag{21.2.13}$$

Now the total change of volume is zero and hence

$$\delta V' + \delta V'' + \int_{(sL)} (\tilde{V}_s - \tilde{v}_s^{(L)})\delta m\, \mathrm{d}\Sigma = 0. \tag{21.2.14}$$

The two contributions (21.2.10) and (21.2.12), combined, then become

$$-p(\delta V' + \delta V'') = p\int_{(sL)} (\tilde{v}_s^{(L)} - \tilde{V}_s)\delta m\, \mathrm{d}\Sigma. \tag{21.2.15}$$

Finally, then, on combining (21.2.6), (21.2.8) and (21.2.15), we may write the equilibrium condition as

$$\delta^1 F = \int_{(sL)} \{(\tilde{f}_s + p\tilde{V}_s) - (\tilde{f}_s^{(L)} + p\tilde{v}_s^{(L)})\}\delta m\, \mathrm{d}\Sigma = 0. \tag{21.2.16}$$

Since δm is an arbitrary function of position on the surface of integration, the expression in curly brackets must vanish. We thus obtain the condition of equilibrium

$$\tilde{f}_s + p\tilde{V}_s = \mu_s^{(L)}, \tag{21.2.17}$$

where $\mu_s^{(L)}$ is the chemical potential of the solid in solution, and $\tilde{f}_s$ and $\tilde{V}_s$, the molar free energy and molar volume of the solid phase, may depend on position on the interface. This famous condition of Gibbs means, however, that the left-hand side of (21.2.17) must have the same value, for equilibrium, at every point on the surface.

21.3. Discussion of the condition (21.2.17)

As an example of the application of this condition, consider a crystal under uniform stress conditions so that $\tilde{f}_s$ and $\tilde{V}_s$ are uniform throughout the crystal. Suppose two solutions are each in contact with a different plane surface of the crystal. If the orientation of these two surfaces is described by the two unit normals $\boldsymbol{N}^{(1)}$ and $\boldsymbol{N}^{(2)}$, it is a condition of equilibrium that $\boldsymbol{N}^{(1)}$ and $\boldsymbol{N}^{(2)}$ are both principal directions of the Cauchy stress tensor $\boldsymbol{\Theta}$ in the crystal. Therefore, (from 21.1.4)

$$\boldsymbol{\Theta N}^{(1)} = -p^{(1)}\boldsymbol{N}^{(1)} \tag{21.3.1}$$

and

$$\boldsymbol{\Theta N}^{(2)} = -p^{(2)}\boldsymbol{N}^{(2)}, \tag{21.3.2}$$

where $p^{(1)}$ and $p^{(2)}$ are the pressures in the two solutions.

If $p^{(1)} = p^{(2)} = p$ then the two solutions must have the same chemical potential $\mu_s^{(L)}$ and, if (21.2.17) is satisfied, equilibrium can be achieved. Thus, in this case, the same solution may be in equilibrium when in contact with these two crystal faces.

In the case of $p^{(1)} \neq p^{(2)}$, we see that two different solutions, one in contact with face 1 and the other in contact with face 2, could both be in equilibrium with the crystal if each satisfies the appropriate condition for the corresponding face.

Another important consideration is whether the above equilibrium between a fluid and a stressed solid is metastable, in the sense that the fluid satisfying these conditions may be supersaturated with respect to the growth of crystals of the solid which are at the hydrostatic pressure of the fluid. In other words, is

$$\mu_s^{(L)} > \mu_s^{(L)}(p), \tag{21.3.3}$$

where $\mu_s^{(L)}(p)$ is the chemical potential of the solid in the solution when it is at equilibrium with the hydrostatically stressed solid at pressure p?

If the inequality (21.3.3) is satisfied then the solution is supersaturated and if a disturbance were to occur or seed crystals were present in the solution, crystals at pressure p would tend to form. We shall examine this inequality in the next section.

It should be noted, finally, that the condition (21.2.17) must apply, for equilibrium, to any stressed state of the solid, whether or not it is regarded as deformed to a finite extent from a given reference state.

21.4. The equilibrium condition (21.2.17) – stable or metastable?

Let us consider a non-hydrostatically stressed solid in contact, and at equilibrium, with a solution of the solid. The temperature T will be uniform throughout both the solid and solution. Let the pressure of the fluid be p. Let us choose the reference state for the solid as that state in which it is under the hydrostatic pressure p and at temperature T. We shall consider states of stress of the solid where one of the eigenvalues of $\boldsymbol{\Theta}$ is equal to $-p$, as required by the mechanical conditions of equilibrium.

Now, the left-hand side of (21.2.17) may be written

$$\tilde{f}_s + p\tilde{V}_s = \tilde{v}_s(f + pJ), \tag{21.4.1}$$

where $\tilde{v}_s$ is the reference volume per mole of the solid phase, and $\tilde{V}_s$ is the molar volume at the equilibrium stressed state being studied. f, as we may be reminded, is the Helmholtz free energy *per unit reference volume.* We may, without loss of generality, restrict ourselves to those deformations from the above reference state which obey the geometrical condition $\boldsymbol{V} = \boldsymbol{E}$. In that case

$$J = \det(\boldsymbol{H}) = \det(\boldsymbol{E} + \boldsymbol{e}). \tag{21.4.2}$$

We shall proceed by finding an expression for $f + pJ$. It will be sufficient to determine this expression up to terms of second order only in the e_α.

With the above reference state, we may write, neglecting higher order terms,

$$f(T, \boldsymbol{e}) = f^0(T) - pE_\alpha e_\alpha + \tfrac{1}{2}\hat{c}_{\alpha\beta}e_\alpha e_\beta. \tag{21.4.3}$$

We note that $\hat{\boldsymbol{c}}$ is the isothermal stiffness matrix and f^0 is the free energy density, at temperature T and at the reference state.

Now, up to second order terms,

$$J = 1 + E_\alpha e_\alpha + (e_1e_2 + e_2e_3 + e_3e_1), \tag{21.4.4}$$

and we may write

$$f + pJ = f^0 + p + \tfrac{1}{2}\hat{c}_{\alpha\beta}e_\alpha e_\beta + p(e_1e_2 + e_2e_3 + e_3e_1). \tag{21.4.5}$$

If, using (16.1.4), we express the elements of $\hat{\boldsymbol{c}}$ in terms of those of $\boldsymbol{c}$, it is not difficult to see that (21.4.5) may be expressed as

$$f + pJ = f^0 + p + \tfrac{1}{2}c_{\alpha\beta}e_\alpha e_\beta. \tag{21.4.6}$$

For later reference, an alternative expression for $f + pJ$ is useful. From

(21.4.6), we may prove that, again up to quadratic terms in the e_α,

$$f + pJ = f^0 + p + \tfrac{1}{2}s_{\alpha\beta}(\Theta_\alpha + pE_\alpha)(\Theta_\beta + pE_\beta). \tag{21.4.7}$$

This follows from the fact that, to first order in the e_α,

$$\Theta_\alpha + pE_\alpha = c_{\alpha\beta}e_\beta; \quad e_\alpha = s_{\alpha\beta}(\Theta_\beta + pE_\beta). \tag{21.4.8}$$

We should note that $c_{\alpha\beta}$ and $s_{\alpha\beta}$ are the *isothermal* stiffness and compliance elements of infinitesimal strain theory. It must, however, be emphasised that the equations (21.4.6) and (21.4.7) are derived using finite strain theory, and do not depend on the approximations of infinitesimal strain theory. If (21.4.6) is multiplied by $\tilde{v}_s$, then

$$\tilde{f} + p\tilde{V}_s = \tilde{f}^0 + p\tilde{v}_s + \tfrac{1}{2}\tilde{v}_s c_{\alpha\beta}e_\alpha e_\beta. \tag{21.4.9}$$

Now, from (21.2.17), the left-hand side is that quantity for the solid, *in its non-hydrostatic stressed state*, which must, for equilibrium, be equal to $\mu_s^{(L)}$, the chemical potential of the solid in the solution. From (21.2.17) we see that $\tilde{f}^0 + p\tilde{v}_s$ is that quantity for the solid, *in its hydrostatic reference state*, which must be equal to $\mu_s^{(L)}(p)$, the chemical potential of the solid in the solution when both solid and solution are at equilibrium at the same T and p.

Thus, we may say that the condition of equilibrium, for the non-hydrostatically stressed solid in contact with the solution, is

$$\mu_s^{(L)} - \mu_s^{(L)}(p) = \tfrac{1}{2}\tilde{v}_s c_{\alpha\beta}e_\alpha e_\beta. \tag{21.4.10}$$

We see, from (21.4.7) and (21.4.9), that (21.4.10) may be expressed in the following alternative forms:

$$\mu_s^{(L)} - \mu_s^{(L)}(p) = \tfrac{1}{2}\tilde{v}_s e_\alpha(\Theta_\alpha + pE_\alpha) \tag{21.4.11}$$

$$= \tfrac{1}{2}\tilde{v}_s s_{\alpha\beta}(\Theta_\alpha + pE_\alpha)(\Theta_\beta + pE_\beta). \tag{21.4.12}$$

We have expressed the equilibrium condition in several very useful forms. For we can say that, if the right-hand side of (21.4.10), (21.4.11) or (21.4.12) is positive, then

$$\mu_s^{(L)} > \mu_s^{(L)}(p). \tag{21.4.13}$$

If this is the case, then the equilibrium between the stressed solid and the solution is metastable, in the sense that crystals, at the pressure p of the solution, will tend to grow in the solution. If negative, the equilibrium would be stable and crystals will not grow at pressure p. We shall show that, for a *stable* crystal, (21.4.13) holds and thus the equilibrium will be metastable. This equilibrium could possibly be achieved by having the solution and solid in contact and with no crystal-growing nuclei present. Such nuclei could be dust, minute seed crystals in the solution or crystal defects on the solid surface, which could act as Frank–Read sources. In such a case, if a disturbance were created by, say, dropping a minute seed crystal into the solution, then the system would not be at equilibrium, and the seed crystal, at hydrostatic pressure, would grow at the expense of the non-hydrostatically stressed solid.

In chapters 22 and 23 it will be shown that, for the solid to be a stable phase at constant hydrostatic pressure, the isothermal stiffness matrix $\boldsymbol{c}$, and its reciprocal the isothermal compliance matrix $\mathbf{s}$, must be non-negative matrices. That is, for arbitrary values of $\boldsymbol{w}$, a six-dimensional vector, we must have

$$c_{\alpha\beta}w_\alpha w_\beta \geqslant 0; \qquad s_{\alpha\beta}w_\alpha w_\beta \geqslant 0, \tag{21.4.14}$$

for a stable solid. If the inequality holds, we say that the solid is strongly stable. Thus (21.4.13) holds for such a solid.

To illustrate these results, let us choose the principal axis system of $\boldsymbol{\Theta}$ to be our co-ordinate axis system. In this axis system let

$$\begin{aligned} \Theta_a &= -p_a, \quad a = 1, 2, 3, \\ &= 0, \qquad a = 4, 5, 6. \end{aligned} \tag{21.4.15}$$

Then we may write (21.4.12) as

$$\mu_s^{(L)} - \mu_s^{(L)}(p) = \tfrac{1}{2}\tilde{v}_s s_{\alpha\beta}(p_\alpha - pE_\alpha)(p_\beta - pE_\beta). \tag{21.4.16}$$

$s_{\alpha\beta}$ are the compliance matrix elements referred to the above axis system. p must, of course, be equal to one of the quantities p_1, p_2, p_3.

Suppose we take a case where $p_1 = p_2 = p$, i.e., $\boldsymbol{\Theta}$ has two equal eigenvalues. Equation (21.4.16) then becomes

$$\mu_s^{(L)} - \mu_s^{(L)}(p) = \tfrac{1}{2}\tilde{v}_s(p_3 - p)^2 s_{33}. \tag{21.4.17}$$

If the right-hand side of this equation is positive then the solid/solution system will be metastable. By choosing in (21.4.14) a vector $\boldsymbol{w}$ whose only non-zero component is w_3, we see that s_{33} must be positive for a strongly stable solid. Thus, for such a solid, the solid/solution system is metastable.

If no two eigenvalues of $\boldsymbol{\Theta}$ are equal, and if we take $p_3 = p$, then the right-hand side of (21.4.16) is

$$\tfrac{1}{2}\tilde{v}_s\{(p_1 - p)^2 s_{11} + (p_2 - p)^2 s_{22} + 2(p_1 - p)(p_2 - p)s_{12}\}. \tag{21.4.18}$$

This expression is non-negative, if the discriminant

$$\begin{vmatrix} s_{11} & s_{12} \\ s_{12} & s_{22} \end{vmatrix} \geqslant 0. \tag{21.4.19}$$

We can see that this is a necessary condition for the stability of the solid, if we take $\boldsymbol{w}$ in (21.4.14) as a vector whose only non-zero components are w_1 and w_2. Again, the equilibrium is metastable, as it is in general, if the solid phase is strongly stable.

If some solutions are easily supersaturated, the equilibrium condition (21.2.17) could possibly be realised. This topic is lucidly discussed by Gibbs (1906), pp. 196–197, who makes some reference to experimental work. Little experimental work has been carried out on the effects of non-hydrostatic stress on crystal growth or the solution of crystals, but, with its interest to geochemists and others, it is a potentially interesting field. The author would welcome information concerning experimental work in this field.

Summary of important equations

A. *Fluid/solid mechanical boundary condition*

$$\boldsymbol{\Theta N} = -p\boldsymbol{N}. \tag{21.1.4}$$

B. *Defining equations*

$\tilde{v}_s$ = molar volume of solid at the reference state.

$\tilde{V}_s$ = molar volume of solid at the stressed, i.e., deformed state.

$$\tilde{v}_s^{(L)} = (\partial V^{(L)}/\partial M_s)_{T,p} \tag{21.2.5}$$

= partial molar volume of the solid material in the solution. M_s is the mass of solid (in moles) in the solution.

$\tilde{f}_s$ = molar Helmholtz free energy of the solid.

$$\tilde{f}_s^{(L)} = (\partial F^{(L)}/\partial M_s)_{T,p} \tag{21.2.9}$$

= partial molar free energy of the solid in the solution.

C. *Condition of chemical equilibrium*

$$\tilde{f}_s + p\tilde{V}_s = \mu_s^{(L)}, \tag{21.2.17}$$

where p is the pressure of the fluid and $\mu_s^{(L)}$ is the chemical potential of the solid in solution.

D. *Second-order formulae for the left-hand side of (21.2.17)*

$$\tilde{f}+p\tilde{V}_s=\tilde{f}^0+p\tilde{v}_s+\tfrac{1}{2}\tilde{v}_s c_{\alpha\beta}e_\alpha e_\beta \tag{21.4.9}$$

$$=\tilde{f}^0+p\tilde{v}_s+\tfrac{1}{2}\tilde{v}_s s_{\alpha\beta}(\Theta_\alpha+pE_\alpha)(\Theta_\beta+pE_\beta),$$

where $\tilde{f}^0$ is the molar Helmholtz free energy of the solid at the hydrostatic reference state whose pressure is p.

E. *The metastability of the solution/stressed solid*

1. $\mu_s^{(L)}(p)$ is the chemical potential of the solid in the solution, when the solution is in equilibrium with the solid *at its reference state*. That is,

$$\mu_s^{(L)}(p)=\tilde{f}^0+p\tilde{v}_s.$$

$$\mu_s^{(L)}-\mu_s^{(L)}(p)=\tfrac{1}{2}\tilde{v}_s c_{\alpha\beta}e_\alpha e_\beta \tag{21.4.10}$$

$$=\tfrac{1}{2}\tilde{v}_s e_\alpha(\Theta_\alpha+pE_\alpha) \tag{21.4.11}$$

$$=\tfrac{1}{2}\tilde{v}_s s_{\alpha\beta}(\Theta_\alpha+pE_\alpha)(\Theta_\beta+pE_\beta), \tag{21.4.12}$$

where $c_{\alpha\beta}$ and $s_{\alpha\beta}$ are the *isothermal* stiffness and compliance elements of infinitesimal strain theory. The above equations, however, are not dependent on the approximations of infinitesimal strain theory.

2. In the principal axis system,

$$\mu_s^{(L)}-\mu_s^{(L)}(p)=\tfrac{1}{2}\tilde{v}_s\{(p_1-p)^2 s_{11}+(p_2-p)^2 s_{22}$$

$$+2(p_1-p)(p_2-p)s_{12}\}, \tag{21.4.18}$$

if the co-ordinate axis system is the principal axis system of $\boldsymbol{\Theta}$, and $-p_1$, $-p_2$, $-p_3=-p$ (the fluid pressure), are its eigenvalues.

3. The condition of stability at constant hydrostatic pressure is

$$c_{\alpha\beta}w_\alpha w_\beta\geqslant 0;\qquad s_{\alpha\beta}w_\alpha w_\beta\geqslant 0, \tag{21.4.14}$$

where $\boldsymbol{w}$ is an arbitrary six-dimensional vector.

Examples

1. A solid is subject to homogeneous deformations, which obey the restriction $\boldsymbol{V}=\boldsymbol{E}$. If the reference state is hydrostatic at pressure p, show that, up to terms quadratic in the e_α,

$$u(s,\boldsymbol{e})-u_0(s)-p(1-J)=\tfrac{1}{2}c_{\alpha\beta}e_\alpha e_\beta,$$

$$=\tfrac{1}{2}s_{\alpha\beta}(\Theta_\alpha+pE_\beta)(\Theta_\beta+pE_\beta),$$

$$=\tfrac{1}{2}s_{\alpha\beta}(p_\alpha-pE_\alpha)(p_\beta-pE_\beta),$$

where, in the last expression, $1 \leqslant \alpha \leqslant 3$, $1 \leqslant \beta \leqslant 3$, and where the co-ordinate axis system is the principal axis system of $\mathbf{\Theta}$. The eigenvalues of $\mathbf{\Theta}$ are $-p_1$, $-p_2$, $-p_3$. $c_{\alpha\beta}$ and $s_{\alpha\beta}$ are the *adiabatic* stiffness and compliance elements of infinitesimal strain theory.

2. For hydrostatics, discuss the growth and solution of crystals in relation to the equality or otherwise of the chemical potential of the solid in solution and that of the crystalline solid.

   In particular, show that if a liquid is brought into contact with a solid, and if $\mu_s^{(L)} > \mu_s$, the chemical potential of the solid, then the crystal will tend to grow.

3. In section 2, δM was treated as if it were positive. Satisfy yourself that (21.2.16) would be obtained if δM were negative, and, also, if δm is positive at some points on the surface and negative at other points.

22

The thermodynamic stability of a phase

The conditions of stability of a phase depend on the boundary conditions. This arises from the fact that the property of stability for a phase cannot, in general, be divorced from that of its environment. Thus, necessary, but not sufficient, stability conditions may be derived by considering special cases of boundary conditions. Three cases are considered here: that of the fixed boundary; that of constant hydrostatic pressure; that of dead loading and rigid smooth constraints. In the last case, the constraints correspond, in the homogeneous case, to the geometrical restriction $\boldsymbol{V} = \boldsymbol{E}$, and are important in that the thermodynamic stability may be studied under the condition that mechanical equilibrium is maintained.

To introduce and illustrate the arguments, the stability of a fluid phase is first discussed.

22.1. The stability of a fluid phase

First, we shall discuss the stability of a fluid phase which is enclosed in a rigid adiabatic container. Then, we shall show that we obtain the same condition of stability when the fluid is held at a constant pressure – this will not be the case for solids.

Consider then a variation in the fluid phase, such that the changes in the entropy density s, and in J are δs and δJ, which may be functions of position in the fluid. Since the enclosure is rigid and adiabatic, we must have

$$\delta S = \int \delta s \, \mathrm{d}v = 0, \tag{22.1.1}$$

and

$$\delta V = \int \delta J \, \mathrm{d}v = 0, \tag{22.1.2}$$

where integration is throughout the reference state of the phase.

Now u, the internal energy per unit reference volume, may be regarded as a function of s and J and we may express δu, to second order in δs and

δJ, as follows,

$$\delta u = \frac{\partial u}{\partial s}\delta s + \frac{\partial u}{\partial J}\delta J + \frac{1}{2}\left\{\frac{\partial^2 u}{\partial s^2}(\delta s)^2 + 2\frac{\partial^2 u}{\partial s \partial J}\delta s\, \delta J + \frac{\partial^2 u}{\partial J^2}(\delta J)^2\right\} + \cdots. \tag{22.1.3}$$

Because of the restrictions (22.1.1) and (22.1.2), the general stability condition (19.1.17) will be

$$\delta U = \int \delta u \, \mathrm{d}v \geqslant 0. \tag{22.1.4}$$

On substituting (22.1.3) into this equation, and on introducing two undetermined parameters T_0 and $-p_0$ to take account of (22.1.1) and (22.1.2), we obtain, up to second order terms,

$$\delta U = \int \left\{\left(\frac{\partial u}{\partial s} - T_0\right)\delta s + \left(\frac{\partial u}{\partial J} + p_0\right)\delta J + \delta^2 u\right\} \mathrm{d}v \geqslant 0, \tag{22.1.5}$$

where $\delta^2 u$ is the second order term in (22.1.3). Hence, from (22.1.4), we first require that the equilibrium conditions be satisfied, that is,

$$T_0 = (\partial u/\partial s) = T, \tag{22.1.6}$$

and

$$p_0 = -(\partial u/\partial J) = p. \tag{22.1.7}$$

In words, T and $p = -\partial u/\partial J$ (see example 1) must be uniform throughout the phase. The condition (22.1.5) now reduces to

$$\int \delta^2 u \, \mathrm{d}v \geqslant 0. \tag{22.1.8}$$

From this equation we might be tempted to conclude that for stability, $\delta^2 u$ must be non-negative everywhere. We will be able to prove this, but at this stage it would be correct to draw this conclusion only if (22.1.8) could apply to any arbitrary part of the phase. However, it is true for the whole phase only, because we have taken as the boundary conditions for the region of integration those for a rigid adiabatic enclosure. Clearly any variation which satisfies these conditions for the whole phase may not be assumed to obey these for any arbitrary part of the phase.

To prove that $\delta^2 u$ must, for stability, be non-negative everywhere in the phase, let us imagine the phase to be divided into two parts 1 and 2.

We consider homogeneous, but different, virtual variations in each part, so that the accessory conditions are

$$v_1 \delta J_1 + v_2 \delta J_2 = 0, \tag{22.1.9}$$

$$v_1 \delta s_1 + v_2 \delta s_2 = 0, \tag{22.1.10}$$

where v_1 and v_2 are the reference volumes of the two parts.

Since a fluid phase is homogeneous, the second derivatives of u with respect to the variables s and J have uniform values throughout the phase. Hence, on using (22.1.9) and (22.1.10) to express δs_2 and δJ_2 as $-(v_1/v_2)\delta s_1$ and $-(v_1/v_2)\delta J_1$ respectively, we may write (22.1.8) as

$$\int_1 \delta^2 u_1 \, \mathrm{d}v_1 + (v_1/v_2)^2 \int_2 \delta^2 u_1 \, \mathrm{d}v_2 \geqslant 0, \tag{22.1.11}$$

where

$$\delta^2 u_1 = \tfrac{1}{2}\{(\partial^2 u/\partial s^2)(\delta s_1)^2 + 2(\partial^2 u/\partial s \partial J)\, \delta s_1 \, \delta J_1 + (\partial^2 u/\partial J^2)(\delta J_1)^2\}. \tag{22.1.12}$$

Equation (22.1.11) may now be written, since $\delta^2 u_1$ is independent of position, as

$$v_1 \delta^2 u_1 \{1 + (v_1/v_2)\} \geqslant 0, \tag{22.1.13}$$

or, since the other factors are positive,

$$\delta^2 u_1 \geqslant 0. \tag{22.1.14}$$

Since region 1 may be arbitrarily chosen, we may say that the stability condition is

$$\delta^2 u \geqslant 0, \tag{22.1.15}$$

everywhere in the phase, and for arbitrary virtual changes, since δs_1 and δJ_1 do not themselves have to satisfy the accessory conditions. It is thus a necessary condition of stability that $\delta^2 u$ is a non-negative definite quadratic form.

If, in the above treatment, we were to evaluate higher order terms in δu, we would, for instance, obtain the nth order term as

$$v_1 \delta^n u_1 \{1 - (-v_1/v_2)^{n-1}\}.$$

We may take part 1 as very much smaller than part 2. Each higher order term in δu is then approximately equal to that for region 1 alone. This justifies us in taking (22.1.14) as a stability condition and shows how to consider the stability if $\delta^2 u$ is zero.

Suppose, now, that we consider stability under the constant pressure condition. First, by considering, as we may, variations which satisfy (22.1.1) and (22.1.2), we can prove that T and p must be uniform. We can further consider variations, at constant pressure, which do not satisfy these restrictions, and the stability condition (19.1.17) becomes

$$\delta U - T\,\delta S + p\,\delta V \geqslant 0, \tag{22.1.16}$$

where, since p is constant, $-p\delta V$ is the exact expression for the work done, and as stated above T and p are uniform through the fluid. On expressing δU, δS, and δV in integral form, we arrive at the stability condition (22.1.8). Since δs and δJ are arbitrary functions of position, we see that (22.1.15) is an immediate consequence of (22.1.8). In fact, (22.1.15) is a necessary and sufficient condition for (22.1.8). Thus, although we have enlarged the set of possible variations, we have obtained the same stability condition as in the case of the adiabatic rigid boundary.

To return to the stability condition (22.1.15), we note, from (22.1.3), that $\delta^2 u$ has a cross term in $\delta s\,\delta J$. This cross term may be eliminated by transforming from u to f, the Helmholtz free energy density as follows. We see from (22.1.6), that, to first order,

$$\delta T = u_{ss}\,\delta s + u_{sJ}\,\delta J, \tag{22.1.17}$$

where $u_{ss} = \partial^2 u/\partial s^2$, $u_{sJ} = \partial^2 u/\partial s\,\partial J$. Thus, if this equation is used to eliminate δs in $\delta^2 u$, we obtain

$$\delta^2 u = \tfrac{1}{2}\{(1/u_{ss})(\delta T)^2 + (u_{JJ} - u_{sJ}/u_{ss})(\delta J)^2\}. \tag{22.1.18}$$

Now, we can show that the coefficient of $(\delta J)^2$ is equal to $f_{JJ} = \partial^2 f/\partial J^2 = -(\partial p/\partial J)_T$. For we may write

$$\frac{\partial^2 f}{\partial J^2} = -\frac{\partial(p, T)}{\partial(J, T)} = -\frac{\partial(p, T)}{\partial(J, s)}\Big/\frac{\partial(J, T)}{\partial(J, s)}, \tag{22.1.19}$$

on using the properties of Jacobians which were developed in chapter 11. In the last expression of (22.1.19), the numerator Jacobian is easily seen to be equal to $u_{JJ}u_{ss} - u_{sJ}^2$, while the denominator is equal to $(\partial T/\partial s)_J = u_{ss}$. We may now write (22.1.15) as

$$\begin{aligned}\delta^2 u &= \tfrac{1}{2}\{(\partial s/\partial T)_J(\delta T)^2 + f_{JJ}(\delta J)^2\}\\ &= \tfrac{1}{2}\{(\rho_0 c_v/T)(\delta T)^2 + n^{(T)}(\delta J)^2\} \geqslant 0,\end{aligned} \tag{22.1.20}$$

since

$$(\partial s/\partial T)_J = \rho_0 c_v/T, \tag{22.1.21}$$

and

$$\partial^2 f/\partial J^2 = -(\partial p/\partial J)_T = -v(\partial p/\partial V)_T = n^{(T)}, \qquad (22.1.22)$$

where c_v is the specific heat capacity at constant volume, ρ_0 is the mass per unit reference volume, and $n^{(T)}$ is the isothermal bulk modulus. Since we may choose variations where, first, $\delta T = 0$, $\delta J \neq 0$, and, then, where $\delta T \neq 0$ while $\delta J = 0$, we see that it is a condition of stability that both c_v and $n^{(T)}$ be non-negative. (See also example 2.)

We may express this condition somewhat loosely by saying that, for stable phases, an inflow of heat must not reduce the temperature (non-negative c_v), and an increase of pressure must not increase the volume (non-negative $n^{(T)}$).

Where the virtual variations may include variations in the chemical composition from place to place in a phase, the stability of the phase against chemical changes may be similarly discussed. The restrictions on the virtual changes of density will be linear in the $\delta\rho_i$, the densities of the chemical components, and may, in cases where a chemical reaction is possible, include those of type (20.3.3) and (20.3.4). The restrictions determine the equilibrium condition that T, p, μ_i, $i = 1, 2, \ldots, \kappa$, be uniform through the system. The stability conditions will include some which, expressed loosely, requires that for a stable phase the chemical potential μ_i must not decrease with increasing concentration of component i. For further details, the reader may refer, for example, to Wilson (1957), although he does not use the simple undetermined parameter technique.

Before we leave this section, an interesting point should be made. In our method of deducing the stability condition (22.1.15), we used variations which involved a discontinuous change of density at the surface between regions 1 and 2. Such variations are acceptable for a fluid but not for a solid. Thus, this method may not be used in seeking a stability condition, similar to (22.1.15), which must be satisfied at every point in the solid phase.

22.2. Discussion of stability

In studying the thermodynamic stability of a phase, some of the possible variations chosen may be such that the conditions of mechanical equilibrium are violated during these variations. These conditions require the vanishing of, first, the resultant of all forces acting on the system, and, second, the moments of these forces about any point. For example, in the well-known case of a compressed strut, the condition of moments is violated in variations which involve a shear change of shape, or a rigid body rotation about an axis perpendicular to the forces producing the compression.

To illustrate this point, let us consider an example which may include the above case of a compressed strut. Suppose we have a solid whose shape, under the compressive forces, is that of a cuboid whose edges have lengths a, b and c, respectively, in the x, y and z directions. We shall take the compressive forces as equal and opposite, and acting normally on the two faces of area ab, perpendicular to the z-axis. Each force is uniformly distributed over the face on which it acts, and its magnitude is F. For simplicity, we shall assume that the condition of dead loading applies, that is, under any variation, the force acting on every surface element of the two faces remains constant. The compressed strut case usually discussed would be that where c is very much greater than a or b.

Now consider a variation of shape, in the form of a simple shear, given by

$$\delta\boldsymbol{D} = \begin{bmatrix} 1 & 0 & \delta\beta \\ 0 & 1 & 0 \\ 0 & 0 & 1 \end{bmatrix}. \tag{22.2.1}$$

It is immediately clear that the moment condition of mechanical equilibrium is violated in the variation. It is also clear that the behaviour of the solid is mechanically unstable, since a non-zero turning moment, or torque, of magnitude $Fc\,\delta\beta$ is produced, and the sense of this torque is such as to tend to remove the solid, by rotation, further from its equilibrium position. Of course, if the forces were extensive, instead of compressive, a torque would be produced, but of a sense which would tend to restore the solid to its initial orientation.

In some theoretical treatments of such problems, attempts are made to remove this difficulty by considering only those variations $\delta\boldsymbol{D}$ which do not violate the mechanical equilibrium conditions. Such variations must satisfy the condition that $\delta\boldsymbol{D\Theta}$ be symmetric (see, e.g., Truesdell and Noll, 1965).

However, such a restriction on the allowed variations would, for example, prevent us from studying the effects of such compressive stresses on a solid state phase transition which is accompanied by a simple shear change of shape. An example of this is the ortho- to clino-enstatite phase transition (Coe, 1970; see also chapter 25).

In fact, the above difficulty is simply and completely overcome for any variation at all of the intrinsic elastic strain, homogeneous or inhomogeneous (McLellan, 1976b). Mechanical equilibrium is maintained by the use of smooth geometrical constraints. The reaction forces at these constraints produce a torque balancing that due to the applied forces responsible, as above, for the state of strain of the phase being studied. In the example above, these are the forces F producing the compression of the solid. The particular constraints used (see section 5 of this chapter) are such that homogeneous strain variations are restricted to those which may be described by triangular matrices $\boldsymbol{H}$. Since any homogeneous strain variation differs only by a rigid body rotation from its unique triangular factor $\boldsymbol{H}$, we may, under these constraints, test the stability of the phase for any homogeneous intrinsic strain. Similarly, any inhomogeneous strain variation is equivalent to a unique inhomogeneous strain, which satisfies the constraints, followed by a rigid body rotation. Thus, the stability of any state of a phase may be tested against any possible elastic strain variation, while mechanical equilibrium is maintained throughout.

For fluids another type of possible violation of mechanical equilibrium conditions may be considered. In terms of attempts, in the theory of continua, to treat solids and fluids on the same basis, this difficulty is rather puzzling, although the physical reasons are simple. The particular point arose in a treatment of equilibrium and stability developed by Coleman and Noll (1959). They were seeking a general local condition of stable thermal equilibrium to be applicable to continuous media, solid or fluid. By local condition we mean one such as (22.1.15), which must hold at every point in a medium. Their postulated general local condition of stable equilibrium is now no longer regarded as such, as, for example, Coleman and Noll (1964) have since pointed out that its application to a fluid is not appropriate. We shall discuss the reasons for this conclusion.

When applied to a fluid, their postulated general local condition is equivalent, *for our discussion only*, to the application of the condition (19.1.17) to a fluid and a strongly stable state. That is,

$$\delta U - \delta W - T\delta S > 0, \tag{22.2.2}$$

for variations from a hydrostatic fluid state, under dead load conditions. In other words the forces applied to the exterior surface which produce the pressure for the state in question are to be held constant, during the variation, on every surface element. Suppose the fluid is initially in the shape of a cube of edge length a. Three concurrent edges define the x, y, and z-axes. If the pressure is p, forces uniformly distributed on each face, of magnitude pa^2, must act in equal and opposite pairs on the six faces of the cube. The variation to be considered is one in which the cube is deformed to a cuboid, of edge lengths αa, βa and γa in the x, y, and z directions. The cuboid is to have the same volume as the cube, and, therefore, $\alpha\beta\gamma = 1$. Under the dead load conditions, the forces remain normal to the faces and of magnitude pa^2. However the force per unit area is not p, but is, for example, $p/\beta\gamma$ on the two faces normal to the x-axis. Such a set of forces could produce a uniform diagonal Cauchy stress tensor whose diagonal elements are $-p/\beta\gamma$, $-p/\alpha\gamma$, $-p/\alpha\beta$.

Now, if we apply the condition (22.2.2) to such a variation, carried out isothermally, then since the volume and temperature are unchanged δU and δS are both zero, and the condition reduces to

$$\delta W < 0, \tag{22.2.3}$$

or, since the forces on the faces are constant,

$$pv\{3-(\alpha+\beta+\gamma)\}<0, \tag{22.2.4}$$

where $v = a^3$. Since $\alpha\beta\gamma = 1$, we may write the quantity in curly brackets as

$$3\{(\alpha\beta\gamma)^{\frac{1}{3}}-(\alpha+\beta+\gamma)/3\}, \tag{22.2.5}$$

which is always negative, since the arithmetic mean of three positive quantities (as are α, β, γ) is always greater than the geometric mean. Hence, it is a consequence of the Coleman and Noll postulate that the pressure of a stable fluid must be positive.

The objection to this result is that to prove it the variation must be carried out under the unrealistic dead load conditions, that is, as seen above, the fluid is required to sustain a non-hydrostatic stress. Thus, the condition of mechanical equilibrium is violated, since a fluid at rest cannot sustain such a stress. In this example, the forces remained normal to the surfaces, but we can see that, in general, the force on a surface element cannot be held constant, since if the orientation of the surface element changes the force must adjust itself to remain always normal.

The difficulty of treating solid and fluid alike, in the above respects, is related to the fact that however we deform a fluid its symmetry group remains unaltered. For a fluid this group is the unimodular group, that is, the group of all non-singular $\boldsymbol{D}$ of determinant $+1$. On the other hand it is a simple matter to describe a deformation of a solid which changes its symmetry group to a sub-group, on taking the deformed state as a new reference state. To specify a symmetry group may not be sufficient and the puzzle may be to find what is needed beside.

In the following sections of this chapter we shall study thermodynamic stability by the application of (19.1.17) to the following boundary conditions: the fixed boundary condition, that is, where $\delta\boldsymbol{X} = 0$ everywhere on the boundary (this is sometimes known as the boundary condition of place); constant hydrostatic pressure; dead loading together with the geometrical constraints mentioned above. The condition of dead loading is sometimes known as the condition of traction. The geometrical constraints may be described as a condition of place. These are applied to two points only, or, strictly, three, if we consider, as is usual, a fixed origin in the phase – that is, if translations are forbidden. Thus, the last boundary condition to be studied may be termed a mixed boundary condition of traction and place.

Since variations satisfying the rigid boundary conditions are possible allowed variations for any other boundary conditions, the stability conditions necessary for a rigid boundary condition are also necessary stability conditions for any other boundary condition.

22.3. The stability of a solid phase under the fixed boundary condition

We know that equilibrium for a fluid phase is possible only if the pressure is uniform throughout. A solid phase, on the other hand, can be in equilibrium under conditions of non-uniform stress. For many practical purposes, such equilibrium is stable – for example, a great number of bridges and buildings, subject to inhomogeneous stresses, have stood and served mankind for many centuries.

In this section we shall use the criterion (19.1.17) to consider the thermodynamic stability of a solid phase against variations which satisfy the rigid boundary condition,

$$\delta \boldsymbol{X} = 0, \tag{22.3.1}$$

everywhere on the boundary surface. The variations will be further restricted to those which satisfy the adiabatic condition,

$$\delta S = \int_{v} \delta s \, \mathrm{d}v = 0, \tag{22.3.2}$$

for the entropy S of the whole phase. Integration is over the reference region of the solid.

No external work can be done, nor can any heat transfer take place to or from the surroundings, when these conditions apply. Thus while they are appropriate to a completely isolated system, we are free to consider such variations as possible for a solid under any experimental conditions.

With the above restrictions, (19.1.17) becomes

$$\delta U \geqslant 0. \tag{22.3.3}$$

In order to find an expression for δU, let us regard u, the internal energy per unit reference volume, as a function of s and the nine variables $D_{\alpha\beta}$.

Usually, of course, u is regarded as a function of s and only six strain co-ordinates, but, since these co-ordinates are functions of the elements of $\boldsymbol{D}$, no difficulty arises. Using (6.4.14), we may write

$$\begin{aligned} du &= T\,ds + B_{\alpha\beta}\,dD_{\beta\alpha} \\ &= T\,ds + \text{trace}(\boldsymbol{B}\,d\boldsymbol{D}). \end{aligned} \tag{22.3.4}$$

(We know that the $B_{\alpha\beta}$ are not all independent of each other, because of the requirement that $\boldsymbol{DB}$ be symmetric (see (6.4.15b)). This requirement is equivalent to three relations, and is related to the fact that no work is done if $\delta\boldsymbol{D}$ describes a rigid body rotation. However, this property of $\boldsymbol{B}$ introduces no complications in the following treatment.) We may now write (22.3.3) as

$$\delta U = \int_v \left\{ \frac{\partial u}{\partial s}\,\delta s + \frac{\partial u}{\partial D_\gamma}\,\delta D_\gamma + \frac{1}{2}\left[\frac{\partial^2 u}{\partial s^2}(\delta s)^2 + \frac{\partial^2 u}{\partial s\,\partial D_\gamma}\,\delta s\,\delta D_\gamma \right.\right.$$
$$\left.\left. + \frac{\partial^2 u}{\partial D_\gamma\,\partial D_\mu}\,\delta D_\gamma\,\delta D_\mu \right] + \cdots \right\} dv \geq 0. \tag{22.3.5}$$

In this equation, we are using an abbreviated notation for the elements of $\boldsymbol{D}$, according to the following scheme,

$$\gamma \leftrightarrow \alpha\beta$$

$$\left.\begin{array}{llllllllll} \gamma: & 1 & 2 & 3 & 4 & 5 & 6 & 7 & 8 & 9 \\ \alpha\beta: & 11 & 22 & 33 & 23 & 13 & 12 & 32 & 31 & 21 \end{array}\right\} \tag{22.3.6}$$

The restriction (22.3.2) can be taken care of by introducing, in the usual way, an undetermined multiplier T_0. If we neglect terms in (22.3.5) of cubic or higher order in the variation, (22.3.5) becomes

$$\delta U = \int_v \left\{ \left(\frac{\partial u}{\partial s} - T_0 \right) \delta s + \frac{\partial u}{\partial D_\gamma}\,\delta D_\gamma + \delta^2 u \right\} dv \geq 0. \tag{22.3.7}$$

Hence, first of all, we must have the uniform temperature condition

$$T_0 = \frac{\partial u}{\partial s}. \tag{22.3.8}$$

The other first order term must be non-negative, that is, in unabbreviated notation,

$$\int_v (\partial u/\partial D_{\alpha\beta})\,\delta D_{\alpha\beta}\,dv = \int_v B_{\beta\alpha}\{\partial(\delta X_\alpha)/\partial x_\beta\}\,dv \geq 0, \tag{22.3.9}$$

where the second integrand follows from (22.3.4) and from

$$\delta D_{\alpha\beta} = \partial(\delta X_\alpha)/\partial x_\beta. \tag{22.3.10}$$

Equation (22.3.9) may be written as

$$\int_v \{\partial(B_{\beta\alpha}\delta X_\alpha)/\partial x_\beta\}\, dv - \int_v \delta X_\alpha (\partial B_{\beta\alpha}/\partial x_\beta)\, dv \geqslant 0, \tag{22.3.11}$$

as may be seen by differentiating the product $B_{\beta\alpha}\delta X_\alpha$ in the first integral. We may now show that this first integral vanishes by applying the divergence theorem (5.2.2) to it, with $f_\beta = B_{\beta\alpha}\delta X_\alpha$. This integral is thus converted to a surface integral over σ, the bounding surface of v, the integrand being $n_\beta B_{\beta\alpha}\delta X_\alpha$, where $\boldsymbol{n}$ is the outward drawn unit normal. Since $\delta \boldsymbol{X}$ is zero everywhere over σ, this surface integral is zero. Hence (22.3.11) becomes

$$-\int \delta X_\alpha (\partial B_{\beta\alpha}/\partial x_\beta) dv \geqslant 0. \tag{22.3.12}$$

Therefore, since $\delta \boldsymbol{X}$ is arbitrary *inside* v, we must have

$$\frac{\partial B_{\beta\alpha}}{\partial x_\beta} = 0, \tag{22.3.13}$$

everywhere inside the solid. This is the well known equilibrium condition for the case of no body forces. The reader may consult example 7 of chapter 6, on this matter.

[By the way, it would not be difficult to include the case of body forces in this treatment. This inclusion would result in the subtraction from δU of the mechanical potential energy increment

$$\int_v \psi^b_\alpha \delta X_\alpha \, dv \tag{22.3.14}$$

where ψ^b_α is the body force per unit reference volume ($= Jf^b_\alpha$). Equation (22.3.12) would become

$$-\int_v \delta X_\alpha \left\{ \frac{\partial B_{\beta\alpha}}{\partial x_\beta} + \psi^b_\alpha \right\} dv \geqslant 0. \tag{22.3.15}$$

From this equation, we obtain the equilibrium condition

$$\frac{\partial B_{\beta\alpha}}{\partial x_\beta} + \psi^b_\alpha = 0, \quad \alpha = 1, 2, 3, \tag{22.3.16}$$

which expresses the requirement that the total force on a volume element must be zero (see example 7, chapter 6). This result is equivalent to (5.4.3).]

Since we have shown that the first order terms must vanish, the stability condition (22.3.7) may be written as

$$\int_v \delta^2 u \, dv \geqslant 0. \tag{22.3.17}$$

Now if the second partial derivatives in $\delta^2 u$ were uniform through the phase, and if we could choose an arbitrary but uniform variation, we could reduce (22.3.17) to

$$\delta^2 u \geqslant 0, \tag{22.3.18}$$

at every point in the solid. However (22.3.1) is not consistent with a uniform variation $\delta \boldsymbol{D}$, just as (22.3.2) is inconsistent with a uniform δs. Further, as noted at the end of section 1, we cannot employ a method generalised from that used in the case of a fluid, to derive (22.1.15) from (22.1.8). Thus, even for a uniform phase, we cannot prove that (22.3.18) is a valid necessary stability condition for the rigid boundary condition.

However, it is possible to derive from (22.3.17) local conditions of stability, some of which are related to the specific heat and to the speeds of acoustic waves. But before we discuss this further, we shall show how to transform (22.3.17) to a very useful form. This involves introducing f, the Helmholtz free energy density (per unit reference volume) and generalising the method used in section 1 to obtain (22.1.20). Since, from (22.3.4), $T = (\partial u/\partial s)_{\boldsymbol{D}}$, we may write

$$\begin{aligned}\delta T &= (\partial^2 u/\partial s^2)_{\boldsymbol{D}} \delta s + (\partial^2 u/\partial s \, \partial D_\alpha) \delta D_\alpha \\ &= u_{ss} \delta s + u_{s\alpha} \delta D_\alpha,\end{aligned} \tag{22.3.19}$$

in an obvious notation. This equation is correct to first order, which is sufficiently accurate for our purpose. On solving this equation for δs and substituting in $\delta^2 u$, we obtain

$$2\delta^2 u = (1/u_{ss})(\delta T)^2 + \{u_{\alpha\beta} - (u_{s\alpha} u_{s\beta}/u_{ss})\} \delta D_\alpha \, \delta D_\beta. \tag{22.3.20}$$

We notice, just as in section 1, that cross terms $\delta T \delta D_\alpha$ do not occur. Now

$$1/u_{ss} = (\partial s/\partial T)_{\boldsymbol{D}} = (\rho_0 c_{\boldsymbol{e}}/T), \tag{22.3.21}$$

because a measurement of specific heat at constant $\boldsymbol{D}$ is the same as measuring it at constant $\boldsymbol{e}$. Further, we can show that the terms quadratic in the δD_α in (22.3.20) are equal to $(\partial^2 f/\partial D_\alpha \, \partial D_\beta) \, \delta D_\alpha \, \delta D_\beta$. We do this by using the properties of Jacobians. Let us define, for this occasion only, $\boldsymbol{b} = \boldsymbol{B}^T$, so that $b_\alpha (= \partial u/\partial D_\alpha = \partial f/\partial D_\alpha)$ is conjugate to D_α. (We note from

(22.3.4) that it is $B_{\alpha\beta}$ which is conjugate to $D_{\beta\alpha}$!). We may then write

$$\frac{\partial^2 f}{\partial D_\alpha \partial D_\beta} = \left(\frac{\partial b_\beta}{\partial D_\alpha}\right)_T = \frac{\partial(b_\beta, \bar{D}_\alpha, T)}{\partial(D_\alpha, \bar{D}_\alpha, T)}$$

$$= \frac{\partial(b_\beta, \bar{D}_\alpha, T)}{\partial(D_\alpha, \bar{D}_\alpha, s)} \bigg/ \frac{\partial(D_\alpha, \bar{D}_\alpha, T)}{\partial(D_\alpha, \bar{D}_\alpha, s)}, \tag{22.3.22}$$

where $\bar{D}_\alpha$ denotes the ordered set obtained by deleting $\bar{D}_\alpha$ from the set $D_1, D_2, \ldots, D_9$, and where the cancelling rule (11.2.5) for Jacobians has been used. On rearranging the variables, (22.3.22) may be expressed as

$$\frac{\partial^2 f}{\partial D_\alpha\, \partial D_\beta} = \frac{\partial(b_\beta, T, \bar{D}_\alpha)}{\partial(D_\alpha, s, \bar{D}_\alpha)} \bigg/ \frac{\partial(T, \mathbf{D})}{\partial(s, \mathbf{D})}. \tag{22.3.23}$$

In the numerator Jacobian, we see that the eight variables in the set $\bar{D}_\alpha$ are common to both sets of variables. Written as a 10×10 determinant, it thus has eight diagonal elements equal to unity, with all other elements in the corresponding rows and columns zero. Hence it reduces to the following 2×2 principal minor

$$\begin{vmatrix} (\partial b_\beta/\partial D_\alpha) & (\partial T/\partial D_\alpha) \\ (\partial b_\beta/\partial s) & (\partial T/\partial s) \end{vmatrix} = \begin{vmatrix} u_{\alpha\beta} & u_{s\alpha} \\ u_{s\beta} & u_{ss} \end{vmatrix}. \tag{22.3.24}$$

Since the denominator Jacobian of (22.3.23) is equal to $(\partial T/\partial s)_{\mathbf{D}} = u_{ss}$, we have finally shown that (22.3.20) may be written as

$$2\delta^2 u = (\rho_0 c_\mathbf{e}/T)(\delta T)^2 + (\partial^2 f/\partial D_\alpha\, \partial D_\beta)\, \delta D_\alpha\, \delta D_\beta. \tag{22.3.25}$$

By considering variations for which $\delta T \neq 0$, and $\delta \mathbf{D} = 0$, we see from (22.3.17), and using (22.3.25), that a necessary stability condition is

$$\rho_0 c_\mathbf{e}/T \geqslant 0. \tag{22.3.26}$$

If we choose $\delta T = 0$, $\delta \mathbf{D} \neq 0$, then it is a necessary condition that

$$\int_v (\partial^2 f/\partial D_\alpha\, \partial D_\beta)\, \delta D_\alpha\, \delta D_\beta\, \mathrm{d}v \geqslant 0. \tag{22.3.27}$$

If, in (22.3.17), we choose variations such that $\delta s = 0$, $\delta \mathbf{D} \neq 0$, we obtain another condition, namely,

$$\int_v (\partial^2 u/\partial D_\alpha\, \partial D_\beta)\, \delta D_\alpha\, \delta D_\beta\, \mathrm{d}v \geqslant 0. \tag{22.3.28}$$

Equations (22.3.27) and (22.3.28) could be termed, respectively, the isothermal and adiabatic stability conditions. In these inequalities, the δD_α are arbitrary except that they must be such that the rigid boundary condition (22.3.1) is obeyed. We may write these two stability conditions in one equation as follows,

$$\int_v A_{\alpha\beta\mu\nu}\,\delta D_{\alpha\beta}\,\delta D_{\mu\nu}\,\mathrm{d}v \geqslant 0 \qquad (22.3.29)$$

where

$$A_{\alpha\beta\mu\nu} = (\partial B_{\beta\alpha}/\partial D_{\mu\nu})_s^* \quad or \quad (\partial B_{\beta\alpha}/\partial D_{\mu\nu})_T^*. \qquad (22.3.30)$$

That is, $A_{\alpha\beta\mu\nu}$ is a kind of adiabatic or isothermal stiffness quantity.

We have still obtained only one local stability condition, that for the specific heat, (22.3.26). A further local stability condition may be derived from (22.3.29). This condition is that

$$A_{\alpha\beta\mu\nu} m_\alpha m_\mu n_\beta n_\nu \geqslant 0, \qquad (22.3.31)$$

at every point in the phase, where $\boldsymbol{m}$ and $\boldsymbol{n}$ are completely arbitrary three dimensional vectors. These two necessary conditions, i.e., isothermal and adiabatic, are known as the Hadamard conditions or inequalities. We shall not give the details of this derivation here. The reader is referred to Truesdell and Noll (1965), who describe fully an elegant and rigorous derivation, due to Noll, of these inequalities.

We shall discuss these conditions a little further. First, as has been pointed out in section 2, stability conditions derived from the rigid boundary conditions are necessary stability conditions for any other boundary conditions. Thus the Hadamard inequalities have a very general application and are very important results. Further, although this has not so far been emphasised, they apply to an inhomogeneous or homogeneous deformation state of the solid. Both the treatment in this section and Noll's derivation are general, in the sense that the state under study may be inhomogeneous. The quantities $A_{\alpha\beta\mu\nu}$ may thus be functions of position. If we now take the state under study as the reference state, we may prove, using a result like (16.2.4), that

$$(\partial^2 f/\partial D_{\alpha\beta}\,\partial D_{\mu\nu})^* = A_{\alpha\beta\mu\nu} = \mathring{c}_{\alpha\beta\mu\nu} + \Theta_{\beta\nu}^*\delta_{\alpha\mu}, \qquad (22.3.32)$$

where, since we have differentiated f, $\boldsymbol{A}$ and $\mathring{\boldsymbol{c}}$ are isothermal matrices. An identical relation between the adiabatic matrices is obtained by differentiating u. At a given point in the phase $\boldsymbol{\Theta}^*$, $\mathring{\boldsymbol{c}}$, and $\boldsymbol{A}$ are all measured at the local reference state determined by the state under study, and, if this latter state is inhomogeneous, are functions of position.

It was noted at the end of the examples of chapter 17 that the phase velocities for the three possible modes of an acoustic wave are given, for a wave direction described by the unit vector $\boldsymbol{n}$, by the eigenvalues of the following 3×3 matrix

$$\Omega_{\alpha\nu} = \mathring{c}_{\alpha\beta\mu\nu}n_\beta n_\mu + (\Theta^*_{\beta\mu}n_\beta n_\mu)$$
$$= (\partial B_{\beta\alpha}/\partial D_{\mu\nu})^* n_\beta n_\mu$$
$$= A_{\alpha\beta\mu\nu}n_\beta n_\mu \qquad (22.3.33)$$

The adiabatic form of $\boldsymbol{\Omega}$ is considered appropriate for acoustic propagation, and, in this case, we see that the Hadamard adiabatic inequality is equivalent to the condition that the adiabatic form of $\boldsymbol{\Omega}$ must be a non-negative definite matrix at every point of the phase. (It should also be noted that the isothermal form of $\boldsymbol{\Omega}$ must be non-negative definite everywhere.)

The eigenvalues of $\boldsymbol{\Omega}$, which are proportional to the square of the speeds of propagation, must thus be non-negative. Hence we arrive at the familiar condition that the wave speeds must be real for a stable phase.

We emphasise again that these stability conditions are termed necessary conditions. They are not sufficient conditions. This is not only because they are derived under particular boundary conditions, but because they are obtained by using only a limited set of variations. For example, in Noll's proof of the Hadamard inequality, the variations used were limited to those of the form

$$\delta\boldsymbol{X} = \boldsymbol{m}\phi(\boldsymbol{x}), \qquad (22.3.34)$$

where $\boldsymbol{m}$ is an arbitrary fixed vector, and $\phi(\boldsymbol{x})$ a scalar function, arbitrary except that it is zero everywhere on the boundary surface. This variation is not, of course, the most general arbitrary variation satisfying the rigid boundary condition.

Finally we may comment on the evaluation of the quantities in (22.3.32). As is commonly the case, the solid phase may be physically and chemically homogeneous at some standard reference state. A convenient one is a zero stress state. f and u would then possibly be known as functions of strain variables, such as the elements of $\boldsymbol{D}$ or $\boldsymbol{\eta}$, where these are referred to the standard reference state. When the state under study is inhomogeneous, and when we have taken it as the reference state, we must find f and u at every point in terms of the strain variables referred to the reference state at that point (the local reference state). The method of finding f and u as functions of strain variables referred to an arbitrary reference state, when they are correspondingly known for a standard reference state, is illustrated in the examples at the end of this chapter. Once f and u have been so determined, the isothermal and adiabatic forms of the matrix $\boldsymbol{A}$ may be obtained by differentiation.

22.4. The stability of a solid phase under constant hydrostatic pressure

It is a simple matter to imagine a physical situation in which the hydrostatic pressure of a solid remains constant. It may be suspended in a fluid (which, of course, does not interact chemically with the solid), suitably acted on by a smooth piston, to which a constant force is applied. Or, it may be immersed in a fluid body so large that the pressure remains effectively constant for small volume changes of the solid.

One might ask: why not generalise this section by considering the solid to be held so that it is under a uniform Cauchy stress tensor which is held constant? This would require, however, that the stress vector $\boldsymbol{t}$ (see chapter 5) be prescribed everywhere on the boundary of the solid. We cannot do this, since the deformed positions $\boldsymbol{X}+\delta\boldsymbol{X}$ cannot be known in advance for any arbitrary variation. The same applies to the direction and area of any surface element, so that it is clearly impossible, in general, to prescribe $\boldsymbol{t}$ in advance, to suit this requirement. For a discussion of such boundary conditions see Truesdell and Noll (1965, p. 125 *et seq.*).

We shall apply the fundamental condition of thermodynamic stability, (19.1.17), to adiabatic variations of the phase under constant hydrostatic pressure. The state under study will be taken as the reference state, and it need not be assumed that the material is homogeneous. Consider an arbitrary adiabatic variation $\delta\boldsymbol{X}$, in which the triangular factor of the deformation matrix, $\boldsymbol{E}+\delta\boldsymbol{D}$, at any point, is given by $\boldsymbol{E}+\delta\boldsymbol{h}$, where $\delta\boldsymbol{h}$ will be a function of position. The work done in this variation is given by

$$\begin{aligned}\delta W &= -p\int_v \delta J \,\mathrm{d}v\\ &= -p\int_v \{\delta h_1+\delta h_2+\delta h_3+\delta h_2\delta h_3+\delta h_3\delta h_1+\delta h_1\delta h_2\}\,\mathrm{d}v,\end{aligned} \tag{22.4.1}$$

to second order terms only. This result follows from the fact that

$$\begin{aligned}1+\delta J &= \det(\boldsymbol{E}+\delta\boldsymbol{D}) = \det(\boldsymbol{E}+\delta\boldsymbol{h})\\ &= (1+\delta h_1)(1+\delta h_2)(1+\delta h_3).\end{aligned} \tag{22.4.2}$$

On introducing an undetermined parameter T_0, for the adiabatic condition (22.3.2), the stability condition (19.1.17) reduces to

$$\int_v \{[(\partial u/\partial h_\alpha)^* + pE_\alpha]\delta h_\alpha + [(\partial u/\partial s)^* - T_0]\delta s + \delta^2 u + p\delta^2 J\}\,\mathrm{d}v \geqslant 0, \tag{22.4.3}$$

where $\delta^2 J$ denotes the second order terms in the integrand of (22.4.1). As before, the first order terms must vanish, so that (22.4.3) reduces to

$$\int_v (\delta^2 u + p\delta^2 J)\, dv \geqslant 0. \tag{22.4.4}$$

Since, in $\delta^2 u$, the partial derivatives of u with respect to the variables h_α are evaluated at the reference state, where $\boldsymbol{D} = \boldsymbol{V} = \boldsymbol{E}$, they may formally be written in terms of $\boldsymbol{e}$ instead of $\boldsymbol{h}$. Thus

$$(\partial^2 u/\partial h_\alpha \partial h_\beta)^* = (\partial^2 u/\partial e_\alpha \partial e_\beta)^* = \hat{c}^{(s)}_{\alpha\beta}. \tag{22.4.5}$$

On substituting this in (22.4.4), and on comparison of the coefficients of $\delta h_\alpha \delta h_\beta$ with (16.1.5), we see that (22.4.4) may be written as

$$\int_v \{(\partial^2 u/\partial s^2)^*(\delta s)^2 + (\partial^2 u/\partial e_\alpha \partial s)^* \delta h_\alpha\, \delta s + c^{(s)}_{\alpha\beta}\, \delta h_\alpha\, \delta h_\beta\} \geqslant 0, \tag{22.4.6}$$

where $\boldsymbol{c}^{(s)}$ is the adiabatic 6×6 stiffness matrix of infinitesimal strain theory.

If we eliminate δs in favour of δT, we may transform (22.4.6) to

$$\int \{(\rho_0 c_{\boldsymbol{e}}/T)(\delta T)^2 + c^{(T)}_{\alpha\beta}\, \delta h_\alpha\, \delta h_\beta\}\, dv \geqslant 0, \tag{22.4.7}$$

where $\boldsymbol{c}^{(T)}$ is the isothermal 6×6 stiffness matrix of infinitesimal strain theory. To perform the required transformation of $\delta^2 u$, we follow the same method as in section 3, apart from the fact that we are using the six strain variables h_α instead of the nine D_α.

We may now obtain the local condition (22.3.26), the isothermal condition

$$\int_v c^{(T)}_{\alpha\beta}\, \delta h_\alpha\, \delta h_\beta\, dv \geqslant 0, \tag{22.4.8}$$

and the adiabatic condition

$$\int_v c^{(s)}_{\alpha\beta}\, \delta h_\alpha\, \delta h_\beta\, dv \geqslant 0. \tag{22.4.9}$$

If we have an inhomogeneous phase, $\boldsymbol{c}^{(s)}$ and $\boldsymbol{c}^{(T)}$ are not independent of position, and we cannot prove that they must be non-negative definite for stability. Nor can we require that the stability conditions (22.4.8) and (22.4.9) apply to any part of the phase. We cannot do this because of the restriction, on (19.1.17), that T must have the same value at every point on the boundary surface of the region considered where heat is exchanged with the surroundings. For arbitrary variations this condition clearly cannot be satisfied for any inner part of the solid.

However, in many cases of importance in the study of materials, homogeneous phases are under consideration. In such cases, $\boldsymbol{c}^{(s)}$ and $\boldsymbol{c}^{(T)}$ are independent of position. Since the boundary condition of constant pressure permits us to choose $\delta \boldsymbol{h}$ as an arbitrary homogeneous variation, we can prove that $\boldsymbol{c}^{(s)}$ and $\boldsymbol{c}^{(T)}$ must be non-negative definite, for a stable homogeneous phase.

A final point concerning the method: why not describe the variation by $\delta \boldsymbol{e}$ in the beginning? This would require that any arbitrary deformation $\delta \boldsymbol{X}$ produce a deformation $\delta \boldsymbol{D}$ which is triangular at every point. This is clearly not possible for inhomogeneous variations. However, we can start with an arbitrary $\delta \boldsymbol{X}$, which determines $\delta \boldsymbol{D}$ at every point, and we can factorise $\boldsymbol{E}+\delta \boldsymbol{D}$ uniquely to obtain its triangular factor.

22.5. The stability of a phase under dead loading with smooth constraints

In section 2 we discussed the difficulty that in variations of a solid under dead loading, the mechanical equilibrium condition of moments may be violated. We shall now describe and use smooth constraints on the solid. Throughout any variation, the reaction forces produced will maintain mechanical equilibrium.

The constraints to be applied to the solid are such that variations are subject to the following two restrictions.

(1): $$\delta X_2 = \delta X_3 = 0, \tag{22.5.1}$$

at a given reference state point a whose co-ordinates are $(\alpha, 0, 0)$, where $\alpha \neq 0$.

(2): $$\delta X_3 = 0, \tag{22.5.2}$$

at a given reference state point b of co-ordinates $(0, \beta, 0)$, where $\beta \neq 0$.

We shall also, as usual, take a fixed origin. These constraints mean that in any deformation the image point A of a will remain on the x_1 axis, while the image point B of b will remain on the x_1x_2 plane.

We see that any inhomogeneous deformation whatsoever may be uniquely related to an inhomogeneous deformation which satisfies these constraints. Consider an inhomogeneous deformation which does not satisfy either constraint. Then A, the image point of a, will not lie on the x_1 axis, nor will B, the image of b, lie on the x_1x_2 plane. It is obvious that we may follow such an inhomogeneous deformation by a unique rigid body rotation which will bring A on to the x_1-axis and B on to the x_1x_2 plane. The product of the deformation and the rotation thus satisfies the constraints.

Another way of stating this property is that any deformation, inhomogeneous or not, may be uniquely factorised into the product of a deformation, satisfying the above constraints, and a rigid body rotation.

Thus, even if we limit ourselves to variations which satisfy the above two constraints we can test the thermodynamic stability of a solid against any intrinsic elastic strain variation whatsoever.

How do these smooth constraints maintain the mechanical equilibrium condition of moments? Since they are smooth constraints, the reaction force exerted at the point A (image of a) will be at right angles to the x_1 axis, that is, normal to the allowed displacements $\delta \boldsymbol{X}$. Such a reaction can produce a torque vector with components in the x_2 and x_3 directions. The reaction force at B must be normal to the allowed displacements and will have a component only in the x_3 direction. This reaction force can produce a torque vector with components in the x_1 and x_2 directions. The total torque vector produced by these two reaction forces may be in any direction at all, and can balance any torque vector produced by the forces loading the solid. Thus, mechanical equilibrium will be maintained through any variation subject to the above constraints. It is perhaps worth remarking that if only one constraint of the two above were applied, then mechanical equilibrium could not be assured for all variations. Both constraints are thus necessary.

Let us now consider the thermodynamic stability of a solid phase under dead load conditions. To do this we consider adiabatic variations which satisfy the constraints (22.5.1) and (22.5.2). The state under study is taken as the reference state. As in section 4, consider an arbitrary adiabatic variation $\delta \boldsymbol{X}$, in which the variation $\delta \boldsymbol{D}$ of the deformation matrix may be a function of position.

If we have dead loading on every element of the exterior surface of the solid, we see from (6.4.21) that the Boussinesq stress tensor must remain constant in any variation. The work done in the variation is exactly

$$\delta W = \int_v B_{\beta\alpha} \delta D_{\alpha\beta} \, \mathrm{d}v. \tag{22.5.3}$$

The reaction forces of the constraints, being normal to the displacements, do no work. Thus, the stability condition (19.1.17) will, on following the methods used in the previous sections, reduce to

$$\int_v \delta^2 u \, \mathrm{d}v \geqslant 0. \tag{22.5.4}$$

This leads to the local specific heat condition (22.3.26), to the isothermal condition

$$\int_v (\partial^2 f/\partial D_\alpha\, \partial D_\beta)^* \,\delta D_\alpha\, \delta D_\beta \,\mathrm{d}v \geq 0, \tag{22.5.5}$$

and to the adiabatic condition

$$\int_v (\partial^2 u/\partial D_\alpha\, \partial D_\beta)^* \,\delta D_\alpha\, \delta D_\beta \,\mathrm{d}v \geq 0, \tag{22.5.6}$$

for variations subject to the constraints.

If the phase is homogeneous at the state in question, then the partial derivatives in the above two equations are independent of position. The boundary conditions permit arbitrary homogeneous variations subject to the smooth constraints. It is a simple matter to verify that for homogeneous deformations these constraints correspond to the geometrical restriction $\boldsymbol{V} = \boldsymbol{E}$. As discussed in chapter 13, we may then describe homogeneous deformations, subject to this restriction, by a triangular matrix $\boldsymbol{h}$ which is equal to $\boldsymbol{e}$ (see (13.1.3 and 4)).

Since, for example,

$$(\partial^2 u/\partial D_\alpha\, \partial D_\beta)^* = (\partial^2 u/\partial e_\alpha\, \partial e_\beta)^* = \hat{c}^{(s)}_{\alpha\beta}, \tag{22.5.7}$$

for α and β from $1, 2, \ldots, 6$, we may therefore say that, for the stability of a homogeneous phase, the 6×6 stiffness matrices $\hat{\boldsymbol{c}}^{(s)}$ and $\hat{\boldsymbol{c}}^{(T)}$ must be non-negative definite.

Summary of important equations

Necessary stability conditions:

A. *Fluid*

$$c_v \geq 0, \qquad n^{(T)} \geq 0$$

B. *Fixed boundary condition*

$$\rho_0 c_{\boldsymbol{e}}/T \geq 0. \tag{22.3.26}$$

$$\int_v (\partial^2 f/\partial D_\alpha\, \partial D_\beta)^* \,\delta D_\alpha\, \delta D_\beta \,\mathrm{d}v \geq 0. \tag{22.3.27}$$

$$\int_v (\partial^2 u/\partial D_\alpha\, \partial D_\beta)^* \,\delta D_\alpha\, \delta D_\beta \,\mathrm{d}v \geq 0. \tag{22.3.28}$$

Hadamard inequalities, to be satisfied at every point,

$$A_{\alpha\beta\mu\nu} m_\alpha m_\mu n_\beta n_\nu \geq 0. \tag{22.3.31}$$

for arbitrary three-component vectors $\boldsymbol{m}$ and $\boldsymbol{n}$.

$$A_{\alpha\beta\mu\nu}=(\partial^2 f/\partial D_{\alpha\beta}\,\partial D_{\mu\nu})^*=(\partial B_{\beta\alpha}/\partial D_{\mu\nu})^*_T,$$

or

$$=(\partial^2 u/\partial D_{\alpha\beta}\,\partial D_{\mu\nu})^*=(\partial B_{\beta\alpha}/\partial D_{\mu\nu})^*_s. \qquad (22.3.30)$$

C. *Constant pressure condition*

$$c_{\boldsymbol{e}}\geqslant 0,\ \text{at every point} \qquad (22.3.26)$$

$$\int_v c^{(T)}_{\alpha\beta}\,\delta h_\alpha\,\delta h_\beta\,\mathrm{d}v\geqslant 0, \qquad (22.4.8)$$

$$\int_v c^{(s)}_{\alpha\beta}\,\delta h_\alpha\,\delta h_\beta\,\mathrm{d}v\geqslant 0, \qquad (22.4.9)$$

for arbitrary variations $\delta\boldsymbol{X}$; $\boldsymbol{E}+\delta\boldsymbol{h}$ is the unique triangular factor of $\delta\boldsymbol{D}=\partial(\delta\boldsymbol{X})/\partial\boldsymbol{x}$. For a homogeneous phase $\boldsymbol{c}^{(T)}$, $\boldsymbol{c}^{(s)}$ must be non-negative definite.

D. *Dead-load conditions with smooth constraints*

$$c_{\boldsymbol{e}}\geqslant 0. \qquad (22.3.26)$$

$$\int_v (\partial^2 f/\partial D_\alpha\,\partial D_\beta)^*\,\delta D_\alpha\,\delta D_\beta\,\mathrm{d}v\geqslant 0. \qquad (22.5.5)$$

$$\int_v (\partial^2 u/\partial D_\alpha\,\partial D_\beta)^*\,\delta D_\alpha\,\delta D_\beta\,\mathrm{d}v\geqslant 0, \qquad (22.5.6)$$

for arbitrary variations subject to the constraints (22.5.1) and (22.5.2). For a homogeneous phase $\hat{\boldsymbol{c}}^{(s)}$ and $\hat{\boldsymbol{c}}^{(T)}$ must be non-negative definite.

Examples

1. In hydrostatics, u is a function of s and J; $U=uv_0$, and so $U(S,V)=u(S/v_0,V/v_0)v_0$. Hence show that

   $T=\partial U/\partial S=\partial u/\partial s;$

   $-p=\partial U/\partial V=\partial u/\partial J.$

2. Use (11.3.36) to show that c_p, the specific heat at constant pressure, must be non-negative for a stable fluid.
3. Suppose we have two reference states, denoted by $\boldsymbol{x}$ and $\boldsymbol{x}'$, such that

   $\boldsymbol{x}'=\boldsymbol{D}_0\boldsymbol{x}.$

If we have another state, defined by $\boldsymbol{X}$, use

$$\frac{\partial X_\alpha}{\partial x_\beta} = \frac{\partial X_\alpha}{\partial x'_\mu}\frac{\partial x'_\mu}{\partial x_\beta}$$

to show that

$$\boldsymbol{D} = \boldsymbol{D}'\boldsymbol{D}_0,$$

where $\boldsymbol{D}$ and $\boldsymbol{D}'$ are the deformation matrices of the state $\boldsymbol{X}$ with respect to the states $\boldsymbol{x}$ and $\boldsymbol{x}'$ respectively.

Examples 4–9 inclusive refer to the situation of example 3.

4. Use (2.3.10) and (6.4.21), together with the fact that the force on a surface element in the state $\boldsymbol{X}$ is given by

$$\mathrm{d}\boldsymbol{F} = \boldsymbol{B}^t\,\mathrm{d}\boldsymbol{\sigma} = \boldsymbol{B}'^t\,\mathrm{d}\boldsymbol{\sigma}',$$

to show that

$$\boldsymbol{B}' = J'(\boldsymbol{D}')^{-1}\boldsymbol{\Theta}.$$

We note that $\boldsymbol{\Theta}$ is independent of the reference state, as it should be. $J' = \det(\boldsymbol{D}')$.

5. If f is the Helmholtz free energy density per unit reference volume, referred to the reference state $\boldsymbol{x}$, and f' that, for the same state $\boldsymbol{X}$, referred to the reference state $\boldsymbol{x}'$, prove that

$$f = J_0 f', \quad \text{where } J_0 = \det(\boldsymbol{D}_0).$$

6. Prove that

$$\boldsymbol{\eta} = \boldsymbol{D}_0^t\boldsymbol{\eta}'\boldsymbol{D}_0 + \boldsymbol{\eta}_0,$$

where $\boldsymbol{\eta}$ and $\boldsymbol{\eta}_0$ are the finite strain tensors of the states $\boldsymbol{X}$ and $\boldsymbol{x}'$ referred to the reference state $\boldsymbol{x}$, and $\boldsymbol{\eta}'$ is the finite strain tensor of state $\boldsymbol{X}$ referred to the reference state $\boldsymbol{x}'$.

7. A crystal has cubic symmetry at a zero stress state. Its free energy density, up to second order terms, is given by (18.5.1) with $p^* = 0$. Take $\boldsymbol{D}_0 = \lambda\boldsymbol{E}$, and show that

$$\boldsymbol{\eta} = \lambda^2\boldsymbol{\eta}' + \boldsymbol{E}(\lambda^2 - 1)/2.$$

If, for brevity, we write $c_{\alpha\beta}$ instead of $\mathring{c}^{(T)}_{\alpha\beta}$, the stiffness elements measured at state $\boldsymbol{x}$, prove that, up to second order terms,

$$\begin{aligned} f'(T, \boldsymbol{\eta}') = \lambda^{-3}\{&f_0(T) + (\tfrac{3}{8})(c_{11} + 2c_{12})(\lambda^2 - 1)^2\} \\ &+ \{(\lambda^2 - 1)/2\lambda\}c_{11}(\eta'_1 + \eta'_2 + \eta'_3) \\ &+ \tfrac{1}{2}\lambda\{c_{12}(\eta'_1 + \eta'_2 + \eta'_3)^2 + (c_{11} - c_{12})(\eta'^2_1 + \eta'^2_2 + \eta'^2_3) \\ &+ c_{44}(\eta'^2_4 + \eta'^2_5 + \eta'^2_6)\}. \end{aligned}$$

Hence show that, as far as f' is concerned, the crystal in the state $\boldsymbol{x}'$ has cubic symmetry.

8. For example 7, what is the pressure at the reference state $\boldsymbol{x}'$? What are the isothermal stiffness matrix elements measured at the state $\boldsymbol{x}'$?
9. For the crystal of example 7, obtain the function f', up to second order terms, for the two cases where

$$(a)\quad \boldsymbol{D}_0 = \begin{bmatrix} a & 0 & 0 \\ 0 & b & 0 \\ 0 & 0 & c \end{bmatrix}, \quad \text{and } (b)\ \boldsymbol{D}_0 = \begin{bmatrix} 1 & e & 0 \\ 0 & 1 & 0 \\ 0 & 0 & 1 \end{bmatrix}.$$

Find $\boldsymbol{\Theta}$ and the stiffness matrix elements for the state $\boldsymbol{x}'$ of (a) and (b).
10. Because the elements $D_{\alpha\beta}$ are partial derivatives $\partial X_\alpha/\partial x_\beta$, they must satisfy the compatibility relations

$$\partial D_{\alpha\beta}/\partial x_\gamma = \partial D_{\alpha\gamma}/\partial x_\beta = \partial^2 X_\alpha/\partial x_\beta\, \partial x_\gamma.$$

(a) Show that these are, of course, satisfied if $\boldsymbol{D}$ is independent of position.
(b) If $\boldsymbol{D}(=\boldsymbol{H})$ is triangular everywhere and is not independent of position, prove that H_{33} must be independent of x_1 and x_2; H_{22}, H_{23} must be independent of x_1; H_{11}, H_{12}, H_{13} may be dependent on all three coordinates x_1, x_2, x_3.

11.(a) Show that homogeneous deformations, subject to the following constraints, must be symmetric:

$$u_x(0, 1, 0) = u_y(1, 0, 0)$$

$$u_y(0, 0, 1) = u_z(0, 1, 0)$$

$$u_z(1, 0, 0) = u_x(0, 0, 1).$$

In these equations, the notation is such that, for example, $u_x(x, y, z)$ denotes the displacement vector component u_x at the reference point $\boldsymbol{x} = x, y, z$.

(b) Design a mechanical system by which these constraints could be applied to a solid. Hence, or otherwise, demonstrate that, if non-zero reaction forces must be applied by this system in order to maintain the moment condition of mechanical equilibrium, these reactions will do work on the solid. (*Hint:* Consider first the two-dimensional case, with the one constraint,

$$u_x(0, 1) = u_y(1, 0).)$$

23

Discussion of the elastic stability conditions

A direct proof is given that $\hat{\boldsymbol{c}}^{(s)}$ must be non-negative definite if $\hat{\boldsymbol{c}}^{(T)}$ is non-negative definite and if $c_{\boldsymbol{e}}$ is non-negative. Some properties of a non-negative definite symmetric matrix are discussed. Another expression of the stability conditions for a homogeneous phase is derived.

23.1. The stiffness matrices

For constant pressure, we have seen that the stability condition for a homogeneous phase requires that the stiffness matrices, isothermal and adiabatic, of infinitesimal strain theory, be non-negative definite. For dead loading under the constraints (22.5.1) and (22.5.2), the stiffness matrices $\hat{\boldsymbol{c}}^{(T)}$ and $\hat{\boldsymbol{c}}^{(s)}$ must be non-negative definite for the stability of a homogeneous phase. Thus the properties of such matrices are of some interest, and we shall discuss these in relation to the stability conditions.

First of all, from the treatment in chapter 22, one would expect that the above condition on the adiabatic stiffness matrix, in the two cases above, would be implied by the conditions on the isothermal stiffness matrix and the specific heat $c_{\boldsymbol{e}}$. We see, in another way, that this is true by considering the relation (16.3.9), between adiabatic and isothermal stiffness matrices. Let us multiply both sides of (16.3.9) by $x_\alpha x_\beta$ where x_α, $\alpha = 1, 2, \ldots, 6$, are the components of an arbitrary six-dimensional vector. Summation over α and β is implied.

We obtain

$$\begin{aligned} x_\alpha C^{(s)}_{\alpha\beta} x_\beta - x_\alpha C^{(T)}_{\alpha\beta} x_\beta &= (T/\rho_0 C_\gamma)(x_\alpha C^{(T)}_{\alpha\mu} M_\mu)(x_\beta C^{(T)}_{\beta\nu} M_\nu) \\ &= (T/\rho_0 C_\gamma)(x_\alpha C^{(T)}_{\alpha\beta} M_\beta)^2. \end{aligned} \qquad (23.1.1)$$

Thus, if $(T/\rho_0 C_\gamma)$ is non-negative and $\boldsymbol{C}^{(T)}$ is non-negative definite, we see that $\boldsymbol{C}^{(s)}$ must also be non-negative definite. This result applies immediately to $\hat{\boldsymbol{c}}^{(s)}$ and $\hat{\boldsymbol{c}}^{(T)}$. For the constant pressure case, if we wish to avoid the approximations of infinitesimal strain theory, we take the special case of (23.1.1), corresponding to the strain co-ordinates $\boldsymbol{e}$ and the stiffness matrix $\boldsymbol{c}$. On using (16.1.5), we may write (23.1.1) as

$$x_\alpha c^{(s)}_{\alpha\beta} x_\beta - x_\alpha c^{(T)}_{\alpha\beta} x_\beta = (T/\rho_0 c_{\boldsymbol{e}})(x_\alpha \hat{c}^{(T)}_{\alpha\beta} \hat{m}_\beta)^2, \qquad (23.1.2)$$

because the difference between $\boldsymbol{c}$ and $\hat{\boldsymbol{c}}$, given, in the hydrostatic case, by (16.1.5), is the same for both the adiabatic and isothermal matrices, and thus it cancels out in the left-hand side of (23.1.1). Hence, we have proved that the non-negative definite condition for the adiabatic stiffness matrix is implied by the conditions that the isothermal stiffness matrix be non-negative definite, and the specific heat $c_{\boldsymbol{e}}$ be non-negative.

23.2. Some properties of a non-negative definite matrix

A non-negative definite symmetric matrix $\boldsymbol{a}$ is one for which the quadratic form

$$x_\alpha a_{\alpha\beta} x_\beta \geqslant 0, \tag{23.2.1}$$

for a vector $\boldsymbol{x}$ which is arbitrary except that the vector, whose components are all zero, is excluded. From the discussion of a positive definite matrix in section 4 of chapter 2, and by particular reference to (2.4.20), we may easily see that for $\boldsymbol{a}$ to be non-negative definite, it is necessary and sufficient that its eigenvalues all be non-negative.

Since $\det(\boldsymbol{a})$ is equal to the product of the eigenvalues of $\boldsymbol{a}$, it is necessary, if $\boldsymbol{a}$ is to be non-negative definite, that

$$\det(\boldsymbol{a}) \geqslant 0. \tag{23.2.2}$$

We may use this to show that every principal minor of $\det(\boldsymbol{a})$ must be non-negative definite. For, if $\boldsymbol{a}$ is an $n \times n$ matrix, then in (23.2.1) we may choose vectors $\boldsymbol{x}$, arbitrary except that $x_{r+1} = x_{r+2} = \cdots x_n = 0$. We may thus prove that the $r \times r$ matrix, whose elements are given by $a_{\alpha\beta}$, where α and β run from 1 to r, must be non-negative definite. If we apply (23.2.2) to this matrix, and, if we let r range from 1 to n, we obtain, as follows, the necessary and sufficient conditions that $\boldsymbol{a}$ be non-negative definite:

$$a_{11} \geqslant 0; \begin{vmatrix} a_{11} & a_{12} \\ a_{21} & a_{22} \end{vmatrix} \geqslant 0; \cdots \begin{vmatrix} a_{11} & \cdots & a_{1n} \\ \vdots & & \vdots \\ a_{n1} & \cdots & a_{nn} \end{vmatrix} \geqslant 0. \tag{23.2.3}$$

We have proved that these conditions are necessary. The reader is referred to Ferrar (1941) for the proof that they are sufficient. We see that (21.4.19) is a special case of the second member of (23.2.3).

An interesting example of the above results is obtained by considering the following expression for the free energy obtained from (21.4.7), for a zero pressure state, using infinitesimal strain theory,

$$f = f^0 + \tfrac{1}{2}\Theta_\alpha e_\alpha = f^0 + \tfrac{1}{2} s_{\alpha\beta}\Theta_\alpha\Theta_\beta. \tag{23.2.4}$$

This expression is correct up to and including quadratic terms. $s_{\alpha\beta}$ is the isothermal compliance matrix and, since this is non-negative definite for a stable phase, we see that $f \geqslant f^0$. So, whether we 'push or pull' a stable solid isothermally from a zero pressure state, the Holmholtz free energy cannot decrease.

We should note that while we may prove that the diagonal elements of a stiffness or compliance matrix must be non-negative for a stable phase, the stability conditions do not imply that the off-diagonal elements must be non-negative. In fact it may be seen from, for example, Clark, 1966, that, for many materials, s_{12} is negative. All of the non-zero stiffness matrix elements quoted in this handbook are positive. However, this is not universally true – for ammonium dihydrogen phosphate (ADP), for instance, s_{12} and c_{12} are both negative (Nye, 1957).

If $\boldsymbol{a}$ is a positive definite matrix the inequality in (23.2.1) holds, and $\boldsymbol{a}$ has positive eigenvalues and is non-singular. In this case, $\boldsymbol{a}$ has a reciprocal $\boldsymbol{a}^{-1}$ whose eigenvalues are the reciprocals of those of $\boldsymbol{a}$ and are therefore positive. Thus, if a stiffness matrix is positive definite then so must be its reciprocal, the corresponding compliance matrix.

Suppose that the state of a phase is changed progressively from a strongly stable towards a weakly stable or unstable state. At least one eigenvalue of the stiffness matrix will tend to zero, and at least one eigenvalue of the compliance matrix will tend to diverge. From the result of example 13, chapter 2, we see that if an eigenvalue of the compliance matrix diverges, then so will some of its elements. In the case of some solid state phase transitions it is observed that as the instability, corresponding to the phase transition, is approached, some compliance elements do diverge. We shall be studying the properties of such phase transitions in later chapters.

23.3. Another form for the elastic stability conditions for a homogeneous phase

We illustrate this form for the case when $\hat{\boldsymbol{c}}^{(T)}$ is the appropriate non-negative definite matrix. We have

$$\det(\hat{\boldsymbol{c}}^{(T)}) = \frac{\partial(T, L_1, L_2, \ldots, L_6)}{\partial(T, e_1, e_2, \ldots, e_6)}. \tag{23.3.1}$$

The conditions (23.2.3) may be written, for strong stability only,

$$\frac{\partial(T, L_1, e_2, \ldots, e_6)}{\partial(T, e_1, e_2, \ldots, e_6)} > 0; \quad \frac{\partial(T, L_1, L_2, e_3, \ldots, e_6)}{\partial(T, e_1, e_2, e_3, \ldots, e_6)} > 0;$$

$$\ldots \frac{\partial(T, L_1, L_2, \ldots, L_6)}{\partial(T, e_1, e_2, \ldots, e_6)} > 0. \tag{23.3.2}$$

The first of these inequalities is, of course, $\hat{c}_{11}^{(T)} > 0$. Now in (23.3.2), let us divide the second inequality by the first, the third by the second and so on. If we apply the cancelling rule (11.2.5) for Jacobians, we obtain the strong stability conditions as follows – the first of which is the first member of (23.3.2),

$$(\partial L_1/\partial e_1)_{T,\bar{e}_1} > 0; \; (\partial L_2/\partial e_2)_{T,L_1,\bar{e}_{12}} > 0;$$
$$(\partial L_3/\partial e_3)_{T,L_1,L_2,\bar{e}_{123}} > 0; \cdots (\partial L_6/\partial e_6)_{T,\bar{L}_6} > 0, \quad (23.3.3)$$

where $\bar{e}_{123}$, for example, is the ordered set obtained from the set $e_1, e_2, \ldots, e_6$, by deleting e_1, e_2, e_3. The conditions (23.3.3) for strong stability have the advantage that they are expressed in terms of single derivatives, each one being, however, a derivative with a different set of variables held constant. These conditions, translated to infinitesimal strain theory, are sometimes seen in the literature, as are analogous conditions for chemical stability (Wilson, 1957). However, while they are elegant in form they are difficult to use. It is usually easier, when considering the stability of a phase, to relate this to the behaviour of the stiffness matrix and its eigenvalues. Also they are limited to the strong stability condition, involving positive definite matrices, because of the fact that since we took quotients of members of (23.3.2), none of them may vanish, except the last member.

Examples

1. Prove that, for a stable crystal of hexagonal symmetry,

$$c_{44} \geqslant 0; \qquad c_{11} \geqslant |c_{12}|; \qquad (c_{11}+c_{12})c_{33} \geqslant 2c_{13}^2.$$

2. Prove that, for a stable crystal of cubic symmetry,

$$c_{44} \geqslant 0; \qquad c_{11} \geqslant c_{12}; \qquad c_{11}+2c_{12} \geqslant 0.$$

3. Prove that a state of a stable phase, arrived at by any infinitesimal adiabatic deformation from a zero pressure hydrostatic state, has an internal energy not less than that of this state.

24

Phase transitions and instability

Some possible processes involved in phase transitions are discussed. Equilibrium conditions for solid state phase transitions are derived for the constant pressure and the dead load conditions. The generalised Clausius–Clapeyron equation is derived for transitions occurring under dead load conditions. Transitions, where the solid becomes thermodynamically unstable at, or very near to, the transition states, are considered, and limiting values of the isothermal compliance elements are obtained. The limiting value of the isothermal compressibility, for a constant pressure instability, is derived. The relation between the instabilities at the constant pressure and the dead load conditions is briefly described.

24.1. Phase transitions

In studying solid state phase transitions it is natural to consider the approach to the transition point as an approach to a thermodynamic instability, that is, to a state where the stability conditions derived in chapter 22 are not satisfied for some variation.

It is well known, however, that the process bringing about the transition can begin before the homogeneous phase in question becomes thermodynamically unstable. This happens, basically, because a local or surface instability occurs first. For example, in melting, it is thought that the melting process begins at a surface of the solid (Tallon, Robinson & Smedley, 1977) when the bulk of the solid is thermodynamically stable. In the liquid–vapour transition, the unstable (metastable) superheated states of the liquid are usually not reached, since instability occurs first at impurities such as dust particles, air bubbles, ionising radiation or particles. In that case the process of evaporation begins locally at the impurity, while the bulk liquid is thermodynamically stable. The transition in this case begins from a thermodynamic state of the bulk liquid which is well in the stable region, and finishes at a stable state of the vapour.

The first order transition of graphite to diamond provides interesting examples of possible processes in which the transition begins and ends at states which lie in the stable regions of the graphite and diamond phases

(Walker and Thrower, 1973). In this transition, it is not difficult with modern techniques to raise the temperature and pressure of the graphite phase so that its state is in the region where diamond is thermodynamically stable. If this is done, the reaction rate of the transition is so low that effectively no conversion occurs. It is possible to increase the reaction rate by adding an impurity catalyst such as nickel. On heating, a molten nickel graphite mixture, saturated with graphite, is produced, while the pure graphite is still stable. The production of this fluid will begin at surfaces of the graphite particles in the specimen. Diamond then crystallises, out of this melt, into a stable state. Thus, an intermediate fluid phase can bring about a transition between stable regions of the two phases.

The conversion of graphite to diamond has also been achieved without the addition of a catalyst. In one experimental procedure, this direct conversion was performed at static pressures exceeding 130 kilobars by flash heating the specimen to temperatures above about 3300 °K. It is believed that this process takes place via a liquid state of carbon, which is more dense than either graphite or diamond. Thermodynamic analysis of the activation volume predicts that such a dense intermediate state would increase the reaction rate. Again, the transition begins and ends at thermodynamically stable states.

On the other hand, there are many solid state phase transitions in which the material has merely to be brought to the transition temperature (usually at atmospheric pressure) for the material to transform rapidly, reversibly, completely and uniformly. If the temperature is restored to a value at which the original phase is stable, the transformation is reversed, rapidly, completely and uniformly. The crystal, in these processes, remains undamaged, and reverts to its original shape. Some examples of such processes are the α–β quartz transition, ferroelectric transitions, e.g., of barium titanate, anti-ferromagnetic transitions, order–disorder transitions in alloys and other crystals, e.g., of ammonium chloride. From the behaviour of the specific heat c_p, transitions of this type are commonly termed λ-transitions. When c_p is plotted graphically, it shows a characteristic lambda shape – a rapid rise on the low temperature side of the transition, followed by a very steep drop at the transition temperature. (For discussions of such processes, see e.g., Kittel, 1971.) It is not quite true that the transformation is so simply reversed, since sometimes a small range of temperature has been discovered in which the two phases may coexist. For example, in the α–β quartz transition, this range is about 1–2 °K at a transition temperature of 847 °K.

In the thermodynamic treatment of solid state phase transitions we shall first be concerned with finding the equilibrium condition between the two phases, under both hydrostatic and non-hydrostatic conditions. Further, if the phase transition occurs in such a way that a thermodynamically unstable state of the phase is at or very close to the transition state, then we shall be able to use the stability theory of chapter 22 to study the limiting values, as the transition and instability is approached, of such thermodynamic quantities as compliance elements, thermal expansion coefficients, electric susceptibilities (relevant to ferroelectric transitions) and specific heats. We shall be able to study these limiting values for the α–β quartz transition and the other rapidly reversible solid state transitions mentioned above – but not for the direct graphite–diamond conversion, unless it is ultimately proved that the graphite becomes thermodynamically unstable very close to the state where the intermediate liquid carbon phase is formed.

How can we find the equilibrium conditions between the two phases? First of all, we know that the onset of a transition (or of an instability) depends on the boundary conditions, that is, on the physical properties of the environment of the specimen. In order to be specific we shall determine the equilibrium conditions for the boundary conditions of constant pressure in the hydrostatic case, and of dead loading.

To determine the equilibrium conditions for one or other of these boundary conditions, it is necessary to find a reversible process between the two phases. In the case of fluids, it is a simple matter to find such a process, in which the transformation, from liquid to vapour, say, is carried out isothermally and reversibly. We simply change the volume slowly, at constant temperatures and pressure. If, in such a process, a small quantity of mass dm is converted from liquid at state c to vapour at state d, the change in the Helmholtz free energy is given by the work done on the specimen. Thus, if $\tilde{V}_c$ and $\tilde{V}_d$ are the specific volumes at the states c and d,

$$\mathrm{d}F = (\tilde{f}_d - \tilde{f}_c)\,\mathrm{d}m = -p(\tilde{V}_d - \tilde{V}_c)\,\mathrm{d}m, \tag{24.1.1}$$

and we therefore obtain the equilibrium condition,

$$\tilde{f}_c + p\tilde{V}_c = \tilde{f}_d + p\tilde{V}_d, \tag{24.1.2}$$

or

$$\tilde{g}_c = \tilde{g}_d. \tag{24.1.3}$$

In the next section, we shall discuss the equilibrium conditions for two phases of a solid state phase transition, at dead loading or constant pressure conditions.

24.2. Equilibrium conditions for solid state phase transitions

In section 1 we showed how to obtain the equilibrium conditions at a transition involving fluid phases. To do this we used a reversible process linking the two phases. For solid phases we shall now describe two possible reversible processes, both of which lead to the same equilibrium condition for the same boundary condition.

For the first such process, let us consider two phases, A and B, of a transition, $A \rightarrow B$, which are at equilibrium together at the transition states a and b respectively. It is reasonable to assume that we may have a specimen in which both phases coexist. These two phases would, in general, be separated by an interfacial zone in which the material would not be homogeneous, nor would it be in the form of phase A or phase B. This inhomogeneous zone would link and accommodate the different structural forms of the two phases. By control of the shape and size of the specimen, a small part of phase A, of mass $\mathrm{d}m$, could be transformed into the other phase B. One effect could be that the interfacial zone would move in the direction from phase B to A. For such an infinitesimal isothermal process, under the dead load condition, the change in the Helmholtz free energy would be

$$\mathrm{d}F = \mathrm{d}m(\tilde{f}_b - \tilde{f}_a) + \mathrm{d}F_i, \tag{24.2.1}$$

where F_i is the Helmholtz free energy of the interfacial zone. Since the process is isothermal, $\mathrm{d}F$ is equal to the work done on the specimen in the process. To evaluate this work we consider the specimen initially to be in the form of phase A at state a, and finally, to be in the form of phase B at state b. The dead loading forces acting on the surface of the specimen are, by definition, such that the force acting on every surface element is constant. We shall assume that in the transition the specimen remains in mechanical equilibrium. The contrary case will be discussed at the end of this section. At the homogeneous state a we assume that these forces are described by a uniform Boussinesq stress tensor. The force on a surface element is given by $\boldsymbol{B}^{\mathrm{t}} \cdot \mathrm{d}\boldsymbol{\sigma}$, where $\mathrm{d}\boldsymbol{\sigma}$ is the reference state of the surface element, and this force, for dead loading, is the constant force acting on $\mathrm{d}\boldsymbol{\Sigma}$, the image surface element of $\mathrm{d}\boldsymbol{\sigma}$, for any deformation occurring during the transformation of phase A to B. The work done in the above infinitesimal isothermal process is given by

$$dW = \int_\sigma \mathrm{d}u_\alpha\, B_{\beta\alpha}\, \mathrm{d}\sigma_\beta \tag{24.2.2}$$

$$= B_{\beta\alpha} \int_v \mathrm{d}D_{\alpha\beta}\, \mathrm{d}v, \tag{24.2.3}$$

since $\boldsymbol{B}$ is uniform over the reference state. To obtain (24.2.3) we have used the theorem (5.2.1).

If we now allow the homogeneous phase A to transform completely to the homogeneous phase B in a series of such infinitesimal steps, we may integrate (24.2.1) and (24.2.3) throughout the process. On equating the results we obtain, since F_i is zero at the beginning and at the end of the process,

$$m\,\Delta\tilde{f}=m(\tilde{f}_b-\tilde{f}_a)=vB_{\beta\alpha}\,\Delta D_{\alpha\beta}, \tag{24.2.4}$$

for the equilibrium condition, where $\Delta\boldsymbol{D}$ is the change in $\boldsymbol{D}$ occurring in the transition and $\boldsymbol{B}$ is the Boussinesq stress tensor for the state a. The quantities m and v are respectively the mass and reference volume of the specimen. Since $\tilde{f}=\tilde{v}f$, where $\tilde{v}$ is the specific reference volume v/m, (24.2.4) may be written as follows,

$$\Delta f=f_b-f_a=B_{\beta\alpha}\,\Delta D_{\alpha\beta}. \tag{24.2.5}$$

Finally, we may, without loss of generality, consider the constraints (22.5.1) and (22.5.2) to be applied. Equation (24.2.5) then becomes (see chapter 22, section 5)

$$\Delta f=L_\alpha\,\Delta e_\alpha, \tag{24.2.6}$$

where $\boldsymbol{L}$ is determined at state a from

$$L_\alpha=B_{\nu\mu}, \tag{24.2.7}$$

where α is the abbreviated symbol for $\mu\nu$ in the scheme (8.2.12). If a is the reference state, $\boldsymbol{B}$, of course, is equal to $\boldsymbol{\Theta}^{(a)}$, the Cauchy stress tensor at the state a, and we may express (24.2.6) as

$$f_b-f_a=\Theta_\alpha^{(a)}\,\Delta e_\alpha. \tag{24.2.8}$$

As an example, if state a is at hydrostatic pressure p, (24.2.8) becomes

$$\Delta f=-p\Delta(e_1+e_2+e_3)\simeq -p\,\Delta J. \tag{24.2.9}$$

The second reversible process which can lead to the equilibrium condition is as follows. This process is possible if the shape of the whole phase could be controlled, so that the transition would proceed uniformly and continuously throughout the whole bulk of the phase. Such a process, for the dead load condition, would clearly result in the same equilibrium condition (24.2.6), as in the first process described.

For transitions at constant pressure we may, by similar considerations, obtain the following well known equilibrium condition for the transition from state a' in phase A to state b' in phase B,

$$f_{b'} - f_{a'} = -p(J_{b'} - J_{a'}), \tag{24.2.10}$$

or

$$g_{a'} = g_{b'}, \tag{24.2.11}$$

where $g = f + pJ$ is the Gibbs free energy per unit reference volume. On multiplying (24.2.11) by $\tilde{v}$, the specific reference volume, it may be expressed in the familiar form

$$\tilde{g}_{a'} = \tilde{g}_{b'}, \tag{24.2.12}$$

where $\tilde{g}$ is the specific (or molar) Gibbs free energy, $\tilde{f} + p\tilde{V}$, $\tilde{V}$ being the specific volume. Here we have put primes on a and b to indicate that these states may not be the same as the states a and b which are linked under dead load conditions.

We must now examine our method of obtaining the equilibrium condition (24.2.6), to see to what sort of phase transitions it will apply. First, it is necessary that the dead load condition can be maintained. This would not be true in general if the transition zone were a liquid intermediate phase, as in the graphite–diamond process described in section 1. However, in principle there is no reason why we should not have graphite and diamond linked by a solid coherent transition zone. The time required to establish it might be very long, but, in principle, it is possible. The method applies to any transition in which we can establish a solid transition zone, and in which the structural changes remain coherent so that the deformations $\boldsymbol{u}$ may be described through the process. That is, the basic lattice determining the shape must remain coherent in the deformations, either in the first or the second of the two reversible processes described. The equilibrium condition will thus apply to a wide range of transitions. Strictly, whether or not it applies to a particular transition will depend on experimental verification.

In the next section we shall discuss the generalised Clausius–Clapeyron equation for phase transitions under the dead load condition.

An important point concerns the boundary conditions used in this section. Whether or not a transition will proceed at a given state of phase A will depend on the environment. For example, suppose A is at a state a' such that, at constant pressure, it will transform to phase B. Suppose, further, that in this transformation the specific volume increases. Let us assume that instead of being held at constant pressure the specimen, at state a', is imbedded in a stable elastic substance. It is clear that, in this environment, the transition will be inhibited and, in order that the transition may proceed, the temperature and pressure must be altered. That is, in this case, the state a' would not be the transition state. The boundary conditions of constant pressure and of dead loading are equivalent to surrounding the specimen by an environment which is neutrally stable, under the conditions considered, to changes of shape and size of the specimen. The boundary conditions we use thus provide specific conditions for studying the properties of a transition. For any other environment the free energy changes in the surrounding material must be taken into account in studying, for example, the equilibrium conditions for coexistence of the two phases at the transition, in that environment.

In deriving the condition (24.2.6), the smooth constraints (22.5.1) and (22.5.2) were considered to be applied. It was taken for granted that in the transition process the phases remained homogeneous, and hence that the change of shape could be described by $\Delta\boldsymbol{e}$, a matrix whose elements are uniform throughout the phase. However, this is clearly not true if non-zero reaction forces must be applied by the smooth constraints in order to maintain the moment condition of mechanical equilibrium. In this case, the reaction forces do no work, but they have the effect of producing an inhomogeneous deformation in the specimen. Suppose that, at the initial state, the specimen is in mechanical equilibrium under the dead load forces applied. At this state no reaction forces are applied and the specimen is homogeneous. If non-zero reaction forces are needed to maintain mechanical equilibrium at the final state (as well as during the transition), they will produce an inhomogeneous deformation in the specimen. These forces act at the origin and at the two constraint points on the x_1 and x_2 axes. Thus, we must write the equilibrium condition as

$$F_b - F_a = B_{\beta\alpha} \int_v \Delta D_{\alpha\beta}\, \mathrm{d}v. \tag{24.2.13}$$

Since state a is homogeneous, while state b is not, this may be written as

$$\int_v f_b\, \mathrm{d}v - v f_a = B_{\beta\alpha} \int_v \Delta D_{\alpha\beta}\, \mathrm{d}v. \tag{24.2.14}$$

This equilibrium condition would depend on the positions of the constraint points. This is another example of the fact that the behaviour of the transition depends on the environment – which includes, in this case, the constraints.

In chapter 28, we study the equilibrium conditions for the coexistence of the two types of Dauphiné twins in a specimen of quartz – an example of the general case of two phases existing in a solid which we have treated in this section. We shall show that the equilibrium condition of coexistence of the two twins, A and B, in one specimen is given by (24.2.5). This equation may be written as

$$g'_a = g'_b, \tag{24.2.15}$$

where g'_a is, for example, the generalised Gibbs free energy density g' for the state a, and where g' is defined as

$$g' = f - B_{\alpha\beta}D_{\beta\alpha}. \tag{24.2.16}$$

If (24.2.15) is not satisfied, the quartz specimen will untwin and we shall show that the direction of untwinning ($A \to B$, or $B \to A$) is determined by a minimum principle for g'.

The discussion of this minimum principle, which is given in chapter 28, is general and may be applied to the general case studied in this section. However the mathematical treatment is more complicated and so it is not given here. The reader is referred to chapter 28 for this treatment and for a discussion of the properties of the thermodynamic function g'.

24.3. The generalised Clausius–Clapeyron equation for coherent phase transitions under dead load

For a first order phase transition, the Clausius–Clapeyron equation relates dT_0/dp, on the transition curve in the two dimensional phase space of hydrostatic states, to $\Delta\tilde{s}$ and $\Delta\tilde{V}$, the changes in specific entropy and volume occurring in the transition. This equation is

$$dT_0/dp = \Delta\tilde{V}/\Delta\tilde{s}. \tag{24.3.1}$$

T_0, of course, is the temperature at which the transition occurs. This result may be generalised to the non-hydrostatic case of a coherent phase transition.

Equation (24.2.6) may be written

$$f_b - L_\alpha e_\alpha^{(b)} = f_a - L_\alpha e_\alpha^{(a)}, \tag{24.3.2}$$

where the states a and b are connected by the constant $\boldsymbol{L}$ path of the transition.

Suppose T_0 is changed to $T_0 + dT_0$, then, in order that the transition conditions may still be satisfied, $\boldsymbol{L}$ must change correspondingly. The strain matrices $\boldsymbol{e}^{(a)}$ and $\boldsymbol{e}^{(b)}$ will also, of course, change. Equation (24.3.2) must be satisfied, and we must have

$$d(f_a - L_\alpha e_\alpha^{(a)}) = d(f_b - L_\alpha e_\alpha^{(b)}). \tag{24.3.3}$$

Since $\mathrm{d}f = -s\,\mathrm{d}T + L_\alpha\,\mathrm{d}e_\alpha$, (24.3.3) reduces to

$$\Delta s\,\mathrm{d}T_0 + \Delta e_\alpha\,\mathrm{d}L_\alpha = 0. \tag{24.3.4}$$

We thus have one restriction on $\mathrm{d}T_0$ and the six components $\mathrm{d}L_\alpha$. If we change L_i by $\mathrm{d}L_i$, leaving all other L_α unchanged, then we find that

$$(\mathrm{d}T_0/\mathrm{d}L_i)_{\bar{L}_i} = -(\Delta e_i/\Delta s), \quad i = 1, 2, \ldots, 6. \tag{24.3.5}$$

This is the generalised Clausius–Clapeyron equation. For a zero stress state, (24.3.4) would be

$$\Delta s\,\mathrm{d}T_0 + \Delta e_\alpha\,\mathrm{d}\Theta_\alpha = 0, \tag{24.3.6}$$

if the state a is the reference state.

24.4. Phase instability

In discussing the stability of a homogeneous phase, the smooth constraints of (22.5.1) and (22.5.2) will usually be considered to be applied. Dead load or constant pressure boundary conditions will be considered. The state of a homogeneous phase will usually then be described by the seven co-ordinates $T, e_1, e_2, \ldots, e_6$, although we shall sometimes use the seven co-ordinates $T, L_1, L_2, \ldots, L_6$. The six strain quantities e_i are the elements of the triangular matrix $\boldsymbol{e}$. It will, however, often be convenient to regard these elements as the elements of a six-dimensional vector, which we shall denote by the same symbol $\boldsymbol{e}$, and which has the six elements $e_1, e_2, \ldots, e_6$. Such a vector will be called the strain vector $\boldsymbol{e}$.

In the following pages, we shall be using the eigenvectors of the stiffness matrices $\hat{\boldsymbol{c}}^{(T)}$ and $\boldsymbol{c}^{(T)}$. These eigenvectors will be chosen to be dimensionless. Since strain elements are dimensionless, these eigenvectors may be regarded as particular strain vectors. To emphasise this point, the symbol $\boldsymbol{e}$ will be used in denoting these eigenvectors. For example, the six (dimensionless) orthonormal eigenvectors of $\hat{\boldsymbol{c}}^{(T)}$ will be denoted by $\hat{\boldsymbol{e}}^{(i)}$, and those of $\boldsymbol{c}^{(T)}$ by $\boldsymbol{e}^{(i)}$, $i = 1, 2, \ldots, 6$.

A homogeneous phase, which is stable under the dead load condition, satisfies the conditions that $\hat{\boldsymbol{c}}^{(T)}$ be non-negative definite and that the specific heat $c_{\boldsymbol{e}}$ be non-negative. If there were at a stable state a six-dimensional vector $\boldsymbol{e}$ such that $\hat{c}^{(T)}_{\alpha\beta} e_\alpha e_\beta$ were zero, this would imply, from the obvious extension of (2.4.20) to six dimensions, that the stiffness matrix has a zero eigenvalue. The corresponding compliance matrix $\hat{\boldsymbol{s}}^{(T)}$ would then have an infinite eigenvalue, and, as a result, some of its elements would be infinite. Since infinite compliance elements and vanishing specific heats are unheard of in experimental studies of stable states, we may conclude that the strong stability conditions are satisfied in a stable phase. That is, $\hat{\boldsymbol{c}}^{(T)}$ is positive definite and c_e is positive, for the dead load condition. Similarly, for the constant pressure condition, in a stable phase $\boldsymbol{c}^{(T)}$ must be positive-definite.

As we change the state of a phase so as to approach a state which is thermodynamically unstable under the dead load condition, $\hat{\boldsymbol{c}}^{(T)}$ will change continuously until it has a zero eigenvalue. Since, even at transitions, a vanishing specific heat is unheard of, we shall assume that c_e remains positive. We shall say that limiting unstable points, such that one eigenvalue of $\hat{\boldsymbol{c}}^{(T)}$ vanishes, lie on an instability surface (for the dead load condition). Such surfaces will define the bounding surfaces, in a phase diagram, of the stable region of the phase in question. In non-hydrostatics, the phase diagram will be in seven-dimensional space defined by seven co-ordinates such as $T, e_1, \ldots, e_6$ or $T, L_1, \ldots, L_6$. An instability surface will be defined, for example, by

$$\hat{\lambda}_i(T, e_1, \ldots, e_6) = 0, \tag{24.4.1}$$

where $\hat{\lambda}_i$ is one of the six eigenvalues of $\hat{\boldsymbol{c}}^{(T)}$. We shall restrict ourselves to cases where only one eigenvalue vanishes, although the more general case may easily be treated. For example, if a phase has high crystal symmetry, two or more eigenvalues may be equal and must vanish together. In the case of a cubic crystal, the stiffness matrix $\boldsymbol{c}$ has two of its eigenvalues both equal to $(c_{11} - c_{12})$. Both of these must remain equal while the phase retains its cubic symmetry. Thus as an instability surface is approached from the stable region both eigenvalues must tend to zero together or remain positive and equal.

If we consider hydrostatic states only, and if stability at constant pressure is considered, the limiting surfaces of instability will be described by

$$\lambda_i(T, p) = 0, \tag{24.4.2}$$

where λ_i is an eigenvalue of $\boldsymbol{c}^{(T)}$, the stiffness matrix of infinitesimal strain theory. Equation (24.4.2) describes a curve in the 2-dimensional hydrostatic sub-space of the seven-dimensional phase space of non-hydrostatics.

If $\hat{\boldsymbol{c}}^{(T)}$ has one vanishing eigenvalue, $\hat{\lambda}_1$, say, there is a vector $\hat{\boldsymbol{e}}^{(1)}$ such that, for the infinitesimal vector $\mathrm{d}\hat{\boldsymbol{e}}^{(1)} = \hat{\boldsymbol{e}}^{(1)}\,\mathrm{d}\varepsilon$, where $\mathrm{d}\varepsilon$ is an infinitesimal quantity, we have

$$\mathrm{d}L_\alpha = \hat{c}^{(T)}_{\alpha\beta}\,\mathrm{d}\hat{e}^{(1)}_\beta = 0, \tag{24.4.3}$$

while, for a strain vector $\boldsymbol{e}'$ which is orthogonal to $\hat{\boldsymbol{e}}^{(1)}$, that is,

$$\hat{e}^{(1)}_\alpha e'_\alpha = 0, \tag{24.4.4}$$

we will have

$$\hat{c}^{(T)}_{\alpha\beta} e'_\alpha e'_\beta > 0. \tag{24.4.5}$$

Thus for isothermal variations, we see that at the dead load instability surface, defined by $\hat{\lambda}_1 = 0$, $\hat{\boldsymbol{e}}^{(1)}$ defines a 'direction' in which the shape can change initially in a neutrally stable manner, while all directions orthogonal to it describe variations of shape which are strongly stable. Similar considerations apply to the constant pressure instability surface defined by the vanishing of λ_1, an eigenvalue of $\boldsymbol{c}^{(T)}$. Here, if $\mathrm{d}\boldsymbol{e}^{(1)} = \boldsymbol{e}^{(1)}\,\mathrm{d}\varepsilon$, where $\boldsymbol{e}^{(1)}$ is the eigenvector corresponding to λ_1,

$$\mathrm{d}\Theta_\alpha = c^{(T)}_{\alpha\beta}\,\mathrm{d}e^{(1)}_\beta = 0, \tag{24.4.6}$$

and so $\boldsymbol{e}^{(1)}$ defines the initially neutrally stable direction, and all directions orthogonal to $\boldsymbol{e}^{(1)}$ describe strongly stable variations.

Thus, if we have transitions in which the whole of the phase uniformly approaches a neutrally stable state, at which state the transition begins, $\hat{\boldsymbol{e}}^{(1)}$ or $\boldsymbol{e}^{(1)}$ will define, under the appropriate boundary conditions, the initial change of shape resulting from the transition. In section 1 we have emphasised that the uniform approach of the whole phase to instability may be interrupted by the transition. However, many examples are known where compliance elements or electric or magnetic susceptibilities (these are generalised stiffness matrix elements appropriate when an electric or magnetic polarisation is introduced as a thermodynamic co-ordinate analogous to the strain elements) diverge at T_0, the transition temperature. Also the microscopic explanations of phase transitions such as the α–β quartz transition, ferroelectric and antiferromagnetic transitions, are based on processes which should occur homogeneously throughout the phase. One such explanation involves the vanishing of the frequency of an optical vibrational mode of the crystal lattice. Optical modes require that there be at least two sub-lattices, each of a different atom or ion. For example, the cubic crystal lattice of NaCl has two cubic sub-lattices, one of sodium ions, one of chlorine ions. The optical mode considered in phase transitions is analogous to that arising from the vibration of the two sub-lattices relative to each other, with all the sodium ions remaining rigidly on their cubic sub-lattice, and the chlorine sub-lattice also remaining rigidly cubic. A vanishing of the frequency of such a mode is obviously a uniform property extending throughout the phase. Thus we could expect that each transition state of the two phases would be at a corresponding thermodynamic instability. They may not exactly coincide – one reason could be that fluctuations in crystal properties may become very large near a transition, and thus induce local instabilities.

Such fluctuations are particularly well known at the critical point of condensation transitions in fluids.

In the next section we shall discuss the limiting values of compliance elements and of the compressibility as an instability surface is approached. Such limiting values will apply to the corresponding transition if the transition states are thermodynamically unstable. If a transition state is close enough to an instability, the limiting values will be a good approximation to those at the transition state.

24.5. Limiting values of the isothermal compliance matrix

Since the mathematics of the surfaces of instability is the same for constant pressure and for dead loading, we shall simplify the notation by studying the properties of a stiffness matrix $\boldsymbol{c}$, whose eigenvalue λ_1 vanishes.

In order to obtain the limiting values of the elements of $\boldsymbol{c}$, we need the following mathematical identity, which applies to a hermitian or real symmetric matrix,

$$c_{\alpha\beta} = \lambda_1 e_\alpha^{(1)} e_\beta^{(1)} + \lambda_2 e_\alpha^{(2)} e_\beta^{(2)} + \cdots + \lambda_6 e_\alpha^{(6)} e_\beta^{(6)}, \tag{24.5.1}$$

that is,

$$\boldsymbol{c} = \sum_{i=1}^{6} \lambda_i \boldsymbol{e}^{(i)} \boldsymbol{e}^{(i)}, \tag{24.5.2}$$

where the $\boldsymbol{e}^{(i)}$, $i = 1, 2, \ldots, 6$, are the orthonormal set of eigenvectors of $\boldsymbol{c}$. This important expression for a hermitian matrix can be proved by showing that the result of the operation of $\boldsymbol{c}$ on each of the $\boldsymbol{e}^{(j)}$, $j = 1, 2, \ldots, 6$, is the same as the result of the operation of the right-hand side of (24.5.2) on the corresponding $\boldsymbol{e}^{(j)}$. This result can immediately be seen to be true following directly, as it does, from the orthonormal nature of the $\boldsymbol{e}^{(i)}$. We can then prove that the operation of $\boldsymbol{c}$ and of the right-hand side of (24.5.2) on any arbitrary vector each results in the same vector. For suppose we express the arbitrary vector $\boldsymbol{e}$ as

$$\boldsymbol{e} = \sum_{j=1}^{6} a_j \boldsymbol{e}^{(j)} = \sum_{j=1}^{6} (\boldsymbol{e} \cdot \boldsymbol{e}^{(j)}) \boldsymbol{e}^{(j)}, \tag{24.5.3}$$

then

$$\boldsymbol{ce} = \sum_{j=1}^{6} \lambda_j a^{(j)} \boldsymbol{e}^{(j)}, \tag{24.5.4}$$

from the eigenvector property of $\boldsymbol{c}$. If $\boldsymbol{e}$ is now operated on by the right-hand side of (24.5.2), the resulting vector is again the right-hand side of (24.5.4). Thus, the two matrices of (24.5.2) must be equal.

Since the eigenvalues of $\boldsymbol{s}$, the reciprocal of $\boldsymbol{c}$, are λ_i^{-1}, with corresponding eigenvectors $\boldsymbol{e}^{(i)}$, we may write

$$s_{\alpha\beta} = \sum_{i=1}^{6} \lambda_i^{-1} e_\alpha^{(i)} e_\beta^{(i)}. \tag{24.5.5}$$

This equation is what is needed for studying the limiting values of the element $s_{\alpha\beta}$.

If, at a surface of instability, λ_1 tends to zero, while the other eigenvalues remain positive, then, *for those elements* $s_{\alpha\beta}$ *for which* $e_\alpha^{(1)} e_\beta^{(1)}$ *is not zero,*

$$s_{\alpha\beta} \to \lambda_1^{-1} e_\alpha^{(1)} e_\beta^{(1)}. \tag{24.5.6}$$

If $e_a^{(1)}$ is a non-zero element of $\boldsymbol{e}^{(1)}$,

$$s_{\alpha\beta}/s_{aa} \to e_\alpha^{(1)} e_\beta^{(1)}/(e_a^{(1)})^2. \tag{24.5.7}$$

Thus, when one eigenvalue vanishes, (24.5.7) determines the ratios of *diverging* compliance elements. Those elements $s_{\alpha\beta}$, such that $e_\alpha^{(1)} e_\beta^{(1)}$ is zero, will not diverge and their limiting values will be given by

$$s_{\alpha\beta} \to \sum_{i=2}^{6} \lambda_i^{-1} e_\alpha^{(i)} e_\beta^{(i)}. \tag{24.5.8}$$

As an example consider a crystal phase of the trigonal crystal class (see example 2). If we take the eigenvector $(1, 1, a, 0, 0, 0)$ to correspond to λ_1 of the above discussion, we see that the diverging elements of $\boldsymbol{s}$ are those for which α and β are restricted to the values 1, 2, 3. The limiting value for this 3×3 sub-matrix $\boldsymbol{s}'$ is given by

$$\boldsymbol{s}'/s_{11} = \begin{bmatrix} 1 & 1 & a \\ 1 & 1 & a \\ a & a & a^2 \end{bmatrix}, \tag{24.5.9}$$

and, for example

$$s_{33}/s_{32} = a. \tag{24.5.10}$$

The elements of $\boldsymbol{s}$ not belonging to this sub-matrix will, if not zero, be determined by the other eigenvalues and eigenvectors of $\boldsymbol{s}$.

For the dead load condition, the results of this section will apply to $\hat{\boldsymbol{c}}^{(T)}$ and $\hat{\boldsymbol{s}}^{(T)}$, while for constant pressure conditions they will apply to $\boldsymbol{c}^{(T)}$ and $\boldsymbol{s}^{(T)}$.

24.6. The limiting value of the isothermal compressibility

The bulk modulus n and its reciprocal, the compressibility κ, are quantities which are relevant to hydrostatic states, since, for example,

$$\kappa^{(T)} = -(\partial V/\partial p)_T V^{-1} = -(\partial J/\partial p)_T. \tag{24.6.1}$$

We may relate $\kappa^{(T)}$ to the isothermal compliance matrix of infinitesimal strain theory. Since

$$\mathrm{d}e_\alpha = s^{(T)}_{\alpha\beta}\,\mathrm{d}\Theta_\beta, \tag{24.6.2}$$

we have, for hydrostatic states,

$$E_\alpha\,\mathrm{d}e_\alpha = \mathrm{d}J = -E_\alpha E_\beta s^{(T)}_{\alpha\beta}\,\mathrm{d}p. \tag{24.6.3}$$

(A similar result holds for adiabatic changes.) Hence

$$\kappa^{(T)} = E_\alpha E_\beta s^{(T)}_{\alpha\beta} \tag{24.6.4}$$

$$= s^{(T)}_{11} + s^{(T)}_{22} + s^{(T)}_{33} + 2(s^{(T)}_{12} + s^{(T)}_{23} + s^{(T)}_{31}), \tag{24.6.5}$$

and

$$n^{(T)} = (E_\alpha E_\beta s^{(T)}_{\alpha\beta})^{-1}. \tag{24.6.6}$$

From (24.5.5) we see that

$$\kappa^{(T)} = \sum_{i=1}^{6} \lambda_i^{-1}(E_\alpha e^{(i)}_\alpha)^2$$

$$= \sum_{i=1}^{6} \lambda_i^{-1}(e^{(i)}_1 + e^{(i)}_2 + e^{(i)}_3)^2. \tag{24.6.7}$$

Thus, at an instability surface at constant pressure, if it is λ_1 which tends to zero, $\kappa^{(T)}$ will diverge, if

$$E_\alpha e^{(1)}_\alpha = e^{(1)}_1 + e^{(1)}_2 + e^{(1)}_3 \neq 0. \tag{24.6.8}$$

In the example, discussed in section 5, of a crystal of the trigonal class, $\kappa^{(T)}$ will diverge, unless $a+2=0$.

In cases of high crystal symmetry, $E_\alpha e^{(1)}_\alpha$ is often identically zero. For example, for a crystal of the cubic class, $E_\alpha e^{(i)}_\alpha$ is identically zero, except for the eigenvector

$$\boldsymbol{e} = (1, 1, 1, 0; 0, 0), \tag{24.6.9}$$

as may readily be shown. This eigenvector corresponds to the eigenvalue $c^{(T)}_{11} + 2c^{(T)}_{12}$ of $\boldsymbol{c}^{(T)}$, and to the eigenvalue $s^{(T)}_{11} + 2s^{(T)}_{12}$ of $\boldsymbol{s}^{(T)}$. Thus, if $c^{(T)}_{11} + 2c^{(T)}_{12}$ tends to zero, $\kappa^{(T)}$ will, in this case, diverge. In fact, $\kappa^{(T)}$ is equal to $3(s^{(T)}_{11} + 2s^{(T)}_{12})$, for a cubic crystal.

24.7. The relation between the instability surfaces at constant pressure and at dead loading

The two stiffness matrices which determine the instability surfaces at constant pressure and at dead load are related, for a state at hydrostatic pressure, by (16.1.5). Thus, for a six-dimensional vector $\boldsymbol{e}$, we may write

$$c^{(T)}_{\alpha\beta}e_\alpha e_\beta - \hat{c}^{(T)}_{\alpha\beta}e_\alpha e_\beta = 2p(e_1e_2+e_2e_3+e_3e_1)$$
$$= p[(e_1+e_2+e_3)^2-(e_1^2+e_2^2+e_3^2)]. \qquad (24.7.1)$$

Suppose that $\hat{\boldsymbol{e}}$ is such that $\hat{c}^{(T)}_{\alpha\beta}\hat{e}_\alpha\hat{e}_\beta$ is zero, that is, the state in question is at the instability surface for dead loading. In that case, $\hat{\boldsymbol{e}}$ is the eigenvector corresponding to the vanishing eigenvalue of $\hat{\boldsymbol{c}}^{(T)}$, and $\mathrm{d}\hat{\boldsymbol{e}} = \hat{\boldsymbol{e}}\,\mathrm{d}\varepsilon$ defines a neutrally stable infinitesimal path under dead loading (see 24.4.3). If p is positive, we see from (24.7.1) that this path will be strongly stable, for constant pressure conditions, if

$$(\hat{e}_1+\hat{e}_2+\hat{e}_3)^2 > \hat{e}_1^2+\hat{e}_2^2+\hat{e}_3^2. \qquad (24.7.2)$$

Similarly, if the state is at the instability surface for constant pressure, and if $\boldsymbol{e}\,\mathrm{d}\varepsilon$ is an eigenvector corresponding to the vanishing eigenvalue of $\boldsymbol{c}^{(T)}$, then the infinitesimal path $\boldsymbol{e}\,\mathrm{d}\varepsilon$ is strongly stable, for the dead load condition, if

$$e_1^2+e_2^2+e_3^2 > (e_1+e_2+e_3)^2. \qquad (24.7.3)$$

Thus, in general, unless there are states at which there are two different neutrally unstable paths – one for constant pressure and one for dead load, the instability surfaces for these two conditions will not have common states. Of course when p is zero, $\hat{\boldsymbol{c}}^{(T)} = \boldsymbol{c}^{(T)}$, and the two instability surfaces will touch there.

Summary of important equations

A. *Equilibrium condition for a transition under the dead load condition*

$$\Delta f = L_\alpha \Delta e_\alpha. \qquad (24.2.6)$$

B. *The generalised Clausius–Clapeyron equation for the dead load condition*

$$\Delta s\,\mathrm{d}T_0 + \Delta e_\alpha\,\mathrm{d}L_\alpha = 0, \qquad (24.3.4)$$

or

$$(\mathrm{d}T_0/\mathrm{d}L_i)_{\bar{L}_i} = -(\Delta e_i/\Delta s), \quad i=1,2,\ldots,6. \qquad (24.3.5)$$

C. *Isothermal stiffness and compliance quantities*

$$c_{\alpha\beta} = \sum_{i=1}^{6} \lambda_i e_\alpha^{(i)} e_\beta^{(i)}, \tag{24.5.1}$$

$$s_{\alpha\beta} = \sum_{i=1}^{6} \lambda_i^{-1} e_\alpha^{(i)} e_\beta^{(i)} \tag{24.5.5}$$

and the isothermal compressibility

$$\kappa^{(T)} = E_\alpha E_\beta s_{\alpha\beta}^{(T)} \tag{24.6.4}$$

$$= \sum_{i=1}^{6} \lambda_i^{-1} (e_1^{(i)} + e_2^{(i)} + e_3^{(i)})^2. \tag{24.6.7}$$

D. *Limiting values as $\lambda_1 \to 0$*

1. If $e_a^{(1)}$, $e_\alpha^{(1)}$, $e_\beta^{(1)}$ are all non-zero,

$$s_{\alpha\beta}/s_{aa} \to e_\alpha^{(1)} e_\beta^{(1)} / (e_a^{(1)})^2. \tag{24.5.7}$$

2. If $e_\alpha^{(1)} e_\beta^{(1)} = 0$,

$$s_{\alpha\beta} \to \sum_{i=2}^{6} \lambda_i^{-1} e_\alpha^{(i)} e_\beta^{(i)}. \tag{24.5.8}$$

3. If $E_\alpha e_\alpha^{(1)}$ is non-zero,

$$\kappa^{(T)} \to \infty.$$

E. *Stability near the transition point under different conditions*

1. A hydrostatic state, on the surface of instability for the dead load condition, will be strongly stable at constant pressure if

$$(\hat{e}_1 + \hat{e}_2 + \hat{e}_3)^2 > \hat{e}_1^2 + \hat{e}_2^2 + \hat{e}_3^2. \tag{24.7.2}$$

2. A hydrostatic state, on the surface of instability for constant pressure, will be strongly stable under the dead load condition if

$$e_1^2 + e_2^2 + e_3^2 > (e_1 + e_2 + e_3)^2. \tag{24.7.3}$$

Examples

1. Consider the stiffness matrix for the tetragonal system.
   (*a*) Show that the vector (1, −1, 0, 0, 0, 0) is an eigenvector, with eigenvalue $(c_{11} - c_{12})$.
   (*b*) Show that a vector (1, 1, a, 0, 0, 0), which is orthogonal to the above vector, is also an eigenvector. Show that there are two of these, with two values a_1 and a_2, such that $a_1 a_2 = -2$, and, thus, that these vectors are orthogonal. Show that the corresponding eigenvalues λ_1 and λ_2 satisfy a quadratic equation.

(*c*) What are the other three eigenvectors and eigenvalues?

2.(*a*) Show that the following scheme of un-normalised eigenvectors and eigenvalues applies to the trigonal classes, $3m$, 32, $\bar{3}m$.

	λ_1	λ_2	λ_3	λ_4	λ_5	λ_6
e_1	1	1	1	$-(a_3/2)$	0	0
e_2	1	1	-1	$(a_3/2)$	0	0
e_3	a_1	$-(2/a_1)$	0	0	0	0
e_4	0	0	a_3	1	0	0
e_5	0	0	0	0	1	$-a_5$
e_6	0	0	0	0	a_5	1

Show that λ_1 and λ_2 are roots of a quadratic equation, as are λ_3, λ_4 and λ_5, λ_6.

(*b*) Show that the scheme for the hexagonal class is the same as in (*a*), but with $a_3 = a_5 = 0$. Hence determine λ_3, λ_4, λ_5, λ_6.

(*c*) In (*a*) and (*b*) state which eigenvectors and eigenvalues contribute to the compressibility.

3. For the trigonal and hexagonal classes show that if $\lambda_1 \to 0$,

$$n^{(T)} \to \lambda_1(2+a^2)/(2+a)^2,$$

and so also tends to zero, so long as a does not tend to -2.

4. Consider the stiffness matrix **c**, for a crystal of the cubic class (see chapter 17). Find those elements of the compliance matrix which diverge when

(*a*) $c_{11} - c_{12} \to 0$, (*b*) $c_{11} + 2c_{12} \to 0$, (*c*) $c_{44} \to 0$.

25

An example of a phase transition involving a simple shear

A structural phase transition, which involves a simple shear change of shape, is discussed. The corresponding instability will be characterised by the vanishing of $\hat{c}_{66}^{(T)}$, for the dead load condition. Since $\hat{c}_{66}^{(T)}$ and $c_{66}^{(T)}$ are equal, they will vanish together. Hence the instability at the dead load condition coincides with that at constant pressure. The case where the dead load corresponds to a simple shearing stress is considered.

25.1. Properties of the transition

Suppose first that the transition takes place, under the dead load condition, from state a to b, respectively, of phases A and B. We shall take the state a as the reference state. Let the homogeneous change of shape be described by

$$\Delta \boldsymbol{D} = \boldsymbol{D} - \boldsymbol{E} = \begin{bmatrix} 0 & t & 0 \\ 0 & 0 & 0 \\ 0 & 0 & 0 \end{bmatrix}. \tag{25.1.1}$$

The condition of equilibrium (24.2.6) is

$$\Delta f = f_b - f_a = L_6 t = \Theta_6^{(a)} t, \tag{25.1.2}$$

since

$$L_\alpha = \Theta_\alpha^{(a)}, \quad \alpha = 1, 2, \ldots, 6, \tag{25.1.3}$$

the Cauchy stress tensor elements at the reference state a.

If the state a is at hydrostatic pressure, (25.1.2) becomes

$$\Delta f = 0. \tag{25.1.4}$$

If the transition temperature T_0 is altered, while the transition state a remains at hydrostatic pressure, the generalised Clausius–Clapeyron equation (24.3.4) becomes

$$\Delta s \, \mathrm{d}T_0 - \Delta(e_1 + e_2 + e_3) \, \mathrm{d}p = 0, \tag{25.1.5}$$

and, if Δs is not zero,

$$\mathrm{d}T_0/\mathrm{d}p = 0, \tag{25.1.6}$$

since the trace of $\Delta \boldsymbol{D}$, which is equal to $\Delta(e_1+e_2+e_3)$, is zero. We therefore see that, to first order at least, T_0 is independent of the pressure.

Consider now the case where a is a non-hydrostatic state and, in particular, where

$$\boldsymbol{\Theta}^{(a)} = \begin{bmatrix} 0 & \tau & 0 \\ \tau & 0 & 0 \\ 0 & 0 & 0 \end{bmatrix}. \tag{25.1.7}$$

Equation (25.1.2) then becomes

$$\Delta f = t\tau. \tag{25.1.8}$$

If T_0 is changed while $\boldsymbol{\Theta}^{(a)}$ remains of the above form of a simple shearing stress, with only $\Theta_6^{(a)}$ non-zero, the generalised Clausius–Clapeyron equation (24.3.4) is

$$\mathrm{d}T_0/\mathrm{d}\tau = -t/\Delta s. \tag{25.1.9}$$

Thus, if Δs is positive, T_0 decreases when τ increases. In other words the application of such a stress will facilitate the transition, in the sense that T_0 will be lowered.

Coe (1970) has discussed this case as a model for the ortho- to clino- inversion of enstatite. His method is consistent with the theory of equilibrium conditions, given in chapter 24, and he obtained the same generalised Clausius–Clapeyron equation (25.1.9).

If the transition is considered to occur at constant pressure from a state a' to a state b', the equilibrium condition (24.2.10) is, in this case,

$$f_{b'} - f_{a'} = 0, \tag{25.1.10}$$

since, as may be seen from (25.1.1), $\Delta J = \Delta V = 0$. Further, the Clausius–Clapeyron equation is

$$\mathrm{d}T_0/\mathrm{d}p = 0, \tag{25.1.11}$$

which is the same for the case where the transition occurs at the dead load condition, from a hydrostatic state a.

25.2. Thermodynamic instability at or near the transition

From the symmetry scheme of chapter 17 for stiffness matrices, we see that $\hat{c}_{66}^{(T)}$ is an eigenvalue of $\hat{\boldsymbol{c}}^{(T)}$ for the orthorhombic, tetragonal, hexagonal and cubic crystal classes. The corresponding eigenvector $\hat{\boldsymbol{e}}$ is one whose only non-zero component is $\hat{\boldsymbol{e}}_6 = t$, say. In other words, the change of shape according to this eigenvector is as given in (25.1.1). Thus the instability leading to the transition of section 1 would result from the vanishing of $\hat{c}_{66}^{(T)}$ for the dead load condition. Similarly, at the constant pressure condition the instability would result from the vanishing of $c_{66}^{(T)}$. In fact, we see from (16.1.4) that $c_{66}^{(T)} = \hat{c}_{66}^{(T)}$. Hence the instabilities at the two boundary conditions will coincide. If the transition states a and b can be considered to be at the surface of instability, the transition will proceed, at the same temperature, for both boundary conditions.

Summary of important equations

A. *Simple shear change of shape*

$$\Delta\boldsymbol{D} = \boldsymbol{D} - \boldsymbol{E} = \begin{bmatrix} 0 & t & 0 \\ 0 & 0 & 0 \\ 0 & 0 & 0 \end{bmatrix} \tag{25.1.1}$$

$$L_\alpha = \Theta_\alpha^{(a)}. \quad \alpha = 1, 2, \ldots, 6. \tag{25.1.3}$$

B. *The equilibrium condition under dead load*

$$\Delta f = L_6 t = \Theta_6^{(a)} t. \tag{25.1.2}$$

1. For a hydrostatic state a, the condition of equilibrium under dead load is, since $\Theta_6^{(a)} = 0$,

$$\Delta f = 0, \tag{25.1.4}$$

and the Clausius-Clapeyron equation is

$$\mathrm{d}T_0/\mathrm{d}p = 0. \tag{25.1.6}$$

2. For

$$\boldsymbol{\Theta}^{(a)} = \begin{bmatrix} 0 & \tau & 0 \\ \tau & 0 & 0 \\ 0 & 0 & 0 \end{bmatrix}, \tag{25.1.7}$$

the equilibrium condition, at dead load, is

$$\Delta f = \tau t, \tag{25.1.8}$$

and the generalised Clausius–Clapeyron equation is

$$\mathrm{d}T_0/\mathrm{d}\tau = -t/\Delta s. \tag{25.1.9}$$

C. *The equilibrium condition at constant pressure*

$$f_{b'} - f_{a'} = 0, \tag{25.1.10}$$

and the Clausius–Clapeyron equation is

$$\mathrm{d}T_0/\mathrm{d}p = 0. \tag{25.1.11}$$

Example

1. Consider the transition of this chapter for the two cases where (a) the state a is at a hydrostatic pressure, (b) the Cauchy stress tensor at state a is given by (25.1.7). Discuss whether the constraints (22.5.1) and (22.5.2) must be applied in order that mechanical equilibrium can be maintained throughout the transition.

26

Limiting values of thermodynamic quantities at an instability

The limiting behaviour of the thermal expansion coefficients, the specific heat and the adiabatic stiffness and compliance matrices is discussed. Limiting relations are derived between those quantities which diverge, in a particular phase, as the instability is approached. It is shown that the limiting behaviour in the low temperature phase of a transition may be completely different from that in the high temperature phase. The Pippard relations are briefly discussed. In the final section we discuss the general relationships between thermodynamic quantities at a transition. To illustrate what is meant by a general relationship, it is shown that if the thermal expansion coefficients remain finite in a phase, as the instability is approached, then the specific heat, c_p, must also remain finite.

26.1. Discussion of limiting values

In chapter 24, we obtained limiting values of the isothermal compliance elements at an instability surface. In this chapter, we shall discuss the limiting behaviour of other thermodynamic quantities, namely, the thermal expansion coefficients, the specific heats and the adiabatic stiffness and compliance matrix elements. We shall obtain limiting relations between those quantities which diverge to infinity at the instability surface.

How do these limiting values and relations apply to the transition? If the transition state of a phase is either at or near to an instability surface, these limiting values will be either equal or very close to the values at the transition state. In fact, one might say that quantitative properties of the homogeneous bulk phase will not, as thermodynamic instability is approached, be significantly affected by the approach to the state at which local or surface instabilities interrupt the passage of the bulk phase, through uniform homogeneous states, to the thermodynamic instability surface. That is to say, they will not be significantly affected until a sudden interruption of this uniform progression occurs, and the transition, triggered off locally or at surfaces, takes place. If fluctuations bring about the transition, their onset will interrupt, in a similarly abrupt way, the progression of the thermodynamic quantities to their limiting values at the thermodynamic instability. The limiting value at a transition should be interpreted in this sense, that is, as its value at the state at which the transition interrupts the above progression. The functional dependence of the quantity on the state variables, as the transition is approached, will be determined rather by the proximity of the instability than by that of the transition state – except, possibly, very close to that state, when an abrupt change in the behaviour may occur.

In a transition $A \to B$, from state a to state b, we should note that there are two instability surfaces involved – one for each phase. In a phase diagram in which the co-ordinates $T, e_1, e_2, \ldots, e_6$ are used, these two surfaces will not, in general, coincide. Just as there is a forbidden region for stable states of liquid or vapour in a phase diagram which uses the co-ordinates T and V, there will, in general, be a forbidden region in the above seven-dimensional diagram. We may thus be concerned, for a particular transition, with two sets of limiting values – one, as the instability surface of phase A is approached, the other, as that of phase B is approached. These two sets need not be the same and, in fact, we shall show that they are significantly different in the case of the α–β quartz transition. For example, the experimental data for the thermal expansion coefficients shows convincingly that these quantities remain finite in the β phase and diverge in the α phase. We shall show that if this experimental fact is accepted, c_p will have the same property – diverging in the α phase, remaining finite in the β phase. This behaviour of c_p explains the characteristic λ shape of the specific heat curve. Similar properties apply to the barium titanate tetragonal–cubic ferroelectric transition, as will be discussed later. Such behaviour appears, in fact, to be characteristic of λ-transitions.

For simplicity, the discussion will be restricted to the zero stress condition. That is, we shall assume that the specimen under study has no forces acting on its exterior surfaces during any changes or variations considered. In a hydrostatic context, this condition would be termed 'zero pressure'. $\boldsymbol{B}$ and p are, thus, always zero, and the two boundary conditions, dead load or constant pressure, both apply. The stiffness matrices $\hat{\boldsymbol{c}}$ and $\boldsymbol{c}$ are equal (see (16.1.4)), as are the thermal expansion coefficients $\hat{\boldsymbol{m}}$ and $\boldsymbol{m}$ (see (14.1.6)). c_p and c_L are also equal (see (15.3.1) and (15.3.2)). Because of (15.4.8), we cannot say that $c_v = c_{\boldsymbol{e}}$.

To simplify the notation, in this zero stress (pressure) condition, we shall drop the superscript (T) and use $\boldsymbol{c}$, $\boldsymbol{s}$, κ, n for the isothermal quantities. m_α, $\alpha = 1, 2, \ldots, 6$, will denote the thermal expansion coefficients, $(\partial e_\alpha/\partial T)_{p=0}$, c_p and $c_{\boldsymbol{e}}$ the specific heats of zero stress (pressure) and at constant strain $\boldsymbol{e}$, respectively.

26.2. The thermal expansion coefficients

First, consider the infinitesimal change, $\mathrm{d}s$, in the entropy which occurs in an isothermal variation. This is given by

$$\mathrm{d}s = (\partial s/\partial e_\alpha)_T \, \mathrm{d}e_\alpha, \tag{26.2.1}$$

where s, the entropy per unit reference volume, is regarded as a function of T and the strain elements e_α. On using the thermodynamic relation of example 7, chapter 11, we may write

$$(\partial s/\partial e_\alpha)_T = -(\partial L_\alpha/\partial T)_{\boldsymbol{e}}. \tag{26.2.2}$$

From the result of example 4, chapter 11, we have

$$(\partial L_\alpha/\partial T)_{\boldsymbol{e}} = -c_{\alpha\beta} m_\beta, \tag{26.2.3}$$

(remember $\hat{\boldsymbol{c}}^{(T)} = \boldsymbol{c}$, $\hat{\boldsymbol{m}} = \boldsymbol{m}$ in the zero stress state!). Thus (26.2.1) may be transformed to

$$\mathrm{d}s = c_{\alpha\beta} m_\beta \, \mathrm{d}e_\alpha. \tag{26.2.4}$$

Now let us choose a particular isothermal variation, such that

$$\mathrm{d}e_\alpha = e_\alpha^{(1)} \, \mathrm{d}\varepsilon, \tag{26.2.5}$$

where $\boldsymbol{e}^{(1)}$ is the normalised eigenvector corresponding to λ_1, the eigenvalue which vanishes at the instability surface; $\mathrm{d}\varepsilon$ is an infinitesimal scalar quantity (not to be confused with the infinitesimal strain tensor). On substituting (26.2.5) in (26.2.4), we obtain

$$\mathrm{d}s = (c_{\alpha\beta} m_\beta e_\alpha^{(1)}) \, \mathrm{d}\varepsilon, \tag{26.2.6}$$

and so

$$\mathrm{d}s/\mathrm{d}\varepsilon = c_{\alpha\beta} m_\beta e_\alpha^{(1)} = \lambda_1 m_\beta e_\beta^{(1)}, \tag{26.2.7}$$

since $\boldsymbol{c}$ is a symmetric matrix. Equation (26.2.7) is to be regarded as the definition of $\mathrm{d}s/\mathrm{d}\varepsilon$, which describes the rate of change of entropy in an isothermal change of shape in the 'direction' (in six dimensions) determined by $\boldsymbol{e}^{(1)}$. We shall be interested in the value of $q_1 = \mathrm{d}s/\mathrm{d}\varepsilon$ as the instability surface is approached. We may express $\boldsymbol{m}$ in terms of the orthonormal set of the eigenvectors of $\boldsymbol{c}$, that is, as

$$\begin{aligned} \boldsymbol{m} &= \mu_\alpha \boldsymbol{e}^{(\alpha)} \\ &= \mu_1 \boldsymbol{e}^{(1)} + \mu_2 \boldsymbol{e}^{(2)} + \cdots + \mu_6 \boldsymbol{e}^{(6)}. \end{aligned} \tag{26.2.8}$$

We should note that $\boldsymbol{e}^{(1)}$ can always be chosen so that $\mu_1 \geqslant 0$. Equation

(26.2.7) becomes

$$q_1 = \mathrm{d}s/\mathrm{d}\varepsilon = \lambda_1\mu_1. \tag{26.2.9}$$

Since we are considering the progression from strongly stable states to the instability surface, λ_1 is $\geqslant 0$, and so q_1 will not be negative. If, for example, q_1 tends to Q_1 at the instability surface then

$$\mu_1 \to Q_1/\lambda_1. \tag{26.2.10}$$

If Q_1 is to be finite then μ_1 must tend to infinity, and in this case some elements of $\boldsymbol{m}$ will diverge. The diverging elements of $\boldsymbol{m}$ may be seen, from (26.2.8), to be those which correspond to non-zero elements of $\boldsymbol{e}^{(1)}$.

Let us define

$$x = \lim_{\lambda_1 \to 0} \{\partial(\ln q_1)/\partial(\ln \lambda_1)\}, \tag{26.2.11}$$

that is,

$$Q_1 = \lim q_1 \propto \lim \lambda_1^x. \tag{26.2.12}$$

Thus Q_1 diverges, is finite, or vanishes, depending respectively on whether x is negative, zero, or positive. Since $\mu_1 = q_1/\lambda_1$, μ_1 will diverge if $x<1$, will remain finite if $x=1$, and will tend to zero if $x>1$.

If μ_1 diverges, the limiting form of $\boldsymbol{m}$ is given by

$$\boldsymbol{m} \to \mu_1\boldsymbol{e}^{(1)} = (q_1/\lambda_1)\boldsymbol{e}^{(1)}. \tag{26.2.13}$$

If m_i and m_j are diverging coefficients, their limiting ratio is

$$m_i/m_j = e_i^{(1)}/e_j^{(1)}. \tag{26.2.14}$$

For axial crystals, such as quartz, there are only three non-zero coefficients $m_1 = m_2$, and m_3. If we take the c-axis as the x_3-axis, we may write

$$m_a/m_c = m_1/m_3 = e_1^{(1)}/e_3^{(1)}. \tag{26.2.15}$$

Thus the limiting ratio m_a/m_c determines the limiting eigenvector, which will then be of the form $e_1^{(1)} = e_2^{(1)}$, $e_3^{(1)} \neq 0$, $e_4^{(1)} = e_5^{(1)} = e_6^{(1)} = 0$. From the symmetry properties of $\boldsymbol{c}$ for such a crystal, such an eigenvector exists, and, in fact, the experimental data for quartz convincingly shows that at the α–β transition, the corresponding eigenvalue vanishes.

We have discussed the possible divergence of the elements of $\boldsymbol{m}$. The volume coefficient of thermal expansion β is defined by

$$\beta = m_1 + m_2 + m_3 = E_\alpha m_\alpha, \tag{26.2.16}$$

and it may possibly diverge. If β diverges, its limiting form will be determined by

$$\beta \to \mu_1(E_\alpha e_\alpha^{(1)}) = \lim\{(q_1/\lambda_1)E_\alpha e_\alpha^{(1)}\}. \tag{26.2.17}$$

For example, in the transition discussed in chapter 25, β will certainly not diverge at the instability surface since $E_\alpha e_\alpha^{(1)}$ is zero.

In the next section we shall discuss the limiting value of c_p, which we shall be able to relate to those of $\boldsymbol{m}$ and β.

26.3. The specific heat c_p

If we consider s as a function of T and the strain elements e_α, we have

$$ds = (\partial s/\partial e_\alpha)\, de_\alpha + (\partial s/\partial T)\, dT. \tag{26.3.1}$$

On using (26.2.2) and (26.2.3) we may write this equation as

$$ds = (\partial s/\partial T)\, dT + c_{\alpha\beta} m_\beta \, de_\alpha. \tag{26.3.2}$$

Therefore

$$c_p/T = (\partial s/\partial T)_{p=0} = c_{\boldsymbol{e}}/T + \tilde{v} c_{\alpha\beta} m_\alpha m_\beta, \tag{26.3.3}$$

where $\tilde{v}$ is the reference volume per unit mass.

Now $c_{\boldsymbol{e}}$ is the specific heat measured at constant strain, i.e., at constant shape and size. There is no evidence, experimental or otherwise, to show that it diverges at a transition. We shall assume that $c_{\boldsymbol{e}}$ does not diverge at a transition. In other treatments of λ-transitions, it is usually assumed that c_v remains finite (see, e.g., Pippard, 1956). These treatments are hydrostatic, and involve the thermodynamic variables p, V, and T only. That is, changes of shape, apart from volume, are not considered, and c_v is then the relevant specific heat measured with the single mechanical coordinate V held constant. The specific heat $c_{\boldsymbol{e}}$ is an irrelevant quantity in hydrostatic thermodynamics as it is impossible, in general, to hold $\boldsymbol{e}$ constant by variation of p alone. In chapter 15, we derived the difference $c_v - c_{\boldsymbol{e}}$, which is given by (15.4.7). If $\boldsymbol{m}$ diverges, then it is an interesting exercise to show that $c_v - c_{\boldsymbol{e}}$ will diverge if $\mu_1 q_1 = \mu_1^2 \lambda_1$ diverges. If there is any reason why c_v or $c_{\boldsymbol{e}}$ must remain finite, this must be based on the fact that when the lattice dimensions of a crystal are held fixed, the vibrational modes, for example, are essentially unaltered. Thus, it is more reasonable to assume that $c_{\boldsymbol{e}}$ remains finite than that c_v does. If we assume that $c_{\boldsymbol{e}}$ remains finite, then we see that c_v may diverge.

On using the expansion (26.2.8) of $\boldsymbol{m}$, (26.3.3) becomes

$$(c_p - c_e)/T\tilde{v} = \sum_{i=1}^{6} (\mu_i^2 \lambda_i). \tag{26.3.4}$$

Therefore c_p will diverge, if μ_1 diverges faster than $\lambda_1^{-\frac{1}{2}}$, and, in this case,

$$(c_p/T\tilde{v}) \to \mu_1^2 \lambda_1 = \lim_{\lambda_1 \to 0} (q_1 \mu_1). \tag{26.3.5}$$

Here, as stated above, we assume that c_e remains finite.

26.4. The adiabatic stiffness matrix at the instability surface

In this section we shall show that the adiabatic stiffness matrix has limiting properties at the instability surface, which are directly related to Q_1, the limiting value of q_1. In fact, we shall show that if $q_1 \to 0$, $\boldsymbol{c}^{(s)}$ and $\boldsymbol{c}$ both have, in the limit, a vanishing eigenvalue corresponding to the same eigenvector $\boldsymbol{e}^{(1)}$. We continue in this section, as elsewhere in this chapter, to denote the isothermal stiffness matrix by $\boldsymbol{c}$.

We shall also show, if Q_1 is finite and non-zero, that $\boldsymbol{c}^{(s)}$ is positive definite up to and at the instability surface. Finally, if Q_1 is infinite, $\boldsymbol{c}^{(s)}$ will have infinite elements at this surface. That is, for some adiabatic variations the material will become infinitely stiff – a possibility that we may reasonably exclude from practical consideration, although perhaps we should not do so absolutely.

From (16.3.9) we have

$$c_{\alpha\beta}^{(s)} = c_{\alpha\beta} + \sigma c_{\alpha\mu} c_{\beta\nu} m_\mu m_\nu, \tag{26.4.1}$$

where

$$\sigma = \tilde{v}T/c_e = T/\rho_0 c_e. \tag{26.4.2}$$

On using (26.2.8) we may write

$$c_{\alpha\beta}^{(s)} = c_{\alpha\beta} + \sigma q_\nu q_\omega e_\alpha^{(\nu)} e_\beta^{(\omega)}, \tag{26.4.3}$$

where

$$q_i = \mu_i \lambda_i, \quad i = 1, 2, \ldots, 6. \tag{26.4.4}$$

(Remember the summation convention applies only to Greek indices.)

Thus

$$c_{\alpha\beta}^{(s)} e_\beta^{(1)} = \lambda_1 e_\alpha^{(1)} + \sigma q_1 q_\nu e_\alpha^{(\nu)}, \tag{26.4.5}$$

since $\boldsymbol{e}^{(1)}$ is an orthonormal eigenvector of $\boldsymbol{c}$. Hence, if $q_1 \to 0$, and $\lambda_1 \to 0$, we see that

$$c^{(s)}_{\alpha\beta}e^{(1)}_{\beta}\to 0. \tag{26.4.6}$$

Therefore, in this case ($Q_1=0$), $\boldsymbol{c}^{(s)}$ and $\boldsymbol{c}$ both have, in the limit, a vanishing eigenvalue λ_1 corresponding to the eigenvector $\boldsymbol{e}^{(1)}$.

If Q_1 is finite and non-zero, we shall now show that $\boldsymbol{c}^{(s)}$ is positive definite, in the limit. Consider an arbitrary six-dimensional vector $\boldsymbol{t}$. Then, from (26.4.3),

$$t_\alpha c^{(s)}_{\alpha\beta}t_\beta=\sum_{i=1}^{6}\lambda_i(t_\alpha e^{(i)}_{\alpha})^2+\sigma q_\nu q_\omega(t_\alpha e^{(\nu)}_{\alpha})(t_\beta e^{(\omega)}_{\beta}). \tag{26.4.7}$$

If we write $T_i=t_\alpha e^{(i)}_{\alpha}$, (26.4.7) becomes

$$t_\alpha c^{(s)}_{\alpha\beta}t_\beta=\lambda_\mu T^2_\mu+\sigma(q_\mu T_\mu)^2. \tag{26.4.8}$$

Now, at least one of the six quantities T_i must be non-zero for a non-null vector $\boldsymbol{t}$, and $\lambda_2,\lambda_3,\ldots,\lambda_6$ are all positive. Therefore, if Q_1 is finite and non-zero, we may conclude from (26.4.8) that $\boldsymbol{c}^{(s)}$ is positive definite in the limit as well as in the stable phase.

We note that if we choose $\boldsymbol{t}=\boldsymbol{e}^{(1)}$, then

$$e^{(1)}_{\alpha}c^{(s)}_{\alpha\beta}e^{(1)}_{\beta}\to\sigma Q_1^2, \tag{26.4.9}$$

that is,

$$(c_{\mathbf{e}}/T\tilde{v})c^{(s)}_{\alpha\beta}e^{(1)}_{\alpha}e^{(1)}_{\beta}\to Q_1^2. \tag{26.4.10}$$

This equation provides a method of determining Q_1. Another method of determining Q_1 follows from (26.2.13) and (26.2.17), from which it may be seen that

$$Q_1=\lim(\lambda_1 m_a/e^{(1)}_{a})=\lim(\lambda_1\beta/E_\alpha e^{(1)}_{\alpha}), \tag{26.4.11}$$

where $e^{(1)}_{a}$ is a non-zero element of $\boldsymbol{e}^{(1)}$, and where the second equality applies only if $E_\alpha e^{(1)}_{\alpha}\neq 0$.

It may be shown that

$$\det(\boldsymbol{c}^{(s)})\to Q_1^2 C, \tag{26.4.12}$$

where C is a positive factor – in fact, on using the result of example 6, chapter 11, and (26.3.4), it can be proved that

$$C=\sigma\lambda_2\lambda_3\lambda_4\lambda_5\lambda_6, \tag{26.4.13}$$

where λ_i, $i=2,\ldots,6$ are the positive eigenvalues of the isothermal stiffness matrix. It is of interest to note that a phase is adiabatically stable at the instability surface, if the neutrally stable path at the instability surface requires a change of entropy ($Q_1 \neq 0$), and is adiabatically unstable if it does not ($Q_1 = 0$).

We have now studied the limiting behaviour of c_p, $\boldsymbol{m}$, β, and $\boldsymbol{c}^{(s)}$, and in chapter 24, that of $\boldsymbol{c}$, $\boldsymbol{s}$, n and κ. We are thus in a position to discuss limiting relations between these quantities. Before doing this, however, we shall briefly discuss the Pippard relations (Pippard, 1956) between these limiting values, for λ-transitions.

26.5. The Pippard relations

Many solid-state transitions are accompanied by very small changes of shape and volume. These changes were often neglected, and such transitions were regarded as second order in the Ehrenfest scheme of classification of transitions. On this classification scheme, a second order transition is such that the Gibbs function and its first derivatives V and S are continuous, while its second derivatives and, therefore, κ and c_p, suffer discontinuous finite jumps.

However it is now recognised that in most solid state transitions, small discontinuities do occur in the shape and entropy of the solid. Thus, they cannot be regarded as true second order transitions. It is a characteristic of the large class of solid state λ-transitions that the specific heat c_p rises rapidly, on the low temperature side, to a very high maximum, whose ultimate height appears to be limited only by the shortcomings of experimental techniques. At the transition point, c_p drops very abruptly to a finite value and remains almost constant in the high temperature phase. This behaviour, together with the small discontinuous changes in shape and entropy, shows that λ-transitions cannot be treated as second order. It is now recognised that no solid state transition, except possibly the super-conducting transition, can be regarded strictly as second order on Ehrenfest's scheme.

We may remark here that where suitable experiments have been carried out, it has been found that the elastic compliances diverge at λ-transitions. Thus we would expect a λ-transition to be associated with a thermodynamic instability.

However, Pippard (1956) discussed the thermodynamics of λ-transitions from a different point of view. He used a cylindrical approximation, empirically justified, for the entropy function near the transition line. That is, he assumed over small ranges in T–p space that the transition line is straight, and that the entropy on this line is a linear function of temperature. Using this approximation he derived what have become known as the Pippard relations, namely,

$$(c_p/T\tilde{V}) = \gamma\beta, \tag{26.5.1}$$

and

$$\beta = \gamma\kappa \tag{26.5.2}$$

where

$$\gamma = (\mathrm{d}p/\mathrm{d}T_0)_\lambda \tag{26.5.3}$$

is the pressure coefficient on the λ-transition line. These relations were in good agreement with experiment, in the sense that linear relations were obtained experimentally between c_p, β and κ, but only for values obtained on the low temperature side of the transitions.

For a true second order transition, the following relations hold (Landau and Lifshitz, 1958, equation 135.11),

$$(\Delta c_p)/T\tilde{V} = \gamma\Delta\beta, \tag{26.5.4}$$

and

$$\Delta\beta = \gamma\Delta\kappa \tag{26.5.5}$$

where, for example, $\Delta\beta$ signifies the change occurring in β during the transition.

Coe and Paterson (1969) generalised the Pippard relations to the case of axial crystals under non-hydrostatic stresses, to obtain

$$c_p/T\tilde{V} = (\mathrm{d}P_a/\mathrm{d}T_0)_\lambda m_a = (\mathrm{d}P_c/\mathrm{d}T_0)_\lambda m_c, \tag{26.5.6}$$

and

$$s_{ij} = (\mathrm{d}T_0/\mathrm{d}P_i)_\lambda m_j = (\mathrm{d}T_0/\mathrm{d}P_j)_\lambda m_i, \tag{26.5.7}$$

where P_a is the uniaxial stress applied in the direction of the a-axis of the crystal (P_a is positive if it is compressive, just as is a pressure). In (26.5.7) ij is any choice of aa, cc and ac. For axial crystals, $m_a (= m_b)$ and m_c are the only non-zero elements of $\boldsymbol{m}$.

From these equations, we obtain the limiting ratios

$$\frac{m_i}{m_j} = \frac{(\mathrm{d}T_0/\mathrm{d}P_i)_\lambda}{(\mathrm{d}T_0/\mathrm{d}P_j)_\lambda} = \frac{\Delta e_i}{\Delta e_j}, \tag{26.5.8}$$

where the right-hand side has been obtained from the Clausius–Clapeyron equation (24.4.5), on noting that $L_i = -P_i$ in infinitesimal strain theory.

In the next section we shall derive relations, analogous to the Pippard relations, between diverging thermodynamic quantities near an instability.

26.6. Limiting relations for diverging quantities for the case when x is zero

In chapter 24 and in sections 26.2 and 26.3 we have derived limiting forms for thermodynamic quantities in cases where they diverge. For convenience, let us gather these here.

$$s_{\alpha\beta} \to \lambda_1^{-1} e_\alpha^{(1)} e_\beta^{(1)}, \tag{24.5.5}$$

$$\kappa \to \lambda_1^{-1} (E_\alpha e_\alpha^{(1)})^2, \tag{24.6.7}$$

$$m_\alpha \to \mu_1 e_\alpha^{(1)}, \tag{26.2.13}$$

$$m_i/m_j \to e_i^{(1)}/e_j^{(1)}, \tag{26.2.14}$$

$$\beta \to \mu_1(E_\alpha e_\alpha^{(1)}), \tag{26.2.17}$$

$$(c_p/T\tilde{v}) \to \mu_1^2 \lambda_1. \tag{26.3.5}$$

In this section we shall assume that x is zero, that is, that Q_1 is finite and non-zero. Then by eliminating μ_1, and λ_1 in the above forms, we shall derive limiting linear relations analogous to the Pippard and generalised Pippard relations.

First, from (26.2.17) and (26.3.5), we obtain the limiting relation

$$c_p/T\tilde{v} = (\mu_1\lambda_1)(E_\alpha e_\alpha^{(1)})^{-1}\beta, \tag{26.6.1}$$

$$= \left(\frac{\mathrm{d}s}{\mathrm{d}J}\right)_N \beta. \tag{26.6.2}$$

Equation (26.6.2) follows from (26.2.9), and from the fact that $(\mathrm{d}s/\mathrm{d}J)_N$, which we here define as the rate of change of entropy, along the neutrally stable path at the instability surface, with respect to J, is given by

$$\left(\frac{\mathrm{d}s}{\mathrm{d}J}\right)_N = \lim\left(\frac{\mathrm{d}s}{\mathrm{d}J}\right) = \lim\left\{\frac{\mathrm{d}s}{\mathrm{d}\varepsilon}(E_\alpha e_\alpha^{(1)})^{-1}\right\}, \tag{26.6.3}$$

since the variation of J along the infinitesimal isothermal path, $\mathrm{d}\varepsilon\,(\boldsymbol{e}^{(1)})$, is

$$\mathrm{d}J = \mathrm{d}\varepsilon\,(E_\alpha e_\alpha^{(1)}). \tag{26.6.4}$$

(26.6.2) may be written as

$$(c_p/T\tilde{v}) = \beta(\mathrm{d}\tilde{s}/\mathrm{d}\tilde{V})_N, \tag{26.6.5}$$

since $\tilde{s} = s\tilde{v}$, and $\tilde{V} = J\tilde{v}$.

This equation may be compared with the Pippard relation (26.5.1), which, on using the Clausius–Clapeyron equation (24.3.1) to eliminate $(\mathrm{d}p/\mathrm{d}T_0)_\lambda$, may be expressed as

$$(c_p/T\tilde{V}) = (\Delta\tilde{s}/\Delta\tilde{V})\beta. \tag{26.6.6}$$

Thus, we see that the Pippard relation differs from (26.6.5) by the appearance of $(\Delta\tilde{s}/\Delta\tilde{V})$, as a coefficient, instead of $(\mathrm{d}\tilde{s}/\mathrm{d}\tilde{V})_N$, the rate óf change of entropy along the neutrally stable path at the instability surface (McLellan, 1976a, 1979).

From (24.6.7) and (26.2.16) we obtain, in the limit,

$$\beta = \kappa(\mu_1\lambda_1)(E_\alpha e_\alpha^{(1)})^{-1}, \tag{26.6.7}$$

and, as above,

$$\beta = \kappa(\mathrm{d}\tilde{s}/\mathrm{d}\tilde{V})_N, \tag{26.6.8}$$

where, of course, κ is the isothermal compressibility. We may compare this equation with the Pippard relation (26.5.2), which may be written as

$$\beta = (\Delta\tilde{s}/\Delta\tilde{V})\kappa. \tag{26.6.9}$$

From (26.2.13) and (26.3.5), we obtain

$$\begin{aligned} c_p/T\tilde{v} &= m_i(\mu_1\lambda_1)(e_i^{(1)})^{-1} \\ &= m_i(\mathrm{d}s/\mathrm{d}e_i)_N, \end{aligned} \tag{26.6.10}$$

since, along the neutrally stable path at the instability surface,

$$\mathrm{d}e_i = \mathrm{d}\varepsilon\,(e_i^{(1)}). \tag{26.6.11}$$

From (24.5.6) and (26.2.13), we may write

$$s_{\alpha\beta} = (\mu_1\lambda_1)^{-1}m_\alpha e_\beta^{(1)} = (\mu_1\lambda_1)^{-1}e_\alpha^{(1)}m_\beta, \tag{26.6.12}$$

and, therefore,

$$s_{\alpha\beta} = (\mathrm{d}e_\alpha/\mathrm{d}s)_N m_\beta = m_\alpha(\mathrm{d}e_\beta/\mathrm{d}s)_N, \tag{26.6.13}$$

and this may be compared with (26.5.7), which, on using the generalised Clausius–Clapeyron equation, (24.3.5), is

$$s_{ij} = (\Delta e_i/\Delta s)m_j = (\Delta e_j/\Delta s)m_i. \tag{26.6.14}$$

Finally from (26.6.10) and (26.6.11), we have, in the limit,

$$\frac{m_\alpha}{m_\beta} = \frac{(\mathrm{d}e_\alpha/\mathrm{d}s)_N}{(\mathrm{d}e_\beta/\mathrm{d}s)_N} = \left(\frac{\mathrm{d}e_\alpha}{\mathrm{d}e_\beta}\right)_N, \tag{26.6.15}$$

which may be compared with (26.5.8).

Thus the present thermodynamic theory may yield linear relations, analogous to the Pippard relations, except that the coefficient in each linear relation is a rate of change along the neutrally stable path at the instability surface, instead of the corresponding ratio of discontinuous increments occurring at the transition. This fact may explain a discrepancy, in the α–β quartz transition, between the observed limiting relation for m_c/m_a and the observed ratio $(\mathrm{d}T_0/\mathrm{d}P_c)_\lambda/(\mathrm{d}T_0/\mathrm{d}P_a)_\lambda$. This point will be discussed in the next chapter.

In the next section, we shall discuss the limiting behaviour of thermodynamic quantities more generally, and, in particular, when x is not zero.

26.7. Possible limiting relations at the instability surface

In the previous section we derived limiting linear relationships between diverging thermodynamic quantities, for the case where the x of (26.2.12) is zero. We will now discuss how we may determine whether or not such quantities diverge and what sort of limiting relationships may occur if x is not zero.

We may write (26.2.8) as

$$\boldsymbol{m} = (q_1/\lambda_1)\boldsymbol{e}^{(1)} + \boldsymbol{m}^{(0)}; \tag{26.7.1}$$

(26.2.16) as

$$\beta = (q_1/\lambda_1)(E_\alpha e_\alpha^{(1)}) + \beta^{(0)}; \tag{26.7.2}$$

and (26.3.4) as

$$(c_p/T\tilde{v}) = (q_1^2/\lambda_1) + (c_p/T\tilde{v})^{(0)}, \tag{26.7.3}$$

where, in these three equations, the quantities with the superscript zero remain finite. We will regard λ_1 as a 'distance' parameter for the approach to the instability surface.

The behaviour of the first terms of the right-hand sides of the above three equations is determined by the value of x defined in (26.2.11). For example, we see that $\boldsymbol{m}$ and β remain finite if λ_1^{x-1} does, that is, if $x \geqslant 1$. If $x < 1$, those elements of $\boldsymbol{m}$, corresponding to non-vanishing elements of $\boldsymbol{e}^{(1)}$, will diverge, as will β if $E_\alpha e_\alpha^{(1)}$ is not zero. The behaviour of $c_p/T\tilde{v}$ is determined by λ_1^{2x-1}, and therefore if $x \geqslant \frac{1}{2}$, it will remain finite, while it will diverge if $x < \frac{1}{2}$. In Table 26.1 are displayed the limiting values of some thermodynamic quantities.

Some interesting points that are evident from Table 26.1 are:

1. if $\boldsymbol{m}$ is finite in the limit, c_p is finite and $Q_1 = 0$;

2. if Q_1 is finite and non-zero, m_α (if $e_\alpha^{(1)} \neq 0$), β (if $E_\alpha e_\alpha^{(1)} \neq 0$), and c_p all diverge;

Table 26.1. *Limiting values at the instability surface*

	Q_1	m_α $e_\alpha^{(1)} \neq 0$	β $E_\alpha e_\alpha^{(1)} \neq 0$	c_p	$e_\alpha^{(1)} e_\beta^{(1)} c_{\alpha\beta}^{(s)}$	$\det \boldsymbol{c}^{(s)}$
Limiting behaviour:	λ_1^x	λ_1^{x-1}	λ_1^{x-1}	λ_1^{2x-1}	σQ_1^2	CQ_1^2
$x\downarrow$						
>1	0	fin.	fin.	fin.	0	0
1	0	fin.	fin.	fin.	0	0
$\frac{1}{2}<x<1$	0	∞	∞	fin.	0	0
$\frac{1}{2}$	0	∞	∞	fin.	0	0
$0<x<\frac{1}{2}$	0	∞	∞	∞	0	0
0	fin.	∞	∞	∞	fin.	fin.
<0	∞	∞	∞	∞	∞	∞

fin., finite and non-zero.

3. if $x \geqslant 1$, c_p, $\boldsymbol{m}$ and β are all finite;
4. if q_1 diverges, that is, if $x<0$, $\det(\boldsymbol{c}^{(s)})$ and $e_\alpha^{(1)} c_{\alpha\beta}^{(s)} e_\beta^{(1)}$ both diverge. The stiffness matrix $\boldsymbol{c}^{(s)}$ would then have some infinite elements. Such infinite stiffnesses are extremely improbable, and the case of negative x may be safely discarded.

When c_p, $\boldsymbol{m}$ (and/or β) diverge, it is possible, from experimental data, to determine whether Q_1 is zero or non-zero. For, from (26.2.13) and (26.3.5), we see that

$$(c_p/T\tilde{v}m_i)e_i^{(1)} \to Q_1, \tag{26.7.4}$$

while, if β diverges, from (26.2.16) and (26.3.5)

$$(c_p/T\tilde{v}\beta)(E_\alpha e_\alpha^{(1)}) \to Q_1. \tag{26.7.5}$$

Hence, if $(c_p/T\tilde{v}m_i)$, where $e_i^{(1)} \neq 0$, tends to a finite non-zero limit, then Q_1 must be non-zero, as it will be if $(c_p/T\tilde{v}\beta)$ tends to a non-zero limit. If the limiting behaviour of $\boldsymbol{c}^{(s)}$ is determined experimentally, then we may determine from (26.4.10) or (26.4.12) whether Q_1 is zero or non-zero. Finally, if the limiting behaviour of $\boldsymbol{c}$ and $\boldsymbol{m}$ are known, (26.4.11) may be used.

For α quartz, as we shall see in more detail in chapter 27, c_p, β, m_a and m_c all diverge. The quantity $(c_p/T\tilde{v}\beta)$ tends to a finite non-zero value, and so we can easily conclude that Q_1 is finite and non-zero. For β quartz, $\boldsymbol{m}$ and β remain finite, and thus, from Table 26.1, we may conclude that c_p remains finite, $Q_1=0$, and $x \geqslant 1$.

From (26.4.12) we see that α quartz remains adiabatically stable when the instability surface is reached, while, from (26.4.5), β quartz becomes neutrally stable to adiabatic variations at the instability surface.

For β quartz, the single experimental observation that $\boldsymbol{m}$ remains finite has important consequences for c_p and $\boldsymbol{c}^{(s)}$.

It is interesting to note that some elements of the adiabatic compliance matrix (or one eigenvalue) will rise rapidly and diverge in β quartz as the temperature is lowered to T_0, while these quantities will remain finite in α quartz. This behaviour is in contrast to that of c_p and of $\boldsymbol{m}$ where the rapid rise and divergence occurs in α quartz as the temperature is increased to T_0, and where $\boldsymbol{m}$ and c_p remain finite in the β quartz. The adiabatic stiffness matrix may be determined from experimental data on the wave speeds of acoustic vibrations (see chapter 17). The above behaviour of $\boldsymbol{c}^{(s)}$ should be of considerable interest in further experimental work, near a λ-transition, on the adiabatic elastic constants.

We may note that, if $0<x<\frac{1}{2}$, c_p, β, m_α and, of course, some elements of the compliance matrix all diverge. However, since in this range of x the quantity Q_1 tends to zero, there will be no linear limiting relations between these diverging quantities as were obtained in 26.6 for the case where x is zero. From the limiting behaviour in Table 26.1, it can be seen that limiting linear relations between the logarithms of the diverging quantities would be obtained, the linear slopes being related to x. For example, the limiting relations

$$\ln(c_p/T\tilde{v}) \to (2x-1)\ln\lambda_1 \tag{26.7.6}$$

and

$$\ln(c_p/T\tilde{v}) \to \frac{2x-1}{x-1}\ln m_\alpha, \tag{26.7.7}$$

would be obtained for diverging quantities in the range $0<x<\frac{1}{2}$.

In the range $0<x<1$, the limiting relation

$$\ln m_\alpha \to (x-1)\ln\lambda_1 \tag{26.7.8}$$

would be obtained for diverging m_α. If β diverges it may be substituted for m_α in the last two equations, while $-\ln\kappa$ or $-\ln s_{\alpha\beta}$, if they diverge, may be substituted for $\ln\lambda_1$ in (26.7.6) or (26.7.8). Such relations could be used to determine experimentally the value of x for a phase approaching a transition.

From Table 26.1, we may deduce, in the range $0\leqslant x<\frac{1}{2}$, the following limiting relation for diverging quantities

$$(c_p/T\tilde{v})s_{\alpha\beta}/m_\alpha m_\beta \to 1, \tag{26.7.9}$$

and, if β and κ diverge, we would obtain the corresponding limiting relation

$$(c_p/T\tilde{v})\kappa/\beta^2 \to 1. \tag{26.7.10}$$

Summary of important equations

A. *Defining equations*

$$\mathrm{d}e_\alpha = e_\alpha^{(1)}\,\mathrm{d}\varepsilon, \tag{26.2.5}$$

then

$$\mathrm{d}s = (c_{\alpha\beta} m_\beta e_\alpha^{(1)})\,\mathrm{d}\varepsilon, \tag{26.2.6}$$

and

$$\mathrm{d}s/\mathrm{d}\varepsilon = c_{\alpha\beta} m_\beta e_\alpha^{(1)} = \lambda_1 m_\beta e_\beta^{(1)}. \tag{26.2.7}$$

$$\boldsymbol{m} = \mu_\alpha \boldsymbol{e}^{(\alpha)},$$

$$= \mu_1 \boldsymbol{e}^{(1)} + \mu_2 \boldsymbol{e}^{(2)} + \cdots + \mu_6 \boldsymbol{e}^{(6)}. \tag{26.2.8}$$

$$q_1 = \mathrm{d}s/\mathrm{d}\varepsilon = \lambda_1 \mu_1 \to Q_1. \tag{26.2.9}$$

$$x = \lim_{\lambda_1 \to 0} \{\partial(\ln q_1)/\partial(\ln \lambda_1)\}, \tag{26.2.11}$$

$$Q_1 = \lim q_1 \propto \lim \lambda_1^x. \tag{26.2.12}$$

B. *If* $\boldsymbol{m}$ *or* β *diverges*

$$\boldsymbol{m} \to (q_1/\lambda_1)\boldsymbol{e}^{(1)}, \tag{26.2.13}$$

$$m_i/m_j \to e_i^{(1)}/e_j^{(1)}, \tag{26.2.14}$$

$$\beta \to (q_1/\lambda_1)(E_\alpha e_\alpha^{(1)}). \tag{26.2.17}$$

C. *If* c_p *diverges*

$$(c_p/T\tilde{v}) \to \mu_1^2 \lambda_1 = \lim(q_1 \mu_1). \tag{26.3.5}$$

D. *The adiabatic stiffness matrix*

$$e_\alpha^{(1)} e_\beta^{(1)} c_{\alpha\beta}^{(s)} \to \sigma Q_1^2, \tag{26.4.9}$$

$$\det(\boldsymbol{c}^{(s)}) \to C Q_1^2, \tag{26.4.12}$$

where

$$C = \sigma \lambda_2 \lambda_3 \lambda_4 \lambda_5 \lambda_6 \tag{26.4.13}$$

and

$$\sigma = \tilde{v} T / c_e = T/\rho_0 c_e. \tag{26.4.2}$$

Example

E. *The Pippard relations*

$$(c_p/T\tilde{V}) = \gamma\beta, \tag{26.5.1}$$

$$\beta = \gamma\kappa \tag{26.5.2}$$

where

$$\gamma = (\mathrm{d}p/\mathrm{d}T_0)_\lambda; \tag{26.5.3}$$

$$c_p/T\tilde{V} = (\mathrm{d}P_a/\mathrm{d}T_0)_\lambda m_a = (\mathrm{d}P_c/\mathrm{d}T_0)_\lambda m_c, \tag{26.5.6}$$

$$s_{ij} = (\mathrm{d}T_0/\mathrm{d}P_i)_\lambda m_j = m_i(\mathrm{d}T_0/\mathrm{d}P_j)_\lambda, \tag{26.5.8}$$

$$m_i/m_j = \Delta e_i/\Delta e_j. \tag{26.5.8}$$

F. *Linear limiting relations for diverging quantities when x is zero*

$$c_p/T\tilde{v} = Q_1(E_\alpha e_\alpha^{(1)})^{-1}\beta, \tag{26.6.1}$$

$$= (\mathrm{d}s/\mathrm{d}J)_N\beta, \tag{26.6.2}$$

$$= (\mathrm{d}\tilde{s}/\mathrm{d}\tilde{V})_N\beta, \tag{26.6.5}$$

$$\beta = (\mathrm{d}\tilde{s}/\mathrm{d}\tilde{V})_N\kappa, \tag{26.6.8}$$

$$c_p/T\tilde{v} = m_i(\mathrm{d}s/\mathrm{d}e_i)_N. \tag{26.6.10}$$

$$s_{\alpha\beta} = (\mathrm{d}e_\alpha/\mathrm{d}s)_N m_\beta = m_\alpha(\mathrm{d}e_\beta/\mathrm{d}s)_N. \tag{26.6.13}$$

Example

1. Use (16.3.11) and (26.3.5) to deduce (26.4.10).

27

The α–β quartz transition

The isothermal stiffness matrix is determined from experimental data, for temperatures near T_0. It is shown that one eigenvalue tends strongly to zero from both sides of the transition. The transition is treated as being associated with a thermodynamic instability. The analysis of the experimental data is consistent with this treatment. Q_1 is shown to be finite and non-zero for the α phase, and zero for the β phase. Experimentally determined estimates are given for limiting values of thermodynamic quantities for both phases. A numerical estimate for $\Delta\tilde{s}$ is obtained using the experimental value of $\Delta\boldsymbol{e}$ and the value of Q_1 for the α phase. This estimate is in good agreement with the experimentally determined value.

27.1. A thermodynamic description of the α–β quartz transition

The low temperature form of quartz is known as α quartz (sometimes as low quartz). This form belongs to the trigonal crystal class and has the point group symmetry 32. When heated at atmospheric pressure, it transforms to β quartz (high quartz) at about 574 °C. β quartz has a higher symmetry than α quartz and belongs to the hexagonal crystal class, with a point group symmetry 622. On cooling, β quartz transforms back to α quartz, usually at a temperature 1 or 2 °C below that at which the α phase transforms on heating.

In the α phase, the transition is heralded well in advance, at temperatures at least 150 °C less than T_0, the transition temperature, when a rapid increase begins to occur in the specific heat c_p, as well as in the coefficients m_a and m_c of thermal expansion along the a and c crystal axes.

In the transition, small discontinuities are observed in the volume, entropy and shape, and these are estimated by Berger *et al.* (1965, 1966) as follows:

$$\Delta\tilde{V} = 0.15 \text{ cm}^3(\text{mole})^{-1}; \qquad \Delta\tilde{s} = 0.15 \text{ cal (mole} \cdot {}^\circ\text{C})^{-1};$$

$$\Delta c/c = \Delta e_c = 0.0012; \qquad \Delta a/a = \Delta e_a = 0.0027. \tag{27.1.1}$$

In (27.1.1), $\Delta\tilde{V}$, Δe_a and Δe_c were determined from X-ray diffraction data.

In the β phase, m_a and m_c remain finite, relatively constant, and small. This is illustrated in fig. 16, which displays the thermal expansions of the lattice constants a and c, as measured by Jay (1933). This figure shows the striking difference between the thermal expansions in the two phases.

According to (24.3.6) the changes in T_0, with axial loading, are given by

$$\left.\begin{aligned} \mathrm{d}T_0/\mathrm{d}P_a &= (\Delta e_a/\Delta s), \\ \mathrm{d}T_0/\mathrm{d}P_c &= (\Delta e_c/\Delta s), \end{aligned}\right\} \tag{27.1.2}$$

where (see chapter 26, section 5) the axial stresses P_a and P_c are positive when compressive. Thus

$$(\mathrm{d}T_0/\mathrm{d}P_a)/(\mathrm{d}T_0/\mathrm{d}P_c) = (\Delta e_a/\Delta e_c). \tag{27.1.3}$$

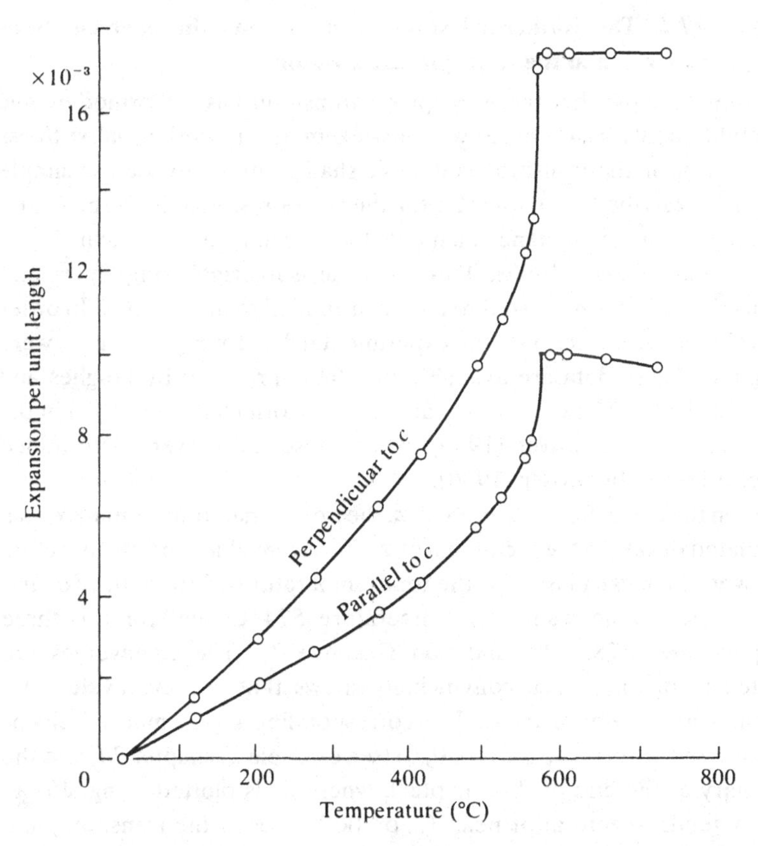

Fig. 16. Thermal expansion of quartz parallel and perpendicular to the c-axis (after Jay, 1933).

If the data of (27.1.1) are used, the rate of change of T_0 with respect to P_a is 2.25 times its rate of change with respect to P_c.

Coe and Paterson (1969) have experimentally tested (27.1.3), by directly measuring the rates of change of T_0. Their results yield the ratio

$$(\mathrm{d}T_0/\mathrm{d}P_a)/(\mathrm{d}T_0/\mathrm{d}P_c) = 2.1 \pm 0.2, \tag{27.1.4}$$

which is in reasonable agreement with the right-hand side of (27.1.3) as measured using X-ray diffraction methods. In their paper Coe and Paterson not only describe their experimental work, but give an excellent account of the experimental difficulties encountered when studying physical properties very close to a phase transition. They also review, comprehensively, other experimental work on thermodynamic properties near the α–β quartz transition.

27.2. The isothermal stiffness matrix and the coefficients of expansion at the α–β quartz transition

In order to show that the α–β quartz transition has a thermodynamic instability as its basic cause, we must examine the isothermal stiffness matrix $\boldsymbol{c}$, near the transition state. We shall continue to use the simple notation described in chapter 26 for the zero stress condition (p. 273).

The adiabatic compliance elements have been measured, near T_0, by Zubov and Firsova (1962). To obtain the isothermal compliance elements the correction of (16.3.6) has been applied to their results. In order to evaluate the correction term, experimental data for m_a, m_c and c_p were required. These data are available in a table provided by Hughes and Lawson (1962). The data for m_a and m_c were originally published in the doctoral thesis of Mayer (1959), while those for c_p were determined experimentally by Moser (1936).

When the correction was applied, $\boldsymbol{c}$, the isothermal stiffness matrix, was calculated by taking the reciprocal of $\boldsymbol{s}$. The eigenvalues and eigenvectors of $\boldsymbol{c}$ were then calculated for the four temperatures, 510, 540, 556, and 571 °C, below the transition temperature 574 °C, and for the three temperatures, 578, 590, and 600 °C above T_0. The eigenvalues are plotted in fig. 17 which convincingly shows that one eigenvalue, λ_1, vanishes at or very near T_0. The corresponding un-normalised eigenvector is of the form $(1, 1, a_1, 0, 0, 0)$ (see example 2, chapter 24, and the summary at the end of this chapter), where a_1 is plotted in fig. 19. λ_1 clearly tends to zero at or near T_0, on both sides of the transition, and particularly strongly on the high temperature β side (McLellan, 1973). The other five eigenvalues show no tendency to vanish, in fact they

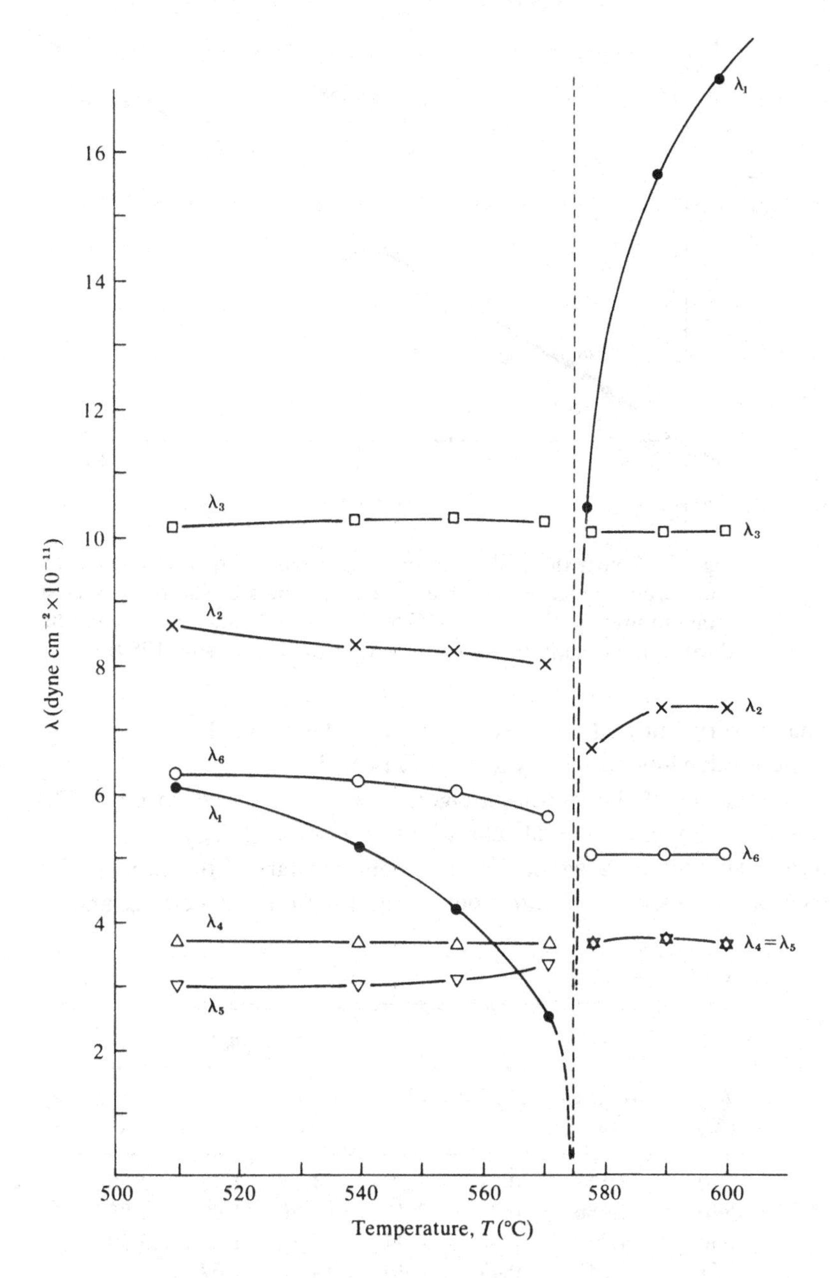

Fig. 17. The eigenvalues of the isothermal stiffness matrix of quartz (McLellan, 1973).

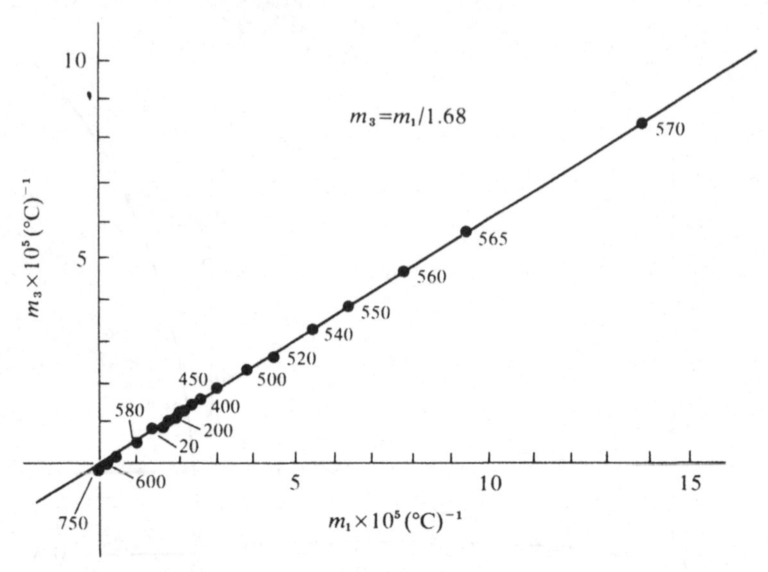

Fig. 18. Linear thermal expansion coefficients of quartz parallel (m_3) and perpendicular (m_1) to the c-axis. Numbers beside the points are temperatures in °C. From 20 °C to 570 °C, the data fits a straight line through the origin: $m_1 = 1.68m_3$ (after Coe & Paterson, 1969).

change very little as T_0 is approached. Thus, there is no doubt that there is a thermodynamic instability at or very near T_0.

The values of the vanishing eigenvalue λ_1 are shown in table 27.1, together with experimental values for the α phase of m_a, m_c, m_a/m_c and $\lambda_1 m_a$. This table shows how surprisingly constant is the ratio m_a/m_c. Figure 18 illustrates, in more detail, the relation between m_a and m_c,

Table 27.1

T (°C)	λ_1 (dyne cm^{-2} $\times 10^{-11}$)	a_1 (°C$^{-1} \times 10^5$)	m_c	m_a	(m_a/m_c)	$\lambda_1 m_a$ (°C^{-1} $\times 10^{-6}$)
510	6.138	0.21	2.39	4.15	1.73	2.55
540	5.186	0.24	3.25	5.50	1.69	2.85
556	4.220	0.36	4.34	7.34	1.69	3.10
571	2.540	0.47	8.40	14.00	1.67	3.50
578	10.35	2.00	0.83	1.9	2.29	β phase
590	15.62	1.16	—	—	—	β phase
600	17.01	1.12	—	—	—	β phase

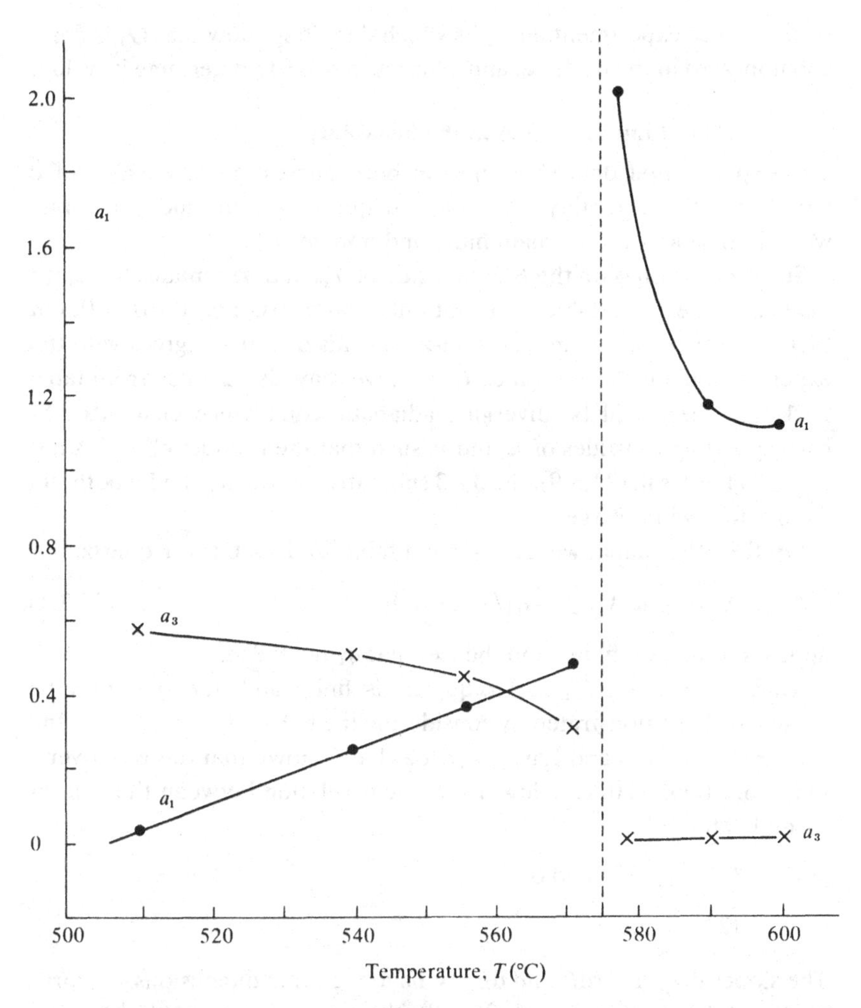

Fig. 19. a_1 and a_3, the eigenvector components for quartz, plotted against temperature; see summary of important equations, chapter 27.

which is accurately linear, in the α phase, over a wide range of temperature up to within a few degrees of T_0, which is as close to T_0 as they have been measured. Independent measurements, by Berger et al. (1966), Jay (1933) and Coenen (1963), show that m_a/m_c is constant through the temperature range 100 °C to 570 °C, and is given by

$$(m_a/m_c) = 1.68 \pm 0.04. \tag{27.2.1}$$

From table 27.1, we may note that $\lambda_1 m_a$, and hence q_1, appears to remain finite and non-zero in the α phase. We shall discuss in the next section other experimental results which definitely show that Q_1 is finite and non-zero in the α phase, and which also serve to determine its value.

27.3. Limiting values at the instability

The experimental data show that in both phases, an eigenvalue of $\boldsymbol{c}$ vanishes at the instability, and that in α quartz, m_a, m_c and c_p diverge, while in β quartz they remain finite and non-zero.

Since λ_1 vanishes on the β quartz side of T_0, and $\boldsymbol{m}$ remains finite, we may conclude, from table 26.1, not only that $q_1 \to 0$, i.e., $Q_1(\beta)=0$, but that c_p must remain finite and non-zero – this of course agrees with the experimental conclusion. Since $Q_1=0$, we may also deduce from table 26.1 that there will be diverging adiabatic compliance elements $s^{(s)}_{\mu\nu}$, corresponding to values of μ and ν, such that the product $e^{(1)}_\mu e^{(1)}_\nu$ is not zero. That is, as in (24.5.9), the 3×3 submatrix in which μ and ν both run from 1 to 3 will diverge.

On the other hand, we have seen in table 27.1 that, for α quartz,

$$\lambda_1 m_a = \mu_1\lambda_1 e^{(1)}_a = q_1/\sqrt{(2+a_1^2)} \tag{27.3.1}$$

appears to remain finite, and, hence, that q_1 does also.

That $Q_1(\alpha)$, i.e., Q_1 for α quartz, is finite and non-zero may be persuasively demonstrated by considering the behaviour of $c_p/T\tilde{v}$, β and κ near T_0. Hughes and Lawson (1962) have shown that there is, over a range of about 170 °C below T_0, a linear relation between the experimental values of

(1) $(c_p/T\tilde{v})$ and β,

(2) β and κ.

The slopes $\mathrm{d}(c_p/T\tilde{v})/\mathrm{d}\beta$ and $\mathrm{d}\beta/\mathrm{d}\kappa$ have the same dimensions. According to (26.6.1), (26.6.5) and (26.6.8), these two slopes should have the same value, given by

$$(\mathrm{d}\tilde{s}/\mathrm{d}\tilde{V})_N = Q_1(\alpha)(E_\mu e^{(1)}_\mu)^{-1}, \tag{27.3.2}$$

for α quartz. According to Hughes and Lawson (1962),

$$\mathrm{d}(c_p/T\tilde{v})/\mathrm{d}\beta = 2.6\times 10^7 \text{ dyne cm}^{-2}(^\circ\mathrm{C})^{-1}, \tag{27.3.3}$$

while, in the same units

$$\mathrm{d}\beta/\mathrm{d}\kappa = 3.6\times 10^7. \tag{27.3.4}$$

Now, we know from experiment that μ_1 diverges, and therefore

$$\boldsymbol{m} \to \mu_1 \boldsymbol{e}^{(1)}, \tag{27.3.5}$$

and

$$m_a/m_c \to a_1^{-1}. \tag{27.3.6}$$

Since the limiting value of m_a/m_c is 1.68 (see (27.2.1)), we may conclude that the limiting value of

$$a_1 = (1.68)^{-1} = 0.59. \tag{27.3.7}$$

Hence the limiting normalised eigenvector corresponding to λ_1 is given by

$$\boldsymbol{e}^{(1)} = (0.65, 0.65, 0.38, 0, 0, 0), \tag{27.3.8}$$

and, thus, in the limit at the instability,

$$E_\mu e_\mu^{(1)} = 1.68. \tag{27.3.9}$$

Therefore, on substituting (27.3.3) and (27.3.9) into (27.3.2), we find that

$$Q_1(\alpha) = 4.40 \times 10^7 \text{ dyne cm}^{-2}(°\text{C})^{-1}. \tag{27.3.10}$$

If we substitute (27.3.4), instead of (27.3.3), in (27.3.2), we obtain, in the same units,

$$Q_1(\alpha) = 6.00 \times 10^7. \tag{27.3.11}$$

There is a 35 per cent discrepancy between these two experimental estimates of $Q_1(\alpha)$. Hughes and Lawson (1962) have shown that there is a very large probable experimental error in the determination of $\mathrm{d}(c_p/T\tilde{v})/\mathrm{d}\beta$, and a very much smaller probable error in that of $\mathrm{d}\beta/\mathrm{d}\kappa$. We shall follow these authors in taking the value of $Q_1(\alpha)$ as that given in (27.3.11). This value is confirmed by a further independent estimate, empirically obtained, as follows.

First, we evaluate, at the temperatures mentioned earlier,

$$\mu_1 = \boldsymbol{m} \cdot \boldsymbol{e}^{(1)} = (2m_a + a_1 m_c)/\sqrt{(2 + a_1^2)}, \tag{27.3.12}$$

and thus we may determine $q_1 = \mu_1 \lambda_1$ at these temperatures. If we calculate $q_1/a_1^{\frac{1}{2}}$, we find that this quantity tends to a finite non-zero value. As may be seen from table 27.2, we may take this limiting value as

$$\lim(q_1/a_1^{\frac{1}{2}}) = 7.90 \times 10^7 \text{ dyne cm}^{-2}(°\text{C})^{-1}. \tag{27.3.13}$$

Hence, if we take the limiting value of a_1 equal to 0.59, as in (27.3.7), we find that

$$Q_1(\alpha) = 6.10 \times 10^7 \text{ dyne cm}^{-2}(°\text{C})^{-1}. \tag{27.3.14}$$

Table 27.2.

T °C	a_1	μ_1 $(\times 10^5)$	λ_1 $(\times 10^{-11})$	q_1 $(\times 10^{-7})$	$q_1/a_1^{\frac{1}{2}}$ $(\times 10^{-7})$
510	0.209	6.65	6.14	4.10	8.90
540	0.238	8.20	5.19	4.25	8.70
556	0.360	11.10	4.22	4.70	7.85
571	0.470	21.30	2.54	5.41	7.90
$m_c/m_a = 0.59$.					
$Q_1(\alpha) = \lim(a_1^{\frac{1}{2}}) \lim(q_1/a_1^{\frac{1}{2}})$				6.10	
$= (E_\alpha e_\alpha^{(1)})(\mathrm{d}\beta/\mathrm{d}\kappa)$				6.00	
$= (E_\alpha e_\alpha^{(1)})\, \mathrm{d}(c_p/T\tilde{v})/\mathrm{d}\beta$				4.40	

α quartz quantities are in c.g.s. units.

27.4. The transition, the instability and the entropy change

In the previous two sections we considered the values of, and the limiting relations between, thermodynamic quantities at the instability. For example, we saw that m_a/m_c is constant in the α phase over a wide range of temperature almost up to T_0. Although the path to the instability is interrupted near T_0 by fluctuations, this constant ratio is clearly the limiting ratio at the instability. However, when the state is only a few degrees from T_0, this ratio may not be constant in that small range, since the thermal expansion may be affected by fluctuations. In this book we shall not deal with fluctuation effects. However, we see from our study of the α–β quartz transition that, to within a few degrees of T_0, the limiting behaviour is that appropriate to the approach to the thermodynamic instability of the bulk homogeneous phase.

The basic microscopic process leading to the α–β quartz transition is considered to be the vanishing of the frequency of a certain optical mode of vibration. While this frequency is non-zero the phase remains stable, since it is only when the frequency vanishes that the lattice structure becomes dynamically unstable (Born and Huang, 1954). If, as T increases, this frequency could be brought to zero in a homogeneous manner, the specimen would reach the instability point. At zero frequency the vibrational mode becomes a static displacement of the silicon and oxygen ions. The ionic displacements arising from this mode are relative to a fixed lattice and centre of mass. They may be described by

$$\boldsymbol{u}_i = A\boldsymbol{\xi}_i, \tag{27.4.1}$$

where $\boldsymbol{u}_i$ is the displacement of the ith ion in the Bravais cell. The vectors

ξ_i, suitably normalised, describe the mode displacement, while A is the amplitude or principal co-ordinate of the mode. At zero frequency the displacement becomes a static one, and the displacement of the ions will be given by (27.4.1). At a particular value of A, and at this value only, the crystal attains the higher hexagonal symmetry of the β phase. To accommodate this displacement the lattice dimensions change slightly, in the transition, by $\Delta \boldsymbol{e}$.

We have said above that only at a single particular value of A in the static displacement, does the crystal attain the higher symmetry form. This point is of general importance, and perhaps needs some explanation. β quartz differs from α quartz in having two perpendicular two-fold symmetry axes at right angles to the common trigonal or hexagonal c-axis, whereas α quartz has only one of these two-fold axes (Bragg and Claringbull, 1965). Thus, for example, the silicon ions in β quartz are equivalent in pairs under the two-fold axis which is absent in α quartz. By equivalent we mean that a rotation of 180° about the two-fold axis would bring one silicon ion of each pair into coincidence with the other member of the pair. In α quartz these ions are inequivalent. In the transition $\alpha \rightarrow \beta$, the static displacement (27.4.1) will produce this equivalence and it is easy to see that these pairs will become equivalent pairs for a unique value A_0 of A. If A is less than or greater than A_0, the symmetry will be that of the lower α form. The amount of the displacement (27.4.1) must be uniquely chosen so that the specimen now has the extra two-fold symmetry axis.

The quantity $A - A_0$ is an example of an order parameter (see Landau and Lifshitz, 1958). It is non-zero in the low symmetry phase, and its value may vary in that phase, while in the high symmetry phase its value must be zero.

Now suppose that in the $\alpha \rightarrow \beta$ transition, $A - A_0$ becomes zero only when the change of shape is given by $\Delta \boldsymbol{e}$. In that case, during the transition the specimen is in the lower symmetry α phase, and it is only at the end point of the transition that the higher symmetry β phase appears. If we assume this, we can estimate $\Delta \tilde{s}$ in terms of $\Delta \boldsymbol{e}$ as follows.

Since the specimen is in the α phase during the transition we may take

$$\Delta s = \int q_1 \, \mathrm{d}\varepsilon \simeq Q_1(\alpha)\, \Delta\varepsilon, \tag{27.4.2}$$

where

$$\Delta\varepsilon = (\boldsymbol{e}^{(1)} \cdot \Delta \boldsymbol{e}). \tag{27.4.3}$$

If we take $\boldsymbol{e}^{(1)}$, $\Delta \boldsymbol{e}$ and $Q_1(\alpha)$, as given by (27.3.8), (27.1.1) and (27.3.10) respectively, we find that

$$\Delta\tilde{s} = 0.14 \text{ cal(mole}\cdot{}^\circ\text{C)}^{-1}, \tag{27.4.4}$$

in good agreement with the value quoted in (27.1.1).

Summary of important equations

A. *$\alpha \to \beta$ quartz transition*

$$\Delta\tilde{V} = 0.15 \text{ cm}^3(\text{mole})^{-1}; \qquad \Delta\tilde{s} = 0.15 \text{ cal(mole}\cdot{}^\circ\text{C)}^{-1};$$

$$\Delta c/c = \Delta e_c = 0.0012; \qquad \Delta a/a = \Delta e_a = 0.0027. \tag{27.1.1}$$

$$\mathrm{d}T_0/\mathrm{d}P_a = (\Delta e_a/\Delta s); \qquad \mathrm{d}T_0/\mathrm{d}P_c = (\Delta e_c/\Delta s), \tag{27.1.2}$$

$$(\mathrm{d}T_0/\mathrm{d}P_a)/(\mathrm{d}T_0/\mathrm{d}P_c) = (\Delta e_a/\Delta e_c). \tag{27.1.3}$$

$$(\mathrm{d}\tilde{s}/\mathrm{d}\tilde{V})_N = Q_1(\alpha)(E_\mu e_\mu^{(1)})^{-1}, \tag{27.3.2}$$

for α quartz.

B. *Eigenvector scheme for α and β quartz*

	λ_1	λ_2	λ_3	λ_4	λ_5	λ_6
e_1	1	1	1	$-(a_3/2)$	0	0
e_2	1	1	-1	$(a_3/2)$	0	0
e_3	a_1	$-(2/a_1)$	0	0	0	0
e_4	0	0	a_3	1	0	0
e_5	0	0	0	0	1	$-a_5$
e_6	0	0	0	0	a_5	1 .

Examples

1. Prove, for α and β quartz, that

$$\boldsymbol{m} = \mu_1\boldsymbol{e}^{(1)} + \mu_2\boldsymbol{e}^{(2)},$$

that is, that $\mu_i = 0$, $i = 3, 4, 5, 6$.

2. Use example 1 to prove, for α quartz, that four eigenvalues $\lambda_i^{(s)}$, of $c^{(s)}$ are each respectively equal to the eigenvalues λ_3, λ_4, λ_5, λ_6 of $\boldsymbol{c}$, and that the corresponding eigenvectors are equal.
3. Use (26.4.12) to prove, for α quartz, that

$$\lambda_1^{(s)}\lambda_2^{(s)} = \sigma Q_1^2\lambda_2.$$

Assume that $\lambda_2^{(s)} \simeq \lambda_2$, and show that at the α quartz instability surface,

$$\lambda_1^{(s)} \simeq 1.2\times 10^{11} \text{ dyne cm}^{-2}.$$

(Take $\tilde{v} = 0.4\ \mathrm{cm^3\ g^{-1}}$; $c_{\boldsymbol{e}} = 0.27\ \mathrm{cal\ {}^\circ K^{-1}\ g^{-1}}$. This value of $c_{\boldsymbol{e}}$ was obtained by calculating $c_{\boldsymbol{e}}$ for the temperatures 510, 540, 556, and 571 °C, by applying the correction equation (11.3.35) to the experimental data for c_p, $\boldsymbol{m}$ and $\boldsymbol{c}$. The value of $c_{\boldsymbol{e}}$ so calculated was substantially the same at all four temperatures.)

28

The thermodynamic theory of the growth of Dauphiné twinning in quartz under stress

The equilibrium condition is deduced for the stable coexistence, under stress, of Dauphiné twins. We shall show that this condition is determined by the principle that a generalised Gibbs free energy be minimised. At equilibrium the generalised Gibbs free energy density of each twin must have the same value. If stresses are applied such that the equilibrium condition is not satisfied, this principle determines which twin will survive the resulting untwinning process. The theory is developed rigorously using finite strain theory. We shall prove that if four relations between the elements of a certain symmetric tensor are satisfied, the stress coexistence condition is satisfied exactly, that is, to any order. The proof of this exact temperature-independent condition employs the method of the integrity basis for invariant functions of the crystal point group.

We shall show that when these temperature-independent coexistence conditions are satisfied, the Helmholtz free energy density for each twin must have the same value. We shall also prove that in the linear infinitesimal elasticity regime, the above minimum principle is equivalent to the empirically enunciated principle of maximum elastic energy storage.

28.1. Description

Natural quartz crystals are frequently twinned. There are three principal types of twin, good descriptions of which are given by Bragg and Claringbull (1965).

Certain properties of a quartz crystal are affected if the crystal consists of a pair of twins. For instance, if Dauphiné twinning, to be discussed in this chapter, occurs, then the usual piezoelectric property of a quartz crystal may not be exhibited.

In Dauphiné twinning, the individual twins may be brought into coincidence, if one twin is rotated by an angle of 180° about the quartz c-axis. This geometrical operation is not a symmetry operation for α quartz, although it is for β quartz. Thus Dauphiné twinning disappears in the β form. Conversely, if a single crystal of β quartz is cooled below T_0, it may split up into Dauphiné twins.

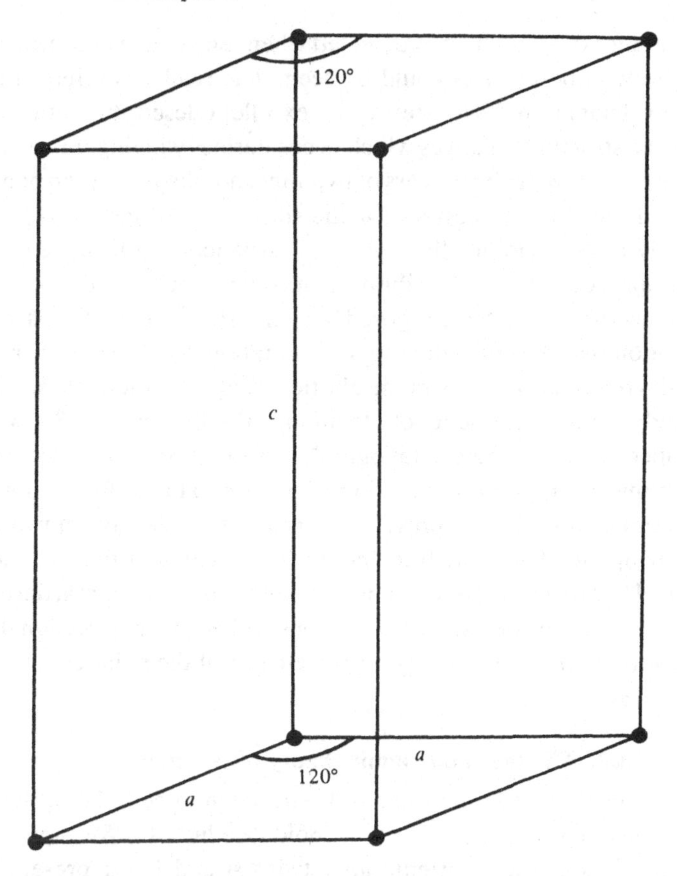

Fig. 20. The unit cell for α quartz.

The unit cell of α quartz is shown in fig. 20. It is a prism with a rhombus as base. The sides of the rhombus are equal to the lattice constant a, and its angle is 120°. This prism is unchanged by a rotation of 180° about the c-axis, and thus it is the unit cell for either orientation in Dauphiné twinning.

An important use of quartz crystals, which depends on the piezoelectric property, is in the stabilisation of radio frequencies in telecommunication equipment. It is necessary that untwinned single crystals be available for this purpose. In the hope of increasing the supply of such crystals, Thomas and Wooster (1951) studied the effect, on twinning, of the application of mechanical and thermal stress. The experimental technique which they used was to apply different stresses, at elevated temperatures of about 400 °C, to quartz plates and bars, cut to different crystallographic orientations. They were able to study the resulting twinning or untwinning.

The presence of twinning was detected when, on etching the surface with acid, the twin boundaries could be seen. The reader is referred to the paper by Thomas and Wooster for an excellent description of the twins and of the structural changes which occur during twinning. In particular, they describe how, in the process of twinning, no silicon–oxygen bonds are broken. It is merely necessary for the silicon and oxygen ions to move relatively to one another through small distances of the same order as their amplitudes of thermal vibration at room temperature.

Thomas and Wooster proposed an empirical theory based on the assumption that the crystallographic orientation developed in a crystal is that which stores the maximum elastic energy. In the next section, a thermodynamic treatment of twinning (McLellan, 1978) will be developed, in which the crystallographic orientation will be determined by minimising a generalised Gibbs function. This will be shown, in infinitesimal strain theory only, to be, rather surprisingly, equivalent to maximizing the Helmholtz free energy. This treatment thus justifies the empirical treatment of Thomas and Wooster. An exact general derivation of the equilibrium coexistence conditions will be given in section 4. This derivation will use the integrity basis elements of the point group 32 (see chapter 18).

28.2. The thermodynamic theory of twinning

For simplicity, we consider the quartz specimen to be a slab which in its reference state is in the shape of a cuboid (see fig. 21). We suppose that the two crystallographic orientations, twins A and B, are present in the specimen, and that the reference state of the zone boundary between the twins is parallel to an external face of the slab. We shall consider

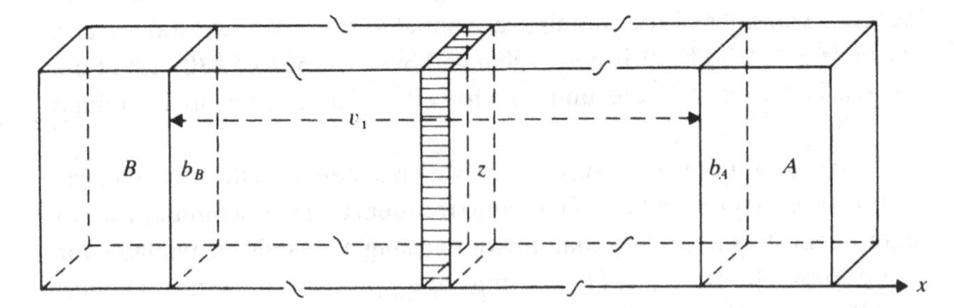

Fig. 21. The twinned specimen referred to the reference state. The reference region of the zone boundary between the twins is denoted by z. The cross-sections b_A and b_B enclose the region v_1; their neighbourhoods are homogeneous and of uniform properties in the deformed state under the applied forces.

variations in which the reference state of the zone boundary moves at right angles to itself, so that its size and properties remain unaltered. In such variations the zone boundary will make no contributions to changes in, for example, the Helmholtz free energy of the specimen.

We shall study the stability of the twinned quartz when it is under dead load conditions which we shall specify shortly. Suppose that there is a small isothermal variation in which the zone boundary moves, as above, so that an amount δn_B, measured in moles, is converted from twin A to twin B. That is,

$$\delta n_B = -\delta n_A. \qquad (28.2.1)$$

The condition of stable equilibrium (19.1.17) becomes, for isothermal variations,

$$\delta F - \delta W \geqslant 0, \qquad (28.2.2)$$

where δW is the work done on the specimen by the forces acting on it under dead load conditions.

We must now specify these forces. This specification is to be made relative to the reference state taken as that for which the specimen is wholly in the form of one of the twins and is acted on by no forces. Since only isothermal variations will be considered, the temperature of the specimen at the reference state may be taken as that of the twinned state under study.

The forces on the specimen are chosen so that the force acting on each and every surface element of the external faces of the specimen is given by (6.4.21), where $\boldsymbol{B}$ has the same value over the external surface, that is, $\boldsymbol{B}$ is uniform over the external faces. In order for dead load conditions to apply, $\boldsymbol{B}$ must not only be uniform over the external surface but constant. The surface forces are thus specified relative to the reference state, while we consider them applied to the twinned state under study. The force, in this state, acting on an external surface element $\mathrm{d}\boldsymbol{\Sigma}$ will be determined by the reference state, $\mathrm{d}\boldsymbol{\sigma}$, of this surface element, according to (6.4.21).

In any variation considered, this force acting on a surface element will be held constant. From (6.4.21), since $\mathrm{d}\boldsymbol{\sigma}$ is a constant vector, this means that $\boldsymbol{B}$, as defined above in terms of the forces acting on external surface elements, will remain constant as well as uniform over the external faces.

The zone boundary between the twins is a narrow region of inhomogeneous structure. Thus, although the above surface force distribution is uniform in the reference state, the stress tensor is not uniform in a macroscopic region about the zone boundary. The Boussinesq stress tensor will be uniform and equal to $\boldsymbol{B}$ in those parts of the specimen which are sufficiently far from the zone boundary. In fact, we may choose the specimen to be as long as is necessary to ensure that this is so.

The twinned state is deformed relative to the reference state.

We now consider the isothermal variation in which a small amount δn_B of twin A is transformed to twin B, in such a way that, referred to its reference state, the zone boundary is translated by a small displacement $\boldsymbol{u}$ which is perpendicular to the zone boundary and parallel to the longest edge of the reference state of the cuboid specimen.

To evaluate the work done in the above variation, we consider the specimen to be divided into three parts by two cross-sections whose reference states are b_B and b_A (see fig. 21). These cross-sections enclose a middle region, of reference state v_1, which includes the zone boundary region z. The cross-sections, b_B and b_A, are chosen to be so far from the zone boundary that they are imbedded deeply in the parts of the specimen which are homogeneous and of uniform properties under the applied forces. Thus, in the neighbourhoods of b_B and b_A the deformation matrices will be uniform and equal, respectively, to the matrices $\boldsymbol{D}^B$ and $\boldsymbol{D}^A$ describing the homogeneous deformations of the parts of the specimen, which as discussed above, are remote from the zone boundary. The Boussinesq stress tensor in the parts of the specimen outside v_1 will likewise be uniform and, in each part, will be equal to the tensor $\boldsymbol{B}$, as defined above in terms of the external surface forces. The part v_1 may be a substantial part of the crystal, in fact, it may be almost all of it.

Since the applied force distribution is uniform in the reference state, the deformation matrix at any point $\boldsymbol{x}$ in the region v_1 will, in the variation, change from $\boldsymbol{D}(\boldsymbol{x})$ to $\boldsymbol{D}(\boldsymbol{x}-\boldsymbol{u})$. The work done by the external surface forces in the variations is given by

$$\delta W=\int_\sigma \mathrm{d}F_\alpha\,\delta X_\alpha=\int_\sigma B_{\beta\alpha}\,\delta X_\alpha\,\mathrm{d}\sigma_\beta=\int_v B_{\beta\alpha}\,\delta D_{\alpha\beta}\,\mathrm{d}v, \qquad (28.2.3)$$

where σ and v are the reference states of the external surface and volume of the specimen. The third integral has been obtained by the use of the divergence theorem (5.2.2). Because $\delta\boldsymbol{D}$ is zero in the homogeneous regions outside v_1, and since $\boldsymbol{B}$, defined by the applied surface forces, is held constant throughout the variation, we may write

$$\delta W=B_{\beta\alpha}\int_{v_1}\{D_{\alpha\beta}(\boldsymbol{x}-\boldsymbol{u})-D_{\alpha\beta}(\boldsymbol{x})\}\,\mathrm{d}v. \qquad (28.2.4)$$

We may choose v_1 and the specimen as large as we like and $\boldsymbol{u}$ as small as we like. Hence, as discussed above, we can take it that $\boldsymbol{D}$ is uniform in the neighbourhoods of b_B and b_A over distances in excess of $|\boldsymbol{u}|$. Using this fact, we may evaluate (28.2.4) to give

$$\delta W = |\boldsymbol{u}|bB_{\beta\alpha}\Delta D_{\alpha\beta} \tag{28.2.5}$$

$$= \delta n_B \tilde{v} B_{\beta\alpha}\Delta D_{\alpha\beta}, \tag{28.2.6}$$

where $\tilde{v}$ is the specific reference volume, b is the cross-sectional area of the reference state of the specimen, and where

$$\Delta \boldsymbol{D} = \boldsymbol{D}^B - \boldsymbol{D}^A \tag{28.2.7}$$

describes the change in shape when twin A is transformed to twin B under the applied forces.

The isothermal variation will be stable if (28.2.2) holds, that is if

$$\delta n_B(f_B - f_A - B_{\beta\alpha}\Delta D_{\alpha\beta}) \geqslant 0, \tag{28.2.8}$$

where f_A and f_B are the Helmholtz free energies of twins A and B, per unit reference volume. Equation (28.2.8) may be written

$$\delta n_B(g'_B - g'_A) \geqslant 0, \tag{28.2.9}$$

where

$$g' = f - B_{\beta\alpha}D_{\alpha\beta} \tag{28.2.10}$$

is a generalised Gibbs free energy density. Thus we see that the condition of stable thermodynamic equilibrium for a twinned state is

$$g'_B - g'_A = 0 \tag{28.2.11}$$

and that the minimum principle for

$$G' = \tilde{v}(n_A g'_A + n_B g'_B) \tag{28.2.12}$$

determines the direction of untwinning, if (28.2.11) is not satisfied. For, if $g'_B - g'_A$ is positive, then, from (28.2.9) we see that the quartz specimen is unstable to variations in which δn_B is negative, and the specimen will untwin and be completely converted to twin A. If δn_B is positive, the corresponding variation is strongly stable. That is, if the zone boundary moves into twin A (δn_B positive), it will, as it were, bounce back and encroach on twin B, until the whole specimen is converted to twin A.

If $g'_B - g'_A$ is negative, the specimen will untwin and be completely converted to twin B. (We can say nothing, theoretically, about the rate of conversion. Thomas and Wooster have shown experimentally that conversion takes place rapidly at temperatures above 400 °C.)

Thus, the stable crystallographic orientation is that which minimises, for isothermal variations under dead load conditions, the generalised Gibbs function G'.

In section 3 we shall discuss the functional form of the generalised Gibbs function g'. In section 4 we shall then derive four temperature-independent coexistence conditions to be satisfied by the stress elements. Then in section 5 we shall be able to discuss the equilibrium condition (28.2.11) and its relation to the conditions to be satisfied by the Helmholtz free energy and the stored elastic energy.

28.3. The functional dependence of g' on the elements of $\boldsymbol{B}$

From (28.2.10) we see that

$$dg' = -s\,dT - D_{\alpha\beta}\,dB_{\beta\alpha}. \tag{28.3.1}$$

Hence

$$D_{\beta\alpha} = -\partial g'/\partial B_{\alpha\beta}, \tag{28.3.2}$$

where g' is to be regarded as a function of T and the elements of $\boldsymbol{B}$.

Under the dead load conditions assumed, $\boldsymbol{B}$ is to be the same in both twins, and $g'_B - g'_A$ is to be evaluated at the common temperature and for the same $\boldsymbol{B}$. For each twin, g' must be invariant to the transformations of the point group 32, and the study of this invariance will depend on how the elements of $\boldsymbol{B}$ appear in g'. In fact, we shall show that just as f and u are functions of only six strain variables, g' is a function of only six stress variables which may be chosen as the six independent elements of the symmetric matrix

$$\boldsymbol{\xi} = \boldsymbol{B}\boldsymbol{B}^{t}. \tag{28.3.3}$$

To prove this statement we need use only the condition that $\boldsymbol{DB}$ is a symmetric matrix, a fact which follows from (6.4.15b). This condition also suffices to show that f is a function of the six independent elements of $\boldsymbol{D}^{t}\boldsymbol{D}$, as the reader may wish to verify.

From (28.3.2) we see that

$$-(\boldsymbol{DB})_{\alpha\beta} = \frac{\partial g'}{\partial B_{\mu\alpha}} B_{\mu\beta}. \tag{28.3.4}$$

Instead of regarding g' as a function of the nine elements of $\boldsymbol{B}$, we introduce as nine new variables, first, the six independent elements of $\boldsymbol{\xi}$ and, second, the three variables defined by

$$\phi_{\alpha} = \delta_{\alpha\beta\gamma} B_{\beta\gamma}, \quad \alpha = 1, 2, 3. \tag{28.3.5}$$

($\delta_{\alpha\beta\gamma}$, the Levi–Civita tensor element, was introduced in section 2, chapter 2.) If, for example, we take $\alpha = 1$, (28.3.5) becomes

$$\phi_1 = B_{23} - B_{32}, \tag{28.3.6}$$

that is, the variables ϕ_α depend on the skew part of $\boldsymbol{B}$. Equation (28.3.4) can now be expressed as

$$-(\boldsymbol{DB})_{\alpha\beta} = \frac{\partial g'}{\partial \xi_{\mu\gamma}} B_{\gamma\alpha} B_{\mu\beta} + \frac{\partial g'}{\partial \phi_\kappa} \delta_{\kappa\mu\alpha} B_{\mu\beta}. \tag{28.3.7}$$

Since μ and γ are both dummy indices and thus can be interchanged, it can be seen that the first term on the right-hand side is symmetric to interchange of α and β, just as $(\boldsymbol{DB})_{\alpha\beta}$ must be. Thus, for $\alpha \neq \beta$, we must have

$$\frac{\partial g'}{\partial \phi_\kappa}(\delta_{\kappa\mu\alpha} B_{\mu\beta} - \delta_{\kappa\mu\beta} B_{\mu\alpha}) = 0. \tag{28.3.8}$$

For example, if $\alpha = 1$, $\beta = 2$, this equation becomes

$$\frac{\partial g'}{\partial \phi_1} B_{31} + \frac{\partial g'}{\partial \phi_2} B_{32} - \frac{\partial g'}{\partial \phi_3}(B_{11} + B_{22}) = 0. \tag{28.3.9}$$

Two further equations may be obtained by setting α, $\beta = 2$, 3 and 3, 1. Since these three equations must be satisfied by, for instance, any symmetric stress tensor $\boldsymbol{B}$, we see that they can be satisfied only if

$$\frac{\partial g'}{\partial \phi_i} = 0, \quad i = 1, 2, 3. \tag{28.3.10}$$

Therefore g' is independent of the skew variables ϕ_i, and may be regarded as a function of T and the six independent elements of $\boldsymbol{\xi}$.

The elements of $\boldsymbol{\xi}$ are quadratic in the elements of $\boldsymbol{\Theta}$, and, in fact, as

$$\boldsymbol{B} \to \boldsymbol{\Theta}, \qquad \boldsymbol{\xi} \to \boldsymbol{\Theta}^2. \tag{28.3.11}$$

In infinitesimal strain theory the generalised Gibbs function, when expanded in powers of the elements of $\boldsymbol{\Theta}$, would, in general, contain odd powers of these elements. It is thus desirable to consider g' as a function of a symmetric stress tensor $\boldsymbol{\chi}$ which is a square root of the symmetric tensor $\boldsymbol{\xi}$, and satisfies the property that as

$$\boldsymbol{B} \to \boldsymbol{\Theta}, \qquad \boldsymbol{\chi} \to \boldsymbol{\Theta}. \tag{28.3.12}$$

To show that we may obtain such a solution $\boldsymbol{\chi}$, of

$$\boldsymbol{\chi}^2 = \boldsymbol{\xi} = \boldsymbol{BB}^{\mathrm{t}}, \tag{28.3.13}$$

we first of all show that $\boldsymbol{\xi}$ is a non-negative definite matrix, i.e., if $\boldsymbol{y}$ is an arbitrary three-component vector, then

$$\boldsymbol{y}\cdot\boldsymbol{\xi}\cdot\boldsymbol{y}=y_\alpha \xi_{\alpha\beta} y_\beta \geqslant 0. \tag{28.3.14}$$

This follows from the fact that if we define a vector $\boldsymbol{x}$ such that

$$x_\mu = y_\alpha B_{\alpha\mu}, \quad \text{i.e., } \boldsymbol{x}=\boldsymbol{B}^t\boldsymbol{y}, \tag{28.3.15}$$

then

$$y_\alpha \xi_{\alpha\beta} y_\beta = y_\alpha B_{\alpha\mu} B_{\beta\mu} y_\beta = x_\mu x_\mu, \tag{28.3.16}$$

and hence this arbitrary quadratic form cannot be negative. In particular, if we choose $\boldsymbol{y}$ such that it is an eigenvector of $\boldsymbol{\xi}$, we can see that the eigenvalues of $\boldsymbol{\xi}$ must not be negative. Suppose that the eigenvalues and corresponding orthonormal eigenvectors are λ_i and $\boldsymbol{v}^i$, $i=1, 2, 3$. Then we can construct a solution $\boldsymbol{\chi}$ of (28.3.13), using the mathematical identity (24.5.2), by taking

$$\boldsymbol{\chi}=\sum_{i=1}^{3}\mu_i \boldsymbol{v}^i\boldsymbol{v}^i, \tag{28.3.17}$$

where

$$\mu_i^2=\lambda_i, \quad i=1,2,3. \tag{28.3.18}$$

In other words, $\boldsymbol{\chi}$ has the same eigenvectors as $\boldsymbol{\xi}$, and the corresponding eigenvalues are the square roots of those of $\boldsymbol{\xi}$. It is left to the reader to verify that this $\boldsymbol{\chi}$ satisfies (28.3.13). If no λ_i is zero, we may construct eight possible square root matrices of $\boldsymbol{\xi}$; if one λ_i is zero we may construct four; while, if two λ_i are zero we may construct two such matrices.

We may choose from these possible $\boldsymbol{\chi}$ the one which satisfies (28.3.12) by selecting that $\boldsymbol{\chi}$ which is closest, as judged by a suitable criterion, to the symmetric stress tensor

$$\boldsymbol{b}=\tfrac{1}{2}(\boldsymbol{B}+\boldsymbol{B}^t). \tag{28.3.19}$$

We may, for instance, choose that $\boldsymbol{\chi}$ for which

$$\sum_{\alpha,\beta}|\chi_{\alpha\beta}-b_{\alpha\beta}| \tag{28.3.20}$$

has the least value. If this criterion is not sufficient to select one single $\boldsymbol{\chi}$, the least value of

$$\sum_{\alpha,\beta}|\chi_{\alpha\beta}^3-b_{\alpha\beta}^3| \tag{28.3.21}$$

could be sought, and so on, to higher odd powers, as necessary. Even powers could possibly be used, but these may not be so effective in separating the possible $\boldsymbol{\chi}$'s, which all have the same $\boldsymbol{\chi}^2$. In practice

$$\boldsymbol{B} \simeq \boldsymbol{\Theta} \simeq \boldsymbol{\chi}, \tag{28.3.22}$$

and it would be possible to choose the correct $\boldsymbol{\chi}$ by inspection.

We can regard g' as a function of T and the six independent elements of $\boldsymbol{\chi}$, and we shall do this in section 4, where we shall derive the stress coexistence conditions for Dauphiné twins.

28.4. The stress conditions of coexistence

If we express g' as a polynomial in powers of the elements of $\boldsymbol{\chi}$, then it may be expressed as a polynomial in the integrity basis elements. As discussed in chapter 18 an integrity basis for a symmetry group is a set of homogeneous polynomials in the elements of $\boldsymbol{\chi}$, in this case, in terms of which all polynomial invariant functions of the elements of $\boldsymbol{\chi}$ may be expressed as a sum of products of the integrity basis elements. For α quartz, g' will be a polynomial expression in the integrity basis elements of the point group 32 (D_3). The coefficients will be functions of temperature only (McLellan, 1974).

There are four such integrity basis elements which change sign under the transformation induced on a second-rank tensor by a rotation of 180° about the c-axis. These are

$$D_3(\boldsymbol{\chi}) = \chi_4(\chi_4^2 - 3\chi_5^2), \tag{28.4.1}$$

$$N_2(\boldsymbol{\chi}) = (\chi_1 - \chi_2)\chi_4 + 2\chi_5\chi_6, \tag{28.4.2}$$

$$N_3(\boldsymbol{\chi}) = \{(\chi_1 - \chi_2)^2 - 4\chi_6^2\}\chi_4 + 4(\chi_2 - \chi_1)\chi_5\chi_6, \tag{28.4.3}$$

$$N_6(\boldsymbol{\chi}) = \chi_5\chi_6\{3(\chi_1 - \chi_2)^2 - 4\chi_6^2\}\{3\chi_4^2 - \chi_5^2\}. \tag{28.4.4}$$

In the above equations it should be emphasised that the abbreviated notation used is as follows, for the symmetric tensor $\boldsymbol{\chi}$,

$$\chi_1 = \chi_{11}; \quad \chi_2 = \chi_{22}; \quad \chi_3 = \chi_{33};$$

$$\chi_4 = \chi_{23}; \quad \chi_5 = \chi_{13}; \quad \chi_6 = \chi_{12}. \tag{28.4.5}$$

The polynomial $D_3(\boldsymbol{\chi})$ is a denominator invariant (McLellan, 1974), which can appear to any order in any invariant polynomial function which is expressed as a polynomial in the integrity basis elements. The other three elements N_2, N_3, N_6, are numerator invariants which can appear singly and to, at most, first order in any term of an invariant polynomial.

As discussed in chapter 18, the most general invariant polynomial is of the form

$$J+\sum_i N_i J_i, \tag{28.4.6}$$

where summation is over all the numerator invariants, and where J and each J_i are polynomials constructed from denominator invariants only.

We may write generally

$$\begin{aligned} g'_B - g'_A = {} & 2s^A_{14}N_2(\chi)\{1+I^+_1(\chi)\} \\ & +(2/3!)\alpha^A D_3(\chi)\{1+I^+_2(\chi)\} \\ & +(2/3!)\beta^A N_3(\chi)\{1+I^+_3(\chi)\} \\ & +(2/6!)\gamma^A N_6(\chi)\{1+I^+_4(\chi)\}. \end{aligned} \tag{28.4.7}$$

In this equation, the $I^+_i(\chi)$, $i=1, 2, 3, 4$ are invariant polynomials of the integrity basis elements of the group 32, expressed in terms of the second rank tensor elements χ_α. The $I^+_i(\chi)$ must have the property that they are invariant to the π rotation about the c-axis. Each such polynomial is an infinite sum of products of the integrity basis elements, the coefficients of each product being temperature dependent, as are the coefficients s^A_{14}, $\alpha^A, \ldots$, of each term in (28.4.7). Besides N_2, N_3, N_6, there are two other numerator invariants, N_a and N_b, of the point group 32 which, however, are invariant to the above π rotation. The right-hand side of (28.4.7) is a polynomial, invariant to the transformations of this point group. Hence it must be of the general form (28.4.6) for such polynomials. By comparison with (28.4.6) we see that the separation of terms in (28.4.7) is unique so long as we require (1) that I^+_1, I^+_3, I^+_4 be constructed using denominator invariants only, and (2) that I^+_2 be constructed according to (28.4.6) but using only N_a, N_b and denominator invariants.

In section 6 it will be shown that s^A_{14} is the isothermal compliance element measured at the zero stress reference state. At this state

$$s^A_{\alpha\beta} = (\partial\eta^A_\alpha/\partial t_\beta)^*_T = (\partial e^A_\alpha/\partial\Theta_\beta)^*. \tag{28.4.8}$$

In the next section we shall discuss the relationships between the derivatives of g' with respect to χ and the elastic stiffness and compliance elements, all evaluated at the zero stress reference state.

Let us define

$$\sigma^A_{\alpha\beta\gamma\ldots} = -(\partial^n g'_A/\partial\chi_\alpha\,\partial\chi_\beta\,\partial\chi_\gamma\cdots)^*. \tag{28.4.9}$$

It may be shown that $D_3(\chi)$ is the only third-order invariant polynomial which contains χ^3_4, while $N_3(\chi)$ is the only third-order invariant which contains the term $\chi_4\chi_6^2$. Therefore

$$\alpha^A = \sigma^A_{444}, \qquad \beta^A = \sigma^A_{466}. \tag{28.4.10}$$

From (28.4.7) we see that the stable equilibrium condition (28.2.11) for a twinned state is satisfied if the following four conditions are satisfied,

$$N_2(\boldsymbol{\chi}) = D_3(\boldsymbol{\chi}) = N_3(\boldsymbol{\chi}) = N_6(\boldsymbol{\chi}) = 0. \tag{28.4.11}$$

This exact result provides a remarkably good example of the use of integrity basis elements.

We see that these coexistence conditions are satisfied, for example, if

$$\text{(i)} \quad \boldsymbol{\chi} = \chi \boldsymbol{E}; \tag{28.4.12}$$

or

$$\text{(ii)} \quad \chi_1 = \chi_2, \qquad \chi_4 = \chi_6 = 0; \tag{28.4.13}$$

or

$$\text{(iii)} \quad \chi_1 = \chi_2, \qquad \chi_6 = 0, \qquad \chi_4 = \pm\sqrt{3}\chi_5, \tag{28.4.14}$$

where $\boldsymbol{E}$ is the unit matrix. If the stress on the specimen were hydrostatic, (28.4.12) would be satisfied.

We should note that, strictly, the conditions (28.4.11) are sufficient, but not necessary, conditions that (28.2.11) be satisfied. In mathematical terms it is possible that the expression in (28.4.7) could vanish for a certain $\boldsymbol{\chi}$ which does not satisfy all the four conditions of (28.4.11).

In such a case this value of $\boldsymbol{\chi}$ would be temperature dependent, since the coefficients of the polynomials in (28.4.7) depend on temperature. The conditions of (28.4.11), on the other hand, are independent of temperature. Thus, if they are satisfied, the twinned state is stable *at all temperatures.* It possibly cannot be proved rigorously, but it is extremely unlikely that there are other temperature-independent values of $\boldsymbol{\chi}$ which satisfy (28.2.11) and do not satisfy the four conditions of (28.4.11).

The coefficients of products, up to third order, of the six variables χ_i in the right-hand side of (28.4.7) may be obtained by using the denominator invariants χ_3 and $\chi_1 + \chi_2$, and the expressions for the three integrity basis elements given in equations (28.4.1–3). If we compare these coefficients with those obtained from a Taylor series for $\Delta g'$, we have, up to third order terms,

$$\Delta g' = g'_B - g'_A = 2s^A_{14}N_2(\boldsymbol{\chi}) + (\sigma^A_{124} + \sigma^A_{114})(\chi_1 + \chi_2)N_2(\boldsymbol{\chi})$$
$$+ 2\sigma^A_{134}\chi_3 N_2(\boldsymbol{\chi}) - \tfrac{1}{4}\sigma^A_{466}N_3(\boldsymbol{\chi}) + \tfrac{1}{3}\sigma^A_{444}D_3(\boldsymbol{\chi}). \tag{28.4.15}$$

The importance of the third-order stability conditions in (28.4.11) can be judged by estimating the order of magnitude of the elements of $\boldsymbol{\chi}$ which make the third-order terms in (28.4.15) comparable with the second-order term. To do this the experimental orders of magnitude of s^A_{14} and the $\sigma^A_{\alpha\beta\gamma}$ are required.

In section 6 it will be shown how to prove that, in unabbreviated suffix notation,

$$\sigma_{\alpha\beta\gamma\delta\mu\nu} = -(\partial^3 g'/\partial\chi_{\alpha\beta}\partial\chi_{\gamma\delta}\partial\chi_{\mu\nu})^*$$
$$= s_{\alpha\beta\omega\delta}s_{\gamma\omega\mu\nu} + s_{\alpha\beta\omega\nu}s_{\mu\omega\gamma\delta} + s_{\mu\nu\alpha\omega}s_{\omega\beta\gamma\delta}$$
$$+ s_{\alpha\beta\alpha'\beta'}s_{\mu\nu\mu'\nu'}s_{\gamma\delta\gamma'\delta'}c^{(3)}_{\alpha'\beta'\gamma'\delta'\mu'\nu'}, \tag{28.4.16}$$

where

$$c^{(3)}_{\alpha'\beta'\gamma'\delta'\mu'\nu'} = (\partial^3 f/\partial\eta_{\alpha'\beta'}\,\partial\eta_{\gamma'\delta'}\,\partial\eta_{\mu'\nu'})^*$$
$$= (\partial^2 t_{\alpha'\beta'}/\partial\eta_{\gamma'\delta'}\,\partial\eta_{\mu'\nu'})^* \tag{28.4.17}$$

is a third order isothermal elastic constant. It must be emphasised that (28.4.16) is valid, as is (28.4.7), only if the reference state is at zero stress.

The third order elastic constants of (28.4.17) have been measured, for quartz at 25 °C, by Thurston, McSkimin and Andreatch (1966). The second order compliances have been measured for α quartz up to the α–β transition temperature, by Zubov and Firsova (1962).

At 25 °C, the second order compliance elements $s_{\alpha\beta\gamma\delta}$ are of the order 10^{-13} cm^2 dyne^{-1}. The third order elastic constants are of the order 10^{12} dyne cm^{-2}. From (28.4.16) and (28.4.17) we see that, if the elements of $\boldsymbol{\chi}$ are of the order 10^{13} dyne cm^{-2} (10 megabars), the third-order terms of (28.4.15) will be comparable to the second-order terms. The second-order condition of (28.4.11) will thus be adequate for most present-day experimental and geological studies. The third-order terms will make a significant contribution at about $\frac{1}{4}$–$\frac{1}{2}$ megabar. They may thus be important to studies at high stresses, if, under such conditions α quartz and Dauphiné twins exist.

For values of the stress tensor such that the strain in each twin may be regarded as infinitesimal to the accuracy desired, the tensor $\boldsymbol{\chi}$ may be taken as equal to $\boldsymbol{\Theta}$. The second-order coexistence condition may then be written

$$(\Theta_1 - \Theta_2)\Theta_4 + 2\Theta_5\Theta_6 = 0. \tag{28.4.18}$$

At 25 °C we see from the order of magnitude of the compliance elements quoted above and from (6.4.15b) and (28.3.13), that the degree of approximation in (28.4.18) would be of the order of 1 per cent for stress elements of the order of 100 kbar. At 500 °C the compliance elements are of the order of 1 (megabar)$^{-1}$, and the degree of approximation would then be of the order of 10 per cent at 100 kbar.

28.5. The equilibrium condition and Helmholtz free energy

For infinitesimal variations, the stress tensor may be taken as linearly related to the corresponding conjugate strain tensor. We may therefore, in the infinitesimal regime, take $\boldsymbol{B}$ to be linearly related to the strain tensor defined by

$$\boldsymbol{d} = \boldsymbol{D} - \boldsymbol{E}. \tag{28.5.1}$$

On using this linear relationship, it can be shown that the Helmholtz free energy density f is given by

$$f = f^0 + \tfrac{1}{2} B_{\alpha\beta} d_{\beta\alpha}, \tag{28.5.2}$$

where f^0 has the same value for each twin. Thus, for the uniform parts of the twins remote from z,

$$\Delta f = f_B - f_A = \tfrac{1}{2} B_{\alpha\beta} \Delta D_{\beta\alpha}, \tag{28.5.3}$$

and, from (28.2.10),

$$\Delta g' = -\tfrac{1}{2} B_{\alpha\beta} \Delta D_{\beta\alpha}. \tag{28.5.4}$$

Hence,

$$\Delta g' = g'_B - g'_A = f_A - f_B = -\Delta f, \tag{28.5.5}$$

and

$$\delta G' = \tilde{v}\, \delta n_B\, \Delta g' = -\delta F, \tag{28.5.6}$$

where F is the Helmholtz free energy of the specimen.

Therefore, in the linear regime, minimising G' is equivalent to maximising the Helmholtz free energy F. If we identify F with the elastic stored energy of Thomas and Wooster (1951), we have given a thermodynamic justification of the principle of maximum stored elastic energy which they enunciated on empirical grounds.

We shall now prove that, if the equilibrium condition (28.2.11) is satisfied under the temperature-independent conditions of (28.4.11), at equilibrium we must have

$$f_A = f_B, \tag{28.5.7}$$

and this is a rigorous result which does not depend on assuming a linear relation between stress and strain.

Since each of the four coexistence conditions of (28.4.11) is homogeneous in the elements of $\boldsymbol{\chi}$, it can be seen that if they are satisfied by a certain tensor $\boldsymbol{\chi}^1$, say, then they are satisfied by $\lambda\boldsymbol{\chi}^1$, where λ is any real number. On referring to (28.3.13) we see that we may say that if the coexistence conditions are satisfied for a certain stress tensor $\boldsymbol{B}^1$, then they are satisfied for $\lambda\boldsymbol{B}^1$, where λ is any real number.

Thus, if (28.2.11) is satisfied by

$$\boldsymbol{B} = \lambda\boldsymbol{B}^1, \tag{28.5.8}$$

we may differentiate (28.2.11) with respect to λ, to obtain

$$\frac{\partial g'_A}{\partial B_{\alpha\beta}} B^1_{\alpha\beta} = \frac{\partial g'_B}{\partial B_{\alpha\beta}} B^1_{\alpha\beta}. \tag{28.5.9}$$

On using (28.3.2) we may write (28.5.9) as

$$B^1_{\alpha\beta} D^A_{\beta\alpha} = B^1_{\alpha\beta} D^B_{\beta\alpha}, \tag{28.5.10}$$

and, on multiplying this equation by λ, we obtain

$$B_{\alpha\beta} D^A_{\beta\alpha} = B_{\alpha\beta} D^B_{\beta\alpha}. \tag{28.5.11}$$

This equation holds when the conditions (28.4.11) are satisfied.

On referring to the definition (28.2.10) of g', we see that when (28.5.11) is satisfied, the Helmholtz free energy densities of the two twins must be equal, as stated in (28.5.7).

Hence we have proved rigorously that the stored elastic energy of either twin must have the same value when the twins are at equilibrium, under the temperature-independent conditions (28.4.11).

28.6. The relationships between the derivatives of $g'(\boldsymbol{\chi})$ and the elastic constants and compliances

From (28.3.2) and the fact that g' is a function of $\boldsymbol{\xi}$, we see that

$$\boldsymbol{D} = -2\boldsymbol{B}^{\mathrm{t}}(\partial g'/\partial\boldsymbol{\xi}). \tag{28.6.1}$$

We know that f may be expressed as a function of $\boldsymbol{\eta}$, and, from (6.4.30b) and (12.2.4), it may be seen that

$$\boldsymbol{B}^{\mathrm{t}} = \boldsymbol{D}(\partial f/\partial\boldsymbol{\eta}). \tag{28.6.2}$$

On comparing these two equations, we see that $-2(\partial g'/\partial\boldsymbol{\xi})$ is the reciprocal matrix of $(\partial f/\partial\boldsymbol{\eta})$, that is, of the stress tensor $\boldsymbol{\Lambda}$. Now, it is commonplace for a stress tensor $\boldsymbol{\Lambda}$ to be specified such that $\det(\boldsymbol{\Lambda})$ is zero.

In other words, $\boldsymbol{\Lambda}$ is then described by a singular matrix. From (6.4.30b) in this case, $\boldsymbol{B}$ will be singular also, since, in practice for real media $\boldsymbol{D}$ must be non-singular. In such a case $(\partial f/\partial \boldsymbol{\eta})$ has no reciprocal. This difficulty must be met by defining (28.6.1), for states where $\boldsymbol{B}$ is singular, in the following way,

$$\boldsymbol{D}' = -2 \lim_{\boldsymbol{B} \to \boldsymbol{B}'} \{\boldsymbol{B}^{\text{t}}(\partial g'/\partial \boldsymbol{\xi})\}, \quad \det \boldsymbol{B} \neq 0. \tag{28.6.3}$$

That is, we must approach a state, for which the stress tensor $\boldsymbol{B}'$ is singular, along a path on which the stress tensor $\boldsymbol{B}$ is non-singular. That such a limit exists may be proved, as follows, by introducing the tensor $\boldsymbol{\chi}$ as in section 5.

In tensor notation we may write (28.6.1) as

$$\boldsymbol{D}^{\text{t}} = -2(\partial g'/\partial \boldsymbol{\chi}) : (\partial \boldsymbol{\chi}/\partial \boldsymbol{\xi}) \cdot \boldsymbol{B}, \tag{28.6.4}$$

that is, in suffix notation,

$$D_{\alpha\beta} = D^{\text{t}}_{\beta\alpha} = -2(\partial g'/\partial \chi_{\gamma\delta})(\partial \chi_{\gamma\delta}/\partial \xi_{\beta\omega})B_{\omega\alpha}. \tag{28.6.5}$$

Now, from (28.3.13), we may show that

$$(\partial \boldsymbol{\xi}/\partial \boldsymbol{\chi}) = \tfrac{1}{2}(\boldsymbol{E} \times \boldsymbol{\chi} + \boldsymbol{\chi} \times \boldsymbol{E}), \tag{28.6.6}$$

where $\boldsymbol{\chi} \times \boldsymbol{E}$, for example, is a direct product of $\boldsymbol{\chi}$ and the unit 3×3 matrix $\boldsymbol{E}$. In suffix notation (28.6.6) may be written as

$$\begin{aligned}(\partial \xi_{\alpha\beta}/\partial \chi_{\mu\nu}) &= \tfrac{1}{2}(\boldsymbol{E} \times \boldsymbol{\chi} + \boldsymbol{\chi} \times \boldsymbol{E})_{\alpha\beta\mu\nu} \\ &= \tfrac{1}{2}(\delta_{\alpha\mu}\chi_{\beta\nu} + \chi_{\alpha\mu}\gamma_{\beta\nu}).\end{aligned} \tag{28.6.7}$$

In (28.6.5) we require the reciprocal of (28.6.6), since

$$(\partial \xi_{\alpha\beta}/\partial \chi_{\gamma\delta})(\partial \chi_{\gamma\delta}/\partial \xi_{\mu\nu}) = \delta_{\alpha\mu}\, \delta_{\beta\nu}, \tag{28.6.8}$$

a well-known mathematical identity for two sets of variables and one which we have used previously, see for example (2.3.4).

To find the reciprocal, in the sense of (28.6.8), we introduce the eigenvalues χ_i, $i = 1, 2, 3$, and the orthonormal set of corresponding eigenvectors $\boldsymbol{u}^i$ of the tensor $\boldsymbol{\chi}$. It can be seen that the right-hand side of (28.6.6) has eigenvalues $\tfrac{1}{2}(\chi_i + \chi_j)$ and corresponding eigenvectors $\boldsymbol{u}^i\boldsymbol{u}^j$, where i and j run from 1 to 3, and $\boldsymbol{u}^i\boldsymbol{u}^j$ is an ordered product, such that

$$(\boldsymbol{u}^i\boldsymbol{u}^j)_{\alpha\beta} = u^i_\alpha u^j_\beta. \tag{28.6.9}$$

We shall now apply the well-known matrix theorem which states that a real symmetric $n \times n$ matrix $\boldsymbol{A}$ can be expressed, as follows, in terms of its eigenvalues A_i and orthonormal set of eigenvectors $\boldsymbol{a}^i$. (This theorem was discussed in chapter 24, section 5.)

$$\boldsymbol{A} = \sum_{i=1}^{n} A_i \boldsymbol{a}^i \boldsymbol{a}^i. \tag{28.6.10}$$

Unless otherwise specified, the summation convention does not apply to latin indices.

Since $(\partial\boldsymbol{\chi}/\partial\boldsymbol{\xi})$ is the reciprocal matrix of $(\partial\boldsymbol{\xi}/\partial\boldsymbol{\chi})$, its eigenvalues will be given by $\{2/(\chi_i+\chi_j)\}$, while its eigenvectors will be the corresponding set $\boldsymbol{u}^i\boldsymbol{u}^j$. Thus, on applying the theorem of (28.6.10), the required tensor is given by

$$(\partial\boldsymbol{\chi}/\partial\boldsymbol{\xi}) = \sum_{i=1}^{3}\sum_{j=1}^{3} \{2/(\chi_i+\chi_j)\}\boldsymbol{u}^i\boldsymbol{u}^j\boldsymbol{u}^i\boldsymbol{u}^j, \tag{28.6.11}$$

as may easily be verified by evaluating $(\partial\boldsymbol{\chi}/\partial\boldsymbol{\xi}):\boldsymbol{u}^k\boldsymbol{u}^l$. Since

$$\boldsymbol{u}^i\boldsymbol{u}^j:\boldsymbol{u}^k\boldsymbol{u}^l = \delta_{ik}\,\delta_{jl}, \tag{28.6.12}$$

from the orthonormal properties of the $\boldsymbol{u}^i$, we see that

$$(\partial\boldsymbol{\chi}/\partial\boldsymbol{\xi}):\boldsymbol{u}^k\boldsymbol{u}^l = \{2/(\chi_k+\chi_l)\}(\boldsymbol{u}^k\boldsymbol{u}^l), \tag{28.6.13}$$

thus verifying the eigenvalue property.

The last step required, to show that a limit to (28.6.1) exists at states where $\boldsymbol{B}$ is singular, is to express $\boldsymbol{B}$ in terms of the χ_i and the $\boldsymbol{u}^i$. Let

$$\boldsymbol{B} = \sum_{(ij)} b_{ij}\boldsymbol{u}^i\boldsymbol{u}^j, \quad \text{and} \quad \boldsymbol{B}^{\mathrm{t}} = \sum_{(ij)} b_{ji}\boldsymbol{u}^i\boldsymbol{u}^j, \tag{28.6.14}$$

where i and j are each summed independently from 1 to 3. From (28.3.13) and (28.6.10) we see that

$$\sum_i \chi_i^2\boldsymbol{u}^i\boldsymbol{u}^i = b_{\alpha\omega}b_{\beta\omega}\boldsymbol{u}^\alpha\boldsymbol{u}^\beta. \tag{28.6.15}$$

On using the orthonormal properties of the $\boldsymbol{u}^i$, we have

$$\delta_{ij}\chi_i^2 = b_{i\omega}b_{j\omega}, \tag{28.6.16}$$

and, in particular,

$$\sum_\omega b_{i\omega}^2 = \chi_i^2. \tag{28.6.17}$$

Thus, if χ_i vanishes so does each $b_{i\omega}$, $\omega = 1, 2, 3$. If we write

$$b_{i\omega} = \chi_i\tilde{b}_{i\omega}, \tag{28.6.18}$$

we see from (28.6.16) that $\tilde{b}_{i\omega}$ is a real orthogonal matrix. If (28.6.13) and

(28.6.14) are substituted into (28.6.4), we find that

$$\boldsymbol{D}^{\mathrm{t}} = -\sum_{(ijk)} \frac{4\tilde{b}_{ij}}{1+(\chi_k/\chi_i)} \frac{\partial g'}{\partial \boldsymbol{\chi}} : \boldsymbol{u}^k \boldsymbol{u}^i \boldsymbol{u}^k \boldsymbol{u}^j, \tag{28.6.19}$$

where the summation of i, j and k is from 1 to 3. Now, if $\boldsymbol{B}$ is singular, one or more of the χ_i are zero. We see from (28.6.19) that the singularity induced in $(\partial g'/\partial \boldsymbol{\xi})$, when $\boldsymbol{B}$ and $\boldsymbol{\xi}$ become singular, has been removed in the complete expression for $\boldsymbol{D}^{\mathrm{t}}$ given in (28.6.19). Thus $(\partial g'/\partial \boldsymbol{\chi})$ is always non-singular. In fact we see that if χ_i vanishes, then, whether χ_k/χ_i tends to a finite or infinite limit, every term in (28.6.19) either vanishes or remains finite and non-zero.

We have thus proved that the limit of (28.6.3) exists.

After this rather long discussion to show how (28.6.1) must be dealt with in practice, we may now use this equation to express the derivatives of $g'(\boldsymbol{\chi})$ in terms of conventional elastic quantities.

First of all, the zero stress reference state is a singular state in the sense above. We can overcome this difficulty by taking a hydrostatic reference state, whose pressure is p^*. We shall later be able to set p^* to zero. At the hydrostatic reference state,

$$\boldsymbol{\chi}^* = \boldsymbol{\Lambda}^* = \boldsymbol{B}^* = \boldsymbol{\Theta}^* = -p^*\boldsymbol{E}, \tag{28.6.20}$$

and

$$\tilde{b}^*_{ij} = \delta_{ij}. \tag{28.6.21}$$

Since $\boldsymbol{D} = \boldsymbol{E}$ at this reference state, we see from (28.6.19) that

$$(\partial g'/\partial \boldsymbol{\chi})^* = -\tfrac{1}{2}\boldsymbol{E}; \tag{28.6.22}$$

from (28.6.1) that

$$(\partial g'/\partial \boldsymbol{\xi})^* = (1/2p^*)\boldsymbol{E}; \tag{28.6.23}$$

and from (28.6.11) that

$$(\partial \boldsymbol{\chi}/\partial \boldsymbol{\xi})^* = -(1/p^*)(\boldsymbol{E}\times\boldsymbol{E}). \tag{28.6.24}$$

By further straightforward analysis, we may show that

$$(\partial^2 \boldsymbol{\chi}/\partial \boldsymbol{\xi}\,\partial \boldsymbol{\xi})^* = (1/p^{*3})(\boldsymbol{E}\times\boldsymbol{E}\times\boldsymbol{E}), \tag{28.6.25}$$

$$(\partial^2 \boldsymbol{\xi}/\partial \boldsymbol{\chi}\,\partial \boldsymbol{\chi})^* = (\boldsymbol{E}\times\boldsymbol{E}\times\boldsymbol{E}), \tag{28.6.26}$$

$$(\partial^2 g'/\partial \boldsymbol{\xi}\,\partial \boldsymbol{\xi})^* = -(1/2p^{*3})(\boldsymbol{E}\times\boldsymbol{E}) + (1/p^{*2})(\partial^2 g'/\partial \boldsymbol{\chi}\,\partial \boldsymbol{\chi}). \tag{28.6.27}$$

The procedure used to relate the derivatives of $g'(\chi)$ to the elastic constants is as follows. From (28.6.1), (28.6.2) and (28.6.22–24), we find expressions for $(\partial D_{\alpha\beta}/\partial B_{\mu\nu})^*$ and $(\partial B_{\mu\nu}/\partial D_{\alpha'\beta'})^*$. From the mathematical identity (28.6.8),

$$\frac{\partial D_{\alpha\beta}}{\partial B_{\mu\nu}}\frac{\partial B_{\mu\nu}}{\partial D_{\alpha'\beta'}}=\delta_{\alpha\alpha'}\,\delta_{\beta\beta'}. \tag{28.6.28}$$

On substituting the expressions for these derivatives into (28.6.28) it is straightforward to prove that, in suffix notation,

$$\sigma_{\alpha\beta\mu\nu}(c_{\mu\nu\alpha'\beta'}-p^*\,\delta_{\mu\alpha'}\,\delta_{\nu\beta'})=\delta_{\alpha\alpha'}\,\delta_{\beta\beta'} \tag{28.6.29}$$

where

$$c_{\mu\nu\alpha\beta}=(\partial^2 f/\partial\eta_{\mu\nu}\,\partial\eta_{\alpha\beta})^* \tag{28.6.30}$$

is the conventional isothermal elastic constant at a hydrostatic reference state. At the zero stress reference state, this elastic constant is equal to $(\partial\Theta_{\mu\nu}/\partial e_{\alpha\beta})^*$, of infinitesimal strain theory, and (28.6.29) shows then that the $\sigma_{\alpha\beta\mu\nu}$ are the conventional isothermal compliance elements.

To obtain the relation (28.4.16) is more tedious. One proceeds by differentiating the identity (28.6.28) with respect to the elements of $\boldsymbol{B}$, to obtain the mathematical identity

$$\frac{\partial^2 D_{\alpha\beta}}{\partial B_{\mu\nu}\,\partial B_{\gamma\delta}}=-\frac{\partial D_{\alpha\beta}}{\partial B_{\alpha'\beta'}}\frac{\partial D_{\mu'\nu'}}{\partial B_{\mu\nu}}\frac{\partial D_{\gamma'\delta'}}{\partial B_{\gamma\delta}}\frac{\partial^2 B_{\alpha'\beta'}}{\partial D_{\mu'\nu'}\,\partial D_{\gamma'\delta'}}. \tag{28.6.31}$$

(A similar mathematical process links the third-order compliance and stiffness elements, as follows [in abbreviated notation]

$$s^{(3)}_{\alpha\beta\gamma}=-s_{\alpha\alpha'}s_{\beta\beta'}s_{\gamma\gamma'}c^{(3)}_{\alpha'\beta'\gamma'}.) \tag{28.6.32}$$

On evaluating the second-order derivatives in (28.6.31) at the hydrostatic reference state, with the help of equations (28.6.22–26), it is possible, after some straightforward analysis, to prove (28.4.16).

Summary of important equations

A. *The condition for a stable variation*

$$\delta n_B=-\delta n_A. \tag{28.2.1}$$

$$\delta n_B(f_B-f_A-B_{\beta\alpha}\Delta D_{\alpha\beta})\geqslant 0. \tag{28.2.8}$$

$$\delta n_B(g'_B-g'_A)\geqslant 0. \tag{28.2.9}$$

$$g'=f-B_{\beta\alpha}D_{\alpha\beta}. \tag{28.2.10}$$

B. *The stable equilibrium condition*

$$g'_B - g'_A = 0. \tag{28.2.11}$$

C. *The properties of the generalised Gibbs free energy g'*

$$dg' = -s\,dT - B_{\alpha\beta}\,dD_{\beta\alpha}. \tag{28.3.1}$$

$$D_{\beta\alpha} = -\partial g'/\partial B_{\alpha\beta}. \tag{28.3.2}$$

$$\boldsymbol{\chi}^2 = \boldsymbol{\xi} = \boldsymbol{B}\boldsymbol{B}^t. \tag{28.3.13}$$

$$\begin{aligned} g'_B - g'_A = {} & 2s^A_{14}N_2(\boldsymbol{\chi})[1 + I_1^+(\boldsymbol{\chi})] \\ & + (2/3!)\alpha^A D_3(\boldsymbol{\chi})[1 + I_2^+(\boldsymbol{\chi})] \\ & + (2/3!)\beta^A N_3(\boldsymbol{\chi})[1 + I_3^+(\boldsymbol{\chi})] \\ & + (2/6!)\gamma^A N_6(\boldsymbol{\chi})[1 + I_4^+(\boldsymbol{\chi})]. \end{aligned} \tag{28.4.7}$$

$\Delta g' = 0$, if

$$N_2(\boldsymbol{\chi}) = D_3(\boldsymbol{\chi}) = N_3(\boldsymbol{\chi}) = N_6(\boldsymbol{\chi}) = 0. \tag{28.4.11}$$

$$\boldsymbol{D} = -2\boldsymbol{B}^t\frac{\partial g'}{\partial \boldsymbol{\xi}}, \qquad \det(\boldsymbol{B}) \neq 0. \tag{28.6.1}$$

$$\boldsymbol{B}^t = \boldsymbol{D}\frac{\partial f}{\partial \boldsymbol{\eta}}. \tag{28.6.2}$$

If $\det(\boldsymbol{B}') = 0$,

$$\boldsymbol{D}' = -2 \lim_{\boldsymbol{B}\to\boldsymbol{B}'} \{\boldsymbol{B}^t(\partial g'/\partial \boldsymbol{\xi})\}, \qquad \det(\boldsymbol{B}) \neq 0. \tag{28.6.3}$$

29

The tetragonal/cubic ferroelectric transition of barium titanate

The elastic and electrical properties of barium titanate are discussed, especially in relation to the tetragonal/cubic ferroelectric transition. Useful thermodynamic relations are derived and applied. It is shown that this transition has the same general behaviour as the α–β quartz transition, although there are some interesting differences – for example, it is possible that the isothermal elastic compliance elements do not diverge, in the cubic phase, as T_0 is approached.

29.1. Elastic and electric properties of crystals

Before we discuss the tetragonal/cubic phase change of barium titanate it will be useful to discuss the thermodynamics of a dielectric crystal.

Such a crystal may be electrically polarised, and its electric polarisation will be determined by $\boldsymbol{P}$, the polarisation vector, that is, the electric dipole moment per unit reference volume.

This polarisation may be induced by an electric field of intensity $\boldsymbol{E}$. For a linear dielectric crystal, $\boldsymbol{E}$ and $\boldsymbol{P}$ are related, as follows, by the dielectric susceptibility tensor $\boldsymbol{\chi}$,

$$P_i = \chi_{ij} E_j. \tag{29.1.1}$$

To assist in distinguishing electrical quantities from elastic quantities, the summation convention will be extended in this chapter, to apply not only to Greek indices, but to i and j. The use of the indices i and j will be reserved for electrical quantities, such as in the above equation, and i and j may each take the values x, y and z. As in the rest of this book, the x, y and z axes are synonymous with the x_1, x_2 and x_3 axes. It will be found that the use of x, y and z for components of electrical quantities, and of the numbers 1 to 6 for strain or stress quantities, in abbreviated notation, will be helpful in interpreting the meaning of such 'mixed' quantities as the piezoelectric coefficients.

Some crystals, for example the tetragonal form of barium titanate, may possess a spontaneous polarisation, $\boldsymbol{P}_s$, which is a 'permanent' polarisation, analogous to a permanent magnet. In the case of barium titanate, $\boldsymbol{P}_s$ disappears on heating the crystal to a temperature above the transition value (120 °C) at which the tetragonal form changes to the cubic form – for which there can be no spontaneous polarisation.

The thermodynamics of such crystals are determined by the equation for the work done, per unit reference volume, in an infinitesimal variation

$$\mathrm{d}'w = L_\mu \,\mathrm{d}e_\mu + E_i \,\mathrm{d}P_i, \tag{29.1.2}$$

where, as stated above, i must be summed over the values x, y, z. There are thus nine thermodynamic co-ordinates, namely the six strain elements, and the three components of $\boldsymbol{P}$. For the thermodynamic functions, we have

$$\left.\begin{aligned} \mathrm{d}u &= T\,\mathrm{d}s + L_\mu \,\mathrm{d}e_\mu + E_i\,\mathrm{d}P_i \\ &\text{and} \\ \mathrm{d}f &= -s\,\mathrm{d}T + L_\mu\,\mathrm{d}e_\mu + E_i\,\mathrm{d}P_i. \end{aligned}\right\} \tag{29.1.3}$$

Hence, for example,

$$s = -\partial f/\partial T; \qquad L_\mu = \partial f/\partial e_\mu; \qquad E_i = \partial f/\partial P_i, \tag{29.1.4}$$

where f is regarded here as a function of T, the six e_μ, and three P_i.

We now have 9×9 stiffness and compliance matrices, which we shall denote by $\boldsymbol{C}$ and $\boldsymbol{S}$. The stiffness matrix $\boldsymbol{C}$ will contain elements of several kinds as follows. First,

$$\begin{aligned} C_{\alpha\beta} &= (\partial^2\psi/\partial e_\alpha\,\partial e_\beta)_{\bar{e}_{\alpha\beta},\boldsymbol{P}} = (\partial L_\alpha/\partial e_\beta)_{\bar{e}_\beta,\boldsymbol{P}} \\ &= c^{\boldsymbol{P}}_{\alpha\beta}, \end{aligned} \tag{29.1.5}$$

where

$$\psi(\boldsymbol{e}, \boldsymbol{P}) = u(s, \boldsymbol{e}, \boldsymbol{P}) \quad \text{or} \quad f(T, \boldsymbol{e}, \boldsymbol{P}). \tag{29.1.6}$$

If ψ is taken as u, we obtain the adiabatic stiffness matrix, while if it is taken as f we obtain the isothermal stiffness matrix. Equation (29.1.5) gives the elastic 6×6 stiffness matrix and, from the functional form of ψ, this matrix is correctly to be regarded as measured at constant $\boldsymbol{P}$, as indicated by the superscript in the last term.

The second kind of element is

$$C_{ij} = (\partial^2\psi/\partial P_i\,\partial P_j)_{\boldsymbol{e},\bar{P}_{ij}} = (\partial E_i/\partial P_j)_{\boldsymbol{e},\bar{P}_j} \tag{29.1.7}$$

and these determine the relation between $\boldsymbol{E}$ and $\boldsymbol{P}$, where $\boldsymbol{e}$, the shape, is held constant – that is, the crystal is then said to be clamped.

The third kind of element is

$$\begin{aligned} C_{\alpha i} &= (\partial^2\psi/\partial e_\alpha\,\partial P_i)_{T,\bar{e}_\alpha,\bar{P}_i} = (\partial L_\alpha/\partial P_i) \\ &= (\partial E_i/\partial e_\alpha). \end{aligned} \tag{29.1.8}$$

These elements describe the piezoelectric effect, although the usual piezoelectric coefficients used in the literature are the corresponding elements of the compliance matrix $\boldsymbol{S}$.

If we define

$$g' = \psi - e_\alpha L_\alpha - E_i P_i, \tag{29.1.9}$$

the compliance matrix is made up of three types of element as defined in the following three equations.

$$S_{\alpha\beta} = -(\partial^2 g'/\partial L_\alpha\, \partial L_\beta)_{\bar{L}_{\alpha\beta},\boldsymbol{E}} = (\partial e_\alpha/\partial L_\beta)$$
$$= s^{\boldsymbol{E}}_{\alpha\beta}; \tag{29.1.10}$$

$$S_{\alpha i} = -(\partial^2 g'/\partial L_\alpha\, \partial E_j)_{\bar{L}_\alpha \bar{E}_i} = (\partial e_\alpha/\partial E_i)$$
$$= (\partial P_i/\partial L_\alpha) = d_{\alpha i}; \tag{29.1.11}$$

and

$$S_{ij} = -(\partial^2 g'/\partial E_i\, \partial E_j)_{\boldsymbol{L},\bar{E}_{ij}} = (\partial P_i/\partial E_j)$$
$$= \chi^{\boldsymbol{L}}_{ij}. \tag{29.1.12}$$

The elements $d_{\alpha i}$ are piezoelectric coefficients, although, as usually defined in the literature, Θ_α replaces L_α, since infinitesimal strain theory is employed. The third type of compliance matrix, $\chi^{\boldsymbol{L}}_{ij}$, is an element of the 3×3 isothermal susceptibility matrix measured at constant $\boldsymbol{L}$. The 9×9 matrices $\boldsymbol{C}$ and $\boldsymbol{S}$ are reciprocal to each other. If the crystal has a centre of inversion, all elements of the 6×3 sub-matrices $[d_{\alpha i}]$ and $[C_{\alpha i}]$ of $\boldsymbol{S}$ and $\boldsymbol{C}$ respectively are zero. In this case, the 3×3 sub-matrix $[\chi^{\boldsymbol{L}}_{ij}]$ is the reciprocal of the 3×3 sub-matrix $[C_{ij}]$, and the 6×6 elastic stiffness sub-matrix $[s^{\boldsymbol{E}}_{\alpha\beta}]$ is the reciprocal $[c^{\boldsymbol{P}}_{\alpha\beta}]$. However, these pairs of sub-matrices are not reciprocals for crystals without a centre of inversion.

In order to discuss the reciprocal matrices of these square sub-matrices of the 9×9 stiffness and compliance matrices, consider the following mathematical identity,

$$\delta_{\alpha\beta} = \left(\frac{\partial \xi_\alpha}{\partial \eta_\mu}\right)_{\bar{\eta}_\mu} \left(\frac{\partial \eta_\mu}{\partial \xi_\beta}\right)_{\bar{\xi}_\beta}. \tag{29.1.13}$$

where the ξ_α and the η_μ are two sets of n quantities, which are related to each other, that is

$$\xi_\alpha = \xi_\beta(\boldsymbol{\eta}); \qquad \eta_\mu = \eta_\mu(\boldsymbol{\xi}). \tag{29.1.14}$$

From (29.1.13) we see that the matrix $(\partial\xi_\alpha/\partial\eta_\mu)_{\bar{\eta}_\mu}$ is the reciprocal of $(\partial\eta_\mu/\partial\xi_\beta)_{\bar{\xi}_\beta}$. We may use these mathematical properties to show (1) that the reciprocal of $[c^{\boldsymbol{P}}_{\alpha\beta}]$ is given by

$$s^{\boldsymbol{P}}_{\alpha\beta} = (\partial e_\alpha/\partial L_\beta)_{\bar{L}_\beta,\boldsymbol{P}}; \tag{29.1.15}$$

(2) that the reciprocal of the 6×6 elastic compliance matrix $s^{\boldsymbol{E}}_{\alpha\beta}$, is

$$c^{\boldsymbol{E}}_{\alpha\beta} = (\partial L_\alpha/\partial e_\beta)_{\bar{e}_\beta,\boldsymbol{E}}; \tag{29.1.16}$$

and (3) that the reciprocal of the 3×3 susceptibility matrix $\chi^{\boldsymbol{L}}_{ij}$ is

$$C^{\boldsymbol{L}}_{ij} = (\partial E_i/\partial P_j)_{\boldsymbol{L},\bar{P}_j}. \tag{29.1.17}$$

Thus, if an elastic stiffness matrix is measured at constant $\boldsymbol{P}$, its reciprocal elastic compliance matrix must be measured at constant $\boldsymbol{P}$ also. Similarly, if a susceptibility matrix is measured at constant strain $\boldsymbol{e}$, its reciprocal must also be measured at constant $\boldsymbol{e}$ – and so on. These results seem rather obvious, but from a mathematical point of view need proving.

It can be seen, for example, that $c^{\boldsymbol{E}}_{\alpha\beta}$ can be expressed as a second derivative of a thermodynamic function $\psi - E_iP_i$, which must be regarded as a function of the elements of $\boldsymbol{e}$ and $\boldsymbol{E}$. Further, $\chi^{\boldsymbol{e}}_{ij}$ may be expressed as a second derivative of this function, with respect to the variables E_i, E_j. The other matrix elements may also be similarly regarded as second derivatives of some function ϕ obtained by Legendre transformations of ψ. We may tabulate this relation as in table 29.1. In this table we have included elastic stiffness and compliance matrices, measured at constant P_1, E_2 and E_3 as an example of a set of such possible mixed quantities. Such quantities could be of possible use in some experimental conditions. The example cited could be of use, for example, if E_2 and E_3 are zero throughout the experiment.

Table 29.1.

Thermodynamic function, ϕ	Appropriate variables	Matrix elements
$\psi - E_iP_i - e_\alpha L_\alpha$	$\boldsymbol{L}$, $\boldsymbol{E}$	$s^{\boldsymbol{E}}_{\alpha\beta}$, $d_{\alpha i}$, $\chi^{\boldsymbol{L}}_{ij}$.
$\psi - e_\alpha L_\alpha$	$\boldsymbol{L}$, $\boldsymbol{P}$	$s^{\boldsymbol{P}}_{\alpha\beta}$, $(\chi^{\boldsymbol{L}})^{-1}_{ij}$.
$\psi - E_iP_i$	$\boldsymbol{e}$, $\boldsymbol{E}$	$c^{\boldsymbol{E}}_{\alpha\beta}$, $\chi^{\boldsymbol{e}}_{ij}$.
ψ	$\boldsymbol{e}$, $\boldsymbol{P}$	$c^{\boldsymbol{P}}_{\alpha\beta}$, $(\chi^{\boldsymbol{e}})^{-1}_{ij}$.
$\psi - E_2P_2 - E_3P_3$	$\boldsymbol{e}$, P_1, E_2, E_3	$c^{P_1,E_2,E_3}_{\alpha\beta}$.
$\psi - e_\alpha L_\alpha - E_2P_2 - E_3P_3$	$\boldsymbol{L}$, P_1, E_2, E_3	$s^{P_1,E_2,E_3}_{\alpha\beta}$.

In the next section we shall discuss the thermodynamic relations between such quantities as $\boldsymbol{\chi}^{\boldsymbol{e}}$, $\boldsymbol{\chi}^{\boldsymbol{L}}$, $\boldsymbol{s}^{\boldsymbol{P}}$ and $\boldsymbol{s}^{\boldsymbol{E}}$.

29.2. Relationships between matrix elements

The relationships between the adiabatic and isothermal matrix elements can easily be obtained from (16.3.7). In this and the remaining sections of this chapter we shall restrict the discussion to the zero stress condition, as discussed in chapter 26, section 1. For example, we have, for the 9×9 compliance matrix $\boldsymbol{S}$,

$$s_{\alpha\beta}^{s,\boldsymbol{E}} = s_{\alpha\beta}^{T,\boldsymbol{E}} - (T/\rho_0 c_{p,\boldsymbol{E}}) m_\alpha m_\beta, \tag{29.2.1}$$

$$d_{\alpha i}^{(s)} = d_{\alpha i}^{(T)} - (T/\rho_0 c_{p,\boldsymbol{E}}) m_\alpha m_i, \tag{29.2.2}$$

$$\chi_{ij}^{s,\boldsymbol{L}} = \chi_{ij}^{T,\boldsymbol{L}} - (T/\rho_0 c_{p,\boldsymbol{E}}) m_i m_j, \tag{29.2.3}$$

where

$$m_\alpha = (\partial e_\alpha/\partial T)_{\boldsymbol{E},p=0}, \tag{29.2.4}$$

$$m_i = (\partial P_i/\partial T)_{\boldsymbol{E},p=0}. \tag{29.2.5}$$

The specific heat $c_{p,\boldsymbol{E}}$ is to be measured at constant zero pressure and constant $\boldsymbol{E}$.

An important relationship may be deduced which connects the susceptibilities $\boldsymbol{\chi}^{\boldsymbol{e}}$ and $\boldsymbol{\chi}^{\boldsymbol{L}}$ with the compliances $\boldsymbol{s}^{\boldsymbol{P}}$ and $\boldsymbol{s}^{\boldsymbol{E}}$. This result enables us to relate, for example, the susceptibilities, measured when the crystal is clamped ($\boldsymbol{e}$ = constant), and when it is free (constant zero stress), to the compliances measured at constant $\boldsymbol{P}$ and at constant $\boldsymbol{E}$.
Consider

$$\frac{\det(\boldsymbol{\chi}^{\boldsymbol{e}})}{\det(\boldsymbol{\chi}^{\boldsymbol{L}})} = \frac{\partial(\boldsymbol{P},\boldsymbol{e})}{\partial(\boldsymbol{E},\boldsymbol{e})}\Big/\frac{\partial(\boldsymbol{P},\boldsymbol{L})}{\partial(\boldsymbol{E},\boldsymbol{L})}, \tag{29.2.6}$$

$$= \frac{\partial(\boldsymbol{P},\boldsymbol{e})}{\partial(\boldsymbol{P},\boldsymbol{L})}\Big/\frac{\partial(\boldsymbol{E},\boldsymbol{e})}{\partial(\boldsymbol{E},\boldsymbol{L})}, \tag{29.2.7}$$

on using the cancelling rule (11, 2, 5) for Jacobians. Thus, we have proved that

$$\frac{\det(\boldsymbol{\chi}^{\boldsymbol{e}})}{\det(\boldsymbol{\chi}^{\boldsymbol{L}})} = \frac{\det(\boldsymbol{s}^{\boldsymbol{P}})}{\det(\boldsymbol{s}^{\boldsymbol{E}})}. \tag{29.2.8}$$

It can be seen that this is true either for adiabatic quantities or for isothermal quantities.

To illustrate this result, consider the particular form of the generalised compliance matrix $\boldsymbol{S}$ for the tetragonal and cubic phases of barium titanate. The tetragonal phase has the point group symmetry $4mm$, i.e., C_{4v}. $\boldsymbol{S}$ must be invariant to the symmetry operations of this group, and so must be of the following form (see, e.g., Bhagavantam, 1966) for the tetragonal phase.

$$\boldsymbol{S}=\begin{bmatrix} s_{11} & s_{12} & s_{13} & 0 & 0 & 0 & 0 & 0 & d_{1z} \\ & s_{11} & s_{13} & 0 & 0 & 0 & 0 & 0 & d_{1z} \\ & & s_{33} & 0 & 0 & 0 & 0 & 0 & d_{3z} \\ & & & s_{44} & 0 & 0 & 0 & d_{5x} & 0 \\ & & & & s_{44} & 0 & d_{5x} & 0 & 0 \\ & & & & & s_{66} & 0 & 0 & 0 \\ & & & & & & \chi_{xx} & 0 & 0 \\ & & & & & & & \chi_{xx} & 0 \\ & & & & & & & & \chi_{zz} \end{bmatrix}, \tag{29.2.9}$$

where, since $\boldsymbol{S}$ is a symmetric 9×9 matrix, it is not necessary to write down the elements under the principal diagonal. The rows and columns are labelled in the order $1, 2, \ldots, 6, x, y, z$. As defined in section 1, the elements $s_{\alpha\beta}$ of $\boldsymbol{S}$ in (29.2.9) are to be measured at constant $\boldsymbol{E}$, while the susceptibility elements χ_{ij} are to be measured at constant $\boldsymbol{L}$.

For the cubic phase of barium titanate, which has the point group symmetry $m3m$ (or 0_h), $\boldsymbol{S}$ has the form given in (29.2.9), but with the extra symmetry requirements,

$$s_{11}=s_{33}, \qquad s_{12}=s_{13}, \qquad s_{66}=s_{44}, \qquad \chi_{xx}=\chi_{zz}=\chi, \tag{29.2.10}$$

and

$$d_{1z}=d_{3z}=d_{5x}=0. \tag{29.2.11}$$

Because the cubic phase has a centre of inversion, as shown in chapter 9, section 2, all piezoelectric coefficients must be zero.

The vanishing of the piezoelectric coefficients has the immediate consequence that

$$\boldsymbol{s}^{P}=\boldsymbol{s}^{E}; \qquad \boldsymbol{\chi}^{e}=\boldsymbol{\chi}^{L}, \tag{29.2.12}$$

where, for the cubic phase, the last two matrices are multiples of the unit 3×3 matrix.

To prove this, consider

$$s_{\alpha\beta}^{P}=\left(\frac{\partial e_\alpha}{\partial L_\beta}\right)_{\bar{L}_\beta,\boldsymbol{P}}=\frac{\partial(e_\alpha,\bar{L}_\beta,\boldsymbol{P})}{\partial(L_\beta,\bar{L}_\beta,\boldsymbol{P})}$$
$$=\frac{\partial(\boldsymbol{L},\boldsymbol{E})}{\partial(\boldsymbol{L},\boldsymbol{P})}\frac{\partial(e_\alpha,\bar{L}_\beta,\boldsymbol{P})}{\partial(L_\beta,\bar{L}_\beta,\boldsymbol{E})}, \tag{29.2.13}$$

where the last term results from the cancelling rule (11.2.5).

The first Jacobian on the right-hand side has the value $(\det \boldsymbol{\chi}^{L})^{-1}$. For the case where the piezoelectric coefficients are zero, it can be seen that the second Jacobian can be expressed as the product of a 6×6 determinant and of a 3×3 determinant. To see this, it should be noted, from (29.1.11), that

$$d_{\alpha i}=(\partial e_\alpha/\partial E_i)_{\boldsymbol{L},\bar{E}_i}=(\partial P_i/\partial L_\alpha)_{\bar{L}_\alpha,\boldsymbol{E}}. \qquad (29.2.14)$$

The 6×6 determinant, referred to above, is

$$\begin{vmatrix} s^{\boldsymbol{E}}_{\alpha\beta} & s^{\boldsymbol{E}}_{\alpha\bar{\beta}_1} & s^{\boldsymbol{E}}_{\alpha\bar{\beta}_2} & s^{\boldsymbol{E}}_{\alpha\bar{\beta}_3} & s^{\boldsymbol{E}}_{\alpha\bar{\beta}_4} & s^{\boldsymbol{E}}_{\alpha\bar{\beta}_5} \\ 0 & 1 & 0 & 0 & 0 & 0 \\ 0 & 0 & 1 & 0 & 0 & 0 \\ 0 & 0 & 0 & 1 & 0 & 0 \\ 0 & 0 & 0 & 0 & 1 & 0 \\ 0 & 0 & 0 & 0 & 0 & 1 \end{vmatrix}, \qquad (29.2.15)$$

where $\bar{\beta}_\mu$, $\mu=1,2,\ldots,5$, are the numbers 1 to 6 with β excluded. It can immediately be seen that this determinant has the value $s^{\boldsymbol{E}}_{\alpha\beta}$. The 3×3 determinant referred to is simply $\det(\boldsymbol{\chi}^{L})$, and this cancels out with the first Jacobian on the right-hand side of (29.2.13). Thus we have proved the first part of (29.2.12). Since the essential ingredient in this proof is the requirement that the piezoelectric coefficients be all zero, we may note that this equality is true for any crystal with a centre of inversion symmetry, as will be the second equality of (29.2.12). This equality may be proved in an exactly similar manner to the first equality, by interchanging in the above proof, P_i for e_α, E_j for L_β, to obtain

$$\boldsymbol{\chi}^{e}=\boldsymbol{\chi}^{L}, \qquad (29.2.16)$$

for any crystal with zero piezoelectric coefficients.

Now, we may return to the relation (29.2.8). For the cubic phase we see that this relation follows trivially from (29.2.12). For the tetragonal phase it reduces to

$$\frac{\det(\boldsymbol{s}^{P})}{\det(\boldsymbol{s}^{E})}=\frac{(\chi^2_{xx}\chi_{zz})^{e}}{(\chi^2_{xx}\chi_{zz})^{L}}. \qquad (29.2.17)$$

In experimental work on the tetragonal/cubic phase transition of barium titanate, it has been shown that whereas $\boldsymbol{s}^{P}$ remains finite, some elements of $\boldsymbol{s}^{E}$, a submatrix of $\boldsymbol{S}$, diverge. If this is accepted, then the generalised 9×9 compliance matrix $\boldsymbol{S}$ has a diverging eigenvalue. From (29.2.9) and from the discussion in the next section, we may see that the corresponding eigenvector is of the form $(e_1, e_2, e_3, 0, 0, 0, 0, 0, P_z)$. We may conclude from (24.5.8) that χ^{L}_{zz}, d_{1z} and d_{3z} will diverge, while χ^{L}_{xx} will remain finite and approach its value for the cubic phase. Thus, in the transition, because of (29.2.12), χ^{e}_{xx} and χ^{L}_{xx} will tend to equality. Hence, for the tetragonal phase, we see from (29.2.17) that

$$\frac{\det(\boldsymbol{s}^{P})}{\det(\boldsymbol{s}^{E})} \to \frac{\chi^{e}_{zz}}{\chi^{L}_{zz}}, \qquad (29.2.18)$$

at the instability. This limiting relation is consistent with experimental observations (Jona and Shirane, 1962).

In the next section we shall discuss some further thermodynamic properties of the tetragonal/cubic transition of barium titanate.

29.3. The tetragonal phase near the instability

It has been found experimentally that some elastic compliance elements, the piezoelectric coefficient d_{1z}, and the susceptibility χ_{zz} all show a tendency to diverge as the transition approaches. For an excellent summary of the experimental results for barium titanate see Jona and Shirane (1962). (More recent experimental results will be cited in the further discussion.) Thus, we may conclude that the transition occurs as the result of the onset of a nearby thermodynamic instability. We shall, therefore, assume that $\boldsymbol{S}$, the isothermal generalised compliance matrix, has a diverging eigenvalue λ_1^{-1}. The z-axis is chosen as the direction of the polarisation in the tetragonal phase. Since, in the transition, the change of shape is from tetragonal to cubic, and since P_z vanishes, we may conclude that the eigenvector corresponding to the diverging eigenvalue is of the form

$$\boldsymbol{e}_1^{(1)} = (e_1^{(1)}, e_2^{(1)}, e_3^{(1)}, 0, 0, 0, 0, 0, P_z^{(1)}). \qquad (29.3.1)$$

We shall restrict our discussion to the zero stress, zero electric field state.

We may easily generalise the considerations of chapter 26 to this case where we have strain plus polarisation co-ordinates. For example, the specific heat c_p will be that measured at constant zero $\boldsymbol{E}$ and should strictly be denoted by $c_{p,\boldsymbol{E}} = c_{L,\boldsymbol{E}}$.

Megaw (1947) has measured the strain thermal expansion coefficients m_a $(= m_1)$ and m_c $(= m_3)$. She has shown (see fig. 22) that, for the tetragonal phase, these diverge to $+\infty$ and $-\infty$ respectively, while for the cubic phase, the single strain thermal expansion coefficient remains finite. Hatta and Ikushima (1972) have measured the specific heat $c_p(\boldsymbol{E} = 0)$, in the temperature range from room temperature to 450 °K, about the

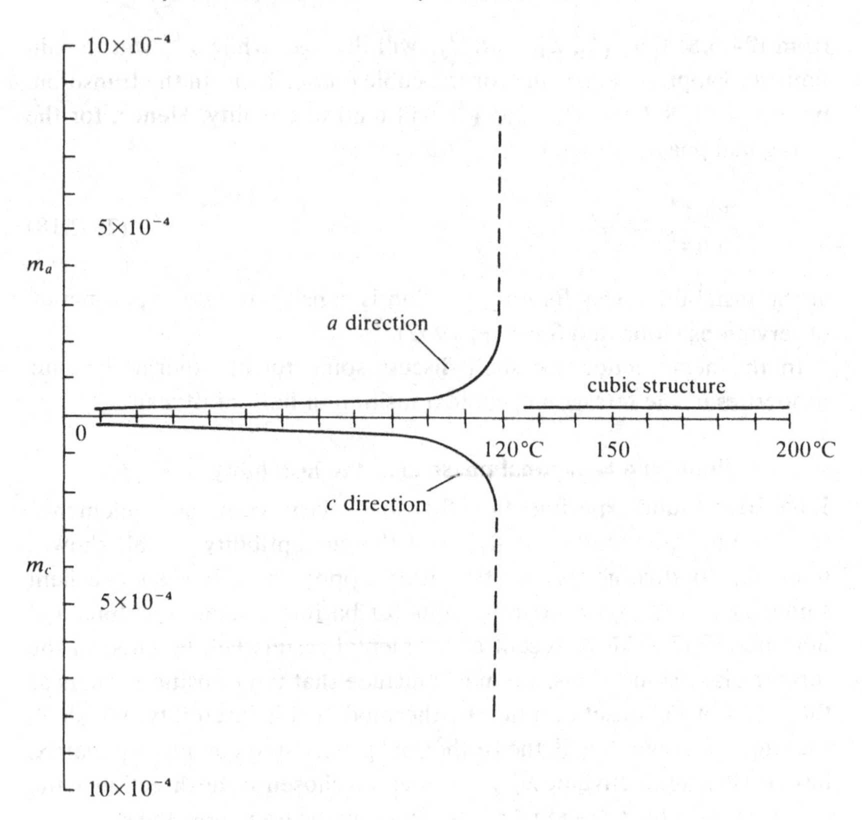

Fig. 22. The linear thermal expansion coefficients for $BaTiO_3$, parallel (m_c) and perpendicular (m_a) to the c-axis (after Megaw, 1947).

transition temperature of 393 °K (120 °C). The specific heat curve has the same characteristic shape as that of quartz about its α–β transition temperature. In other words, the specific heat diverges as the tetragonal phase approaches T_0, while it remains finite as the cubic high temperature phase approaches the temperature T_0.

Thus reference to table 26.1 shows that in the tetragonal phase Q_1 is finite and non-zero, i.e., $Q_1 \propto \lambda_1^x$, where $x = 0$. (The possibility of negative x may be excluded, see chapter 26, section 7, point 4 concerning table 26.1.) The determination of Q_1 would require a knowledge not only of the thermal expansion coefficients m_a, m_c and m_z, but of the isothermal generalised compliance elements near T_0. Unfortunately, the experimental results for these quantities appear in the literature displayed in the form of graphs, making it difficult to obtain accurate estimates.

However, an interesting observation can be made concerning the limiting form of the nine-dimensional eigenvector. First of all, from the data of Megaw (1947), the ratio

$$(m_c/m_a) \to -1.60 \quad \text{(approximately).} \tag{29.3.2}$$

Next, Merz (1949) has shown that

$$P_z \to k e_c^{\frac{1}{2}}, \tag{29.3.3}$$

where k is a constant. e_c is the strain measured from the cubic phase as the standard state; that is, $e_c \to 0$ at the transition. Using this result we see that the thermal expansion coefficient m_z, as defined in (29.2.5), has the following limiting behaviour

$$m_z \to \tfrac{1}{2} k m_c e_c^{-\frac{1}{2}}. \tag{29.3.4}$$

Hence m_z tends to infinity faster, by the factor $e_c^{-\frac{1}{2}}$, than the strain thermal expansion coefficients m_c and m_a. Therefore, to first order in (m_a/m_z) and (m_c/m_z), the normalised eigenvector

$$\boldsymbol{e}^{(1)} \to \left(\frac{m_a}{m_z}, \frac{m_a}{m_z}, \frac{m_c}{m_z}, 0, 0, 0, 0, 0, 1\right), \tag{29.3.5}$$

as may be obtained on using (26.2.13). In the limit, the first three components will tend to zero.

Now, it is known experimentally that elements such as $s_{11}^{\boldsymbol{E}}$, $s_{12}^{\boldsymbol{E}}$, of $\boldsymbol{S}$ appear to diverge (Jona and Shirane, 1962). Hence, we may conclude from (24.5.5) and (29.3.5) that the following quantities

$$\lambda_1^{-1}(m_a/m_z)^2, \qquad \lambda_1^{-1}\{(m_a m_c)/m_z^2\}, \qquad \lambda_1^{-1}(m_c/m_z)^2, \tag{29.3.6}$$

diverge. From (29.3.5) we see that we may take $\mu_1 = m_z$, and, since $\lambda_1\mu_1 = \lambda_1 m_z = Q_1$ which is finite, we may say that the following quantities diverge, that is,

$$\frac{m_a^2}{m_z}, \frac{m_a m_c}{m_z}, \frac{m_c^2}{m_z} \to \infty, \tag{29.3.7}$$

whereas the following quantities, from (29.3.2) and (29.3.4), tend to zero, that is,

$$\frac{m_a}{m_z} \text{ and } \frac{m_c}{m_z} \propto e_c^{\frac{1}{2}} \to 0. \tag{29.3.8}$$

If m_c and m_a behave as m_z^y near the instability, then these two results are consistent with

$$\tfrac{1}{2} < y < 1. \tag{29.3.9}$$

Finally, from (29.3.1) and (24.5.8) we may predict that the piezoelectric coefficients d_{1z} and d_{3z} will diverge. This follows because, for example, from (29.3.5),

$$\lambda_1^{-1} e_a^{(1)} e_z^{(1)} \to \lambda_1^{-1}(m_a/m_z) \to m_a/Q_1. \tag{29.3.10}$$

Therefore, since m_a diverges, so does d_{1z}. Similarly d_{3z} diverges, and in such a way that the ratio

$$d_{1z}/d_{3z} \to (m_a/m_c). \tag{29.3.11}$$

On the other hand, d_{5z} will simply tend to zero, its value in the cubic phase. We may also predict that χ_{zz}^L will diverge, while χ_{xx}^L will not. These predictions are consistent with experiment (Jona and Shirane, 1962), although the limiting relation (29.3.11) has not been verified.

29.4. The cubic phase near the instability

From the experimental data as discussed in the previous section, the single strain thermal expansion coefficient for the cubic phase remains finite and small up to the transition. The specific heat does not diverge in this phase. Hence, from table 26.1, Q_1 is zero, and $x \geq 1$.

The susceptibility χ is, for the cubic phase, an eigenvalue of $\boldsymbol{S}$, and it is well known that, experimentally, χ diverges at the transition. If this were the only eigenvalue of $\boldsymbol{S}$ to diverge, then, from (24.5.8), we could conclude that none of the isothermal elastic compliance coefficients would diverge, in the cubic phase. The experimental data (see Jona and Shirane, 1962) show much more steeply-rising curves for s_{11}^E and $(2s_{12} + s_{66})^E$ on the tetragonal side than on the cubic side of T_0. Thus, it is possible that χ is the only eigenvalue of $\boldsymbol{S}$ to vanish, although experimental work closer to T_0 would be desirable before this could be stated as experimentally confirmed.

In conclusion, it is clear, from this study of barium titanate, that a careful detailed analysis of the experimental data on thermodynamic quantities would be very interesting and useful – as, of course, would data measured nearer T_0.

Summary of important equations

A. *Defining equations*

$$P_i = \chi_{ij} E_j. \tag{29.1.1}$$

$$d'w = L_\mu \, de_\mu + E_i \, dP_i. \tag{29.1.2}$$

$$S_{\alpha\beta} = (\partial e_\alpha / \partial L_\beta)_{\bar{L}_\beta, \boldsymbol{E}} = s_{\alpha\beta}^{\boldsymbol{E}}. \tag{29.1.10}$$

Summary of important equations

$$S_{\alpha i} = (\partial e_\alpha/\partial E_i)_{\boldsymbol{L},\bar{E}_i} = (\partial P_i/\partial L_\alpha)_{\bar{L}_\alpha,\boldsymbol{E}} = \mathrm{d}_{\alpha i}. \tag{29.1.11}$$

$$S_{ij} = (\partial P_i/\partial E_j)_{\boldsymbol{L},\bar{E}_j} = \chi^{\boldsymbol{L}}_{ij}. \tag{29.1.12}$$

B. *A relation between susceptibilities and elastic compliances*

$$\frac{\det(\boldsymbol{s}^{\boldsymbol{P}})}{\det(\boldsymbol{s}^{\boldsymbol{E}})} = \frac{\det(\boldsymbol{\chi}^{\boldsymbol{e}})}{\det(\boldsymbol{\chi}^{\boldsymbol{L}})}. \tag{29.2.8}$$

References

Berger, C., Eyraud, L., Richard, M. & Rivière, R. (1966). *Bull. Soc. Chim., France*, 628.

Berger, C., Richard, M. & Eyraud, L. (1965). *Bull. Soc. Chim., France*, 1491.

Bhagavantam, S. (1966). *Crystal symmetry and physical properties*. New York: Academic Press.

Birss, R. R. (1967). *Electric and magnetic forces*. London: Longmans, Green & Co.

Born, M. & Huang, K. (1954). *The dynamical theory of crystal lattices*. Oxford: Clarendon Press.

Bragg, W. H. & Bragg, W. L. (1933). *The crystalline state*, **1**, London: Bell & Sons.

Bragg, L. & Claringbull, G. F. (1965). *Crystal structures of minerals* (volume 4 of *The crystalline state*). London: Bell & Sons.

Clark, S. P. (ed.) (1966). *Handbook of physical constants*, Memoir 97. The Geological Society of America.

Coe, R. S. (1970). *Contribution to Mineralogy and Petrology*, **26**, 247.

Coe, R. S. & Paterson, M. S. (1969). *J. Geophys. Res.*, **74**, 4921–47.

Coenen, M. (1963). *Silicates Ind.*, **28**, 147.

Coleman, B. D. & Noll, W. (1959). *Arch. Rat. Mech. Anal.*, **4**, 97–128.

Coleman, B. D. & Noll, W. (1964). *Arch. Rat. Mech. Anal.*, **15**, 87–111.

Coxeter, H. S. M. & Moser, W. T. (1972). *Generators and relations for discrete groups*. Berlin: Springer.

Ferrar, W. L. (1941). *Algebra*. Oxford University Press.

Gibbs, J. W. (1906). *The scientific papers of J. Willard Gibbs*, **1**. London: Longmans, Green & Co.

Goldstein, H. (1950). *Classical mechanics*. Cambridge, Mass.: Addison-Wesley.

Hatta, I. & Ikushima, A. (1972). *Phys. Lett.* **40A**, 235–6.

Heine, V. (1963). *Group theory in quantum mechanics*. Oxford: Pergamon Press.

Hughes, A. J. & Lawson, A. W. (1962). *J. Chem. Phys.*, **36**, 2098–2100.

Huntingdon, H. B. (1958). *The elastic constants of crystals*. New York: Academic Press.

Jay, A. H. (1933). *Proc. Roy. Soc.*, **142A**, 237–47.

Jona, F. & Shirane, G. (1962). *Ferroelectric crystals*. Oxford: Pergamon Press.

Killingbeck, J. (1972). *J. Phys. C, Solid State Phys.* **5**, 2497–2502.

Kittel, C. (1971). *Solid state physics*. New York: Wiley.

Landau, L. D. & Lifshitz, E. M. (1958). *Statistical physics*. Oxford: Pergamon Press.

MacDuffee, C. C. (1946). *The theory of matrices*. New York: Chelsea Publishing Co.

Mayer, G. (1959). Doctoral Thesis, University of Paris.

McLellan, A. G. (1973). *Phil. Mag.*, **28**, 1077–85.
McLellan, A. G. (1974). *J. Phys. C, Solid State Phys.*, **7**, 3326–40.
McLellan, A. G. (1976a). *J. Phys. C, Solid State Phys.*, **9**, 939–46.
McLellan, A. G. (1976b). *J. Phys. C, Solid State Phys.*, **9**, 4083–94.
McLellan, A. G. (1978). *J. Phys. C, Solid State Phys.* **11**, 4665–79.
McLellan, A. G. (1979). *J. Phys. C, Solid State Phys.* **12**, 5411–17.
Megaw, H. D. (1947). *Proc. Roy. Soc.*, **189A**, 261–83.
Merz, W. J. (1949). *Phys. Rev.*, **76**, 1221–5.
Moser, H. (1936). *Phys. Zeit.*, **21**, 737–53.
Murnaghan, F. D. (1951). *Finite deformation of an elastic solid.* New York: Wiley.
Nelson, J. B. & Riley, D. P. (1945). *Proc. Phys. Soc.*, **57**, 477.
Nye, J. F. (1957). *Physical properties of crystals.* Oxford University Press.
Pippard, A. B. (1956). *Phil. Mag.* **1**, 473–6.
Tallon, J. L., Robinson, W. H. & Smedley, S. I. (1977). *Nature*, **266**, 337–8.
Thomas, A. L. & Wooster, W. A. (1951). *Proc. Roy. Soc.*, **208A**, 43–62.
Thurston, R. N., McSkimin, H. J. & Andreatch Jr, P. (1966). *J. Appl. Phys.*, **37**, 267–275.
Truesdell, C. & Noll, W. (1965). *Handbuch der Physik*, III/3. Berlin: Springer.
Walker, P. L. & Thrower, P. A. (eds.) (1973). *The chemistry and physics of carbon*, **10**. New York: Dekker.
Wilson, A. H. (1957). *Thermodynamics and statistical mechanics.* Cambridge University Press.
Zemansky, M. W. (1968). *Heat and thermodynamics*, 5th edn. New York: McGraw-Hill.
Zubov, W. G. & Firsova, M. M. (1962). *Soviet Phys. Cryst.* (English trans.), **7**, 374–6.

Index

Index